Computability

Computable Functions, Logic, and the Foundations of Mathematics

3rd edition

Richard L. Epstein

Walter A. Carnielli

Advanced Reasoning Forum

Socorro, New Mexico, USA

ARF

For more information contact:
    Advanced Reasoning Forum
    P. O. Box 635
    Socorro, NM 87801 USA
    <arf@AdvancedReasoningForum.org>

ARF

ISBN 978-0-9815507-2-5

Contents

II COMPUTABLE FUNCTIONS

IV CHURCH'S THESIS, CONSTRUCTIVE MATHEMATICS, and MATHEMATICS AS MODELING

Preface

Why was the theory of computable functions developed before there were any computers?

The formal theory of computable functions and their relation to logic arose as a response to the ferment in the foundations of mathematics at the beginning of this century. The paradoxes of self-reference and the question of how or even whether we are justified in using infinite sets stood at the center of that development, and those paradoxes are no less interesting, nor settled, now. Along with readings from the originators of the subject, the paradoxes and doubts about the infinite serve to motivate the study of the technical mathematics in this book and place the mathematics in its history.

Some mathematicians may prefer a straight mathematical development; for that Part II, *Computable Functions*, and Part III, *Logic and Arithmetic*, will suffice. In Part II we describe the notion of computability, present the Turing machine model, and then develop the theory of partial recursive functions as far as the Normal Form Theorem. In Part III we begin with propositional logic and give an overview of predicate logic and Gödel's theorems, which can serve as a summary for a short course. A full development of the syntactic part of first-order logic and Gödel's theorems then follows. Part I, *The Fundamentals*, can be referred to for notation and basic proof techniques.

Philosophy, however, has been the motive for much of logic and computability. In Part I we give the philosophical background for discussions about the foundations of mathematics while presenting the notions of whole number, function, proof, and real number. Hilbert's paper "On the infinite" sets the stage for the analysis of computability in Part II. In Part IV we consider the significance of the technical work with discussions of Church's Thesis, constructivity in mathematics, and mathematics as modeling.

Many exercises are included, beginning gently in Part I and progressing to a graduate level in the final chapters. The most difficult ones, marked with a dagger †, may be skipped, although all are intended to be read. Solutions to the exercises can be found in the Instructor's Manual (available from the Advanced Reasoning Forum, www.AdvancedReasoningForum.org), which also contains suggestions for course outlines. Sections marked "Optional" are not essential for the technical development of chapters which follow, although they often provide important motivation.

For the second edition: Beyond the addition of a timeline on computability and undecidability written by Epstein, we have confined our changes almost entirely to technical corrections, adding only two new quotes from Gödel (p. 173 and p. 215). One noteworthy change is the replacement of Fermat's Last Theorem by Goldbach's Conjecture as an example of an unsolved arithmetic problem used in several examples; the former has been shown to be true by Andrew Wiles, "Modular elliptic curves and Fermat's Last Theorem", *Annals of Mathematics*, second series, vol. 141, 1995, pp. 443–551. Where other authors have used Fermat's Last Theorem as an example (Arend Heyting on p. 234, Nicholas Goodman on p. 259), a similar substitution of Goldbach's Conjecture would make the same point.

For the third edition: We have added a chapter that gives a very different view of mathematics than in the other articles in the text, viewing mathematics as modeling and not (necessarily) in need of foundations. It is the view that underlies our presentation of the mathematics in this book.

We have made only minor corrections to the body of the text, retaining the same pagination. A few small corrections have been made to the timeline.

The story we tell leaves no room to include a presentation of the semantics of classical predicate logic. That material is now available in a companion to this volume, Epstein's *Classical Mathematical Logic*, Princeton University Press, 2006.

Acknowledgments

We are grateful to the following organizations for their help in the writing of this book: Victoria University of Wellington for a post-doctoral fellowship for Richard Epstein in 1975–1977, during which the notes for a course on recursive function theory for the Philosophy Department were developed that became the basis for this book; the Fundação de Amparo à Pesquisa do Estado de São Paulo, Brazil, which sponsored our collaboration in Brazil and the United States in 1984–1985; the Fulbright Commission for allowing us to continue that collaboration by awarding a fellowship to Richard L. Epstein to work at the University of Campinas, Brazil, from January to June of 1987; and the Alexander von Humboldt Foundation of the Federal Republic of Germany for awarding a fellowship to Walter A. Carnielli for 1988–1989, during which final revisions were made to the text.

We are especially pleased to have this opportunity to thank the many people who have helped us. Max Dickmann, Justus Diller, Benson Mates, Piergiorgio Odifreddi, A.S. Troelstra, and our students Sandra de Amo, Karl Henderscheid, João Meidanis, and Homero Schneider read and suggested many important improvements to the text. Oswaldo Chateaubriand and Peter Eggenberger cleared up much of our confusion about Church's Thesis. The following persons served as reviewers of the text for the first edition and offered many useful suggestions: Herbert Enderton, F. Golshani, Roger Maddux, Mark Mahowald, Bernard Moret, Fred Richman, Rick Smith, Stephen Thomason, and V.J. Vazirani. Finally, we are indebted to our editor for the first edition, John Kimmel, and production editors Nancy Shammas and Bill Bokermann, and to Peter Adams, our editor for the second edition, through whose patience and persistence this became a better text. To all these, and any others whom we may have inadvertently forgotten, our thanks.

Each author wishes to indicate that any mistakes still left in this text are not due to those above who have so generously helped us, but are due entirely to the other author.

For the third edition: We are grateful to Howard Blair, Carlos Augusto Di Prisco, Ricardo Gazoni, Leon Harkleroad, David Isles, and Carolyn Kernberger for their corrections and suggestions for improving the text. We hope that we have introduced no new errors in correcting the old ones.

Richard L. Epstein is also grateful to the following people whose comments greatly helped in the preparation of the timeline: Irving Anellis, Walter Carnielli, William Craig, Martin Davis, Peter Eggenberger, Rogéria Gaudencio, Ivor Grattan-Guinness, Leon Harkleroad, David Isles, Benson Mates, Gregory Moore, Piergiorgio Odifreddi, Jerzy Perzanowski, Clara Helena Sánchez, Stewart Shapiro, Peter Simons, Stanisław Surma, Lesław Szczerba, Jan Wolenski, Jan Zygmunt.

Publishing Acknowledgments

(Full references to the papers and books below may be found in the Bibliography.)

Quotations from Robert J. Baum, *Philosophy and Mathematics*, reprinted with permission of the publisher, Freeman Cooper and Co., San Francisco, California. Copyright © 1973.

Quotations from *Constructive Analysis* by E. Bishop and D. Bridges reprinted with permission of the publisher. Copyright © 1985 Springer–Verlag, Berlin–Heidelberg.

Quotations from "Intuitionism and formalism" by L.E.J. Brouwer reprinted with permission from the Bulletin of the American Mathematical Society. © 1913.

The "Letter to Descartes," by R. C. Buck, reprinted with permission of the publisher, The Mathematical Association of America. Copyright © 1978.

Quotations from "On undecidable propositions of formal mathematical systems," from lectures delivered by Kurt Gödel in 1934. Reprinted with permission of the Institute for Advanced Study, Princeton, literary executors of the estate of Kurt Gödel.

Quotations from "Reflections on Bishop's philosophy of mathematics," by Nicolas D. Goodman, reprinted with permission of the publisher, Springer–Verlag. Copyright © 1981, Lecture Notes in Mathematics, vol. 873.

The Introduction to *Constructive Formalism*, by R. L. Goodstein, reprinted with permission of the publisher, University College, Leicester. Copyright © 1951.

Quotations from *Enumerability, Decidability, Computability*, by Hans Hermes, reprinted with permission of the publisher. Copyright © 1969 Springer–Verlag, Berlin–Heidelberg.

"On the infinite," by David Hilbert, was first delivered on June 4, 1925, before a congress of the Westphalian Mathematical Society in Münster, in honor of Karl Weierstrass. Translated by Erna Putnam and Gerald J. Massey from *Mathematischen Annalen* (Berlin) no. 95 (1926) pp. 161–190. Permission for the translation and inclusion of the article in this volume was kindly granted by the publishers, Springer–Verlag.

Quotations from "Remarks on the notion of standard non-isomorphic natural number series," by David Isles, reprinted with permission of the author. Copyright © 1981.

Quotations from *Introduction to Metamathematics*, by Stephen C. Kleene, reprinted with permission of the publisher, North–Holland. Copyright © 1952.

"Mathematical proofs: the genesis of reasonable doubt," by Gina Bari Kolata, *Science*, vol. 192, pp. 989–990, 4 June 1976, by the A.A.A.S. Reprinted with permission.

"Finite combinatory processes—formulation I," by Emil Post, reprinted with permission of the Association for Symbolic Logic. Copyright © 1936.

Quotations from Joseph R. Shoenfield, *Mathematical Logic*, © 1967, Addison–Wesley Publishing Co., Inc., Reading, Massachusetts, p.214. Reprinted by permission.

Quotations from "On computable numbers, with an application to the Entscheidungsproblem," by Alan M. Turing, published in the *Proceedings of the London Mathematical Society*, ser. 2, vol. 42, 1936–1937. Reprinted by permission of Oxford University Press.

"Is $10^{10^{10}}$ a finite number?," by D. van Dantzig, reprinted with permission of the editors of *Dialectica*. Copyright © 1956.

Quotations from *From Mathematics to Philosophy*, by Hao Wang, reprinted with permission of the publisher, Routledge and Kegan Paul. Copyright © 1974.

A previous version of Chapter 27 appeared at CLE e-Prints, Vol. 8(3), 2001 (Section Logic), /www.cle.unicamp.br/e-prints/vol_8,n_3,2008.html.

I

The FUNDAMENTALS

1 Paradoxes

Much of modern logic came about as a response to problems and paradoxes in the foundations of mathematics. Paradoxes test our intuitions: a contradiction appears, yet the principles that clash are so fundamental we are unwilling to give them up. So we try to resolve the paradox by making clearer distinctions or, perhaps in the end, abandoning or modifying one of the principles.

The difference between truth and falsity and the question of how language reflects upon itself are the themes of the paradoxes of Section A. We will see these paradoxes and themes in one guise or another in the formalization of computable functions, in the study of formal languages, and in reflections about artificial intelligence. Later we will see them reappear not as paradoxes but as tools.

In Section B the principles which apparently come into conflict with our experience concern infinite processes and completed infinities. With this we commence our discussion of how to demarcate the borderline between the finite and the infinite, a question still as central and unsettled as it was in Zeno's time.

A. Self-Referential Paradoxes

If someone says, "I'm over 6 feet tall" or "That's my cup of coffee," he is using self-reference. Self-reference is an apparently essential part of our language which reflects our self-consciousness: without "I" and "my" we can be in the world but not express our knowledge of that fact. Yet the power of self-reference within our language can create puzzling problems.

1. The preoccupation with self-referential problems dates to antiquity. Epimenides the Cretan is reported to have said, "All Cretans are liars." Was he speaking truly?

2. "This sentence is false." Is this true? This is known as the *antinomy of the liar*, or the *liar paradox*, and was first

posed by Eubulides of Miletus, a contemporary of Socrates. So perplexing did it seem to Philetus of Cos (ca. 340–285 B.C.) that on his gravestone was written

> O Stranger: Philetus of Cos am I
> 'Twas the Liar who made me die,
> And the bad nights caused thereby.

> translated by St. George Stock

3. Take three sheets of blank paper.
   a. On the first sheet write, "The sentence on the other side of this is false." On the other side write, "The sentence on the other side of this is true."
   b. On the second sheet write on one side, "The sentence on the other side of this is false." On the other side write, "The sentence on the other side of this is false."
   c. On the third sheet write, "The sentence on the other side of this is true." On the other side write, "The sentence on the other side is false, or God exists."
   Which of these sentences are true? Which are false?

4. In a village there lives a barber who shaves all those and only those villagers who don't shave themselves. Does he shave himself?

5. Consider the set $Z = \{X : X \notin X\}$. Is $Z \in Z$?
   This is *Russell's set theory paradox*.

6. From *The American Mathematical Monthly*, vol. 85, no. 10, 1978

By a strange sequence of events, an undated letter has come to light, asserted to be from Fermat to Descartes. Although the provenience of the letter is clouded we feel it may be of interest to readers of the MONTHLY.

M. René Descartes:
You have argued cogently that he who thinks, is, without regard for the nature of these thoughts. Reflecting upon this, I have found another use for my "method of descent" which I think will interest you.

   Consider: Most people think of themselves from time to time, but we may suppose that there are some selfless people who never think of themselves. Let us hypothesize that I am a person whose sole thoughts are of each of the selfless persons. I will argue that I cannot exist, even though I have thoughts!

   For, either I must be selfless or not selfless. If I am selfless, then at some time my thoughts must turn to myself as one of the selfless persons; but by doing so, I reveal that I am not selfless! On the other hand, if I am not selfless, then I will sometime think of myself. However, since the only object of my thoughts are selfless beings, I myself must have been selfless!

   From this dilemma, I can only conclude that it is inconceivable that I exist.

   I draw the conclusion that my existence depends not only on the fact that I think, but also upon the content of my thoughts.

   May I suggest that you pass this letter on to young Blaise Pascal. He has a

bright mind and wide interests. Perhaps he can clarify the implications of this
for both God and Reality.

Pierre de Fermat

(Translated and communicated by R.C. Buck, who comments: "It would be
interesting if this letter were authentic, and preceded Descartes' 1647 visit to
Pascal and the latter's subsequent drastic change of interests.")

B. Zeno's Paradoxes

Zeno's paradoxes, which we present here, were apparently directed against the
Pythagoreans who thought of space and time as consisting of points and instants.

1. Achilles and the tortoise are going to race. The tortoise is given a head start.

But no matter how swiftly Achilles runs nor how slowly the tortoise crawls, Achilles
can never overtake the tortoise. By the time Achilles reaches the initial position of
the tortoise, the latter will have advanced some short distance; by the time Achilles
covers that distance, the tortoise will have gone a bit further, and this goes on
indefinitely, so Achilles can never catch the tortoise.

This argues that motion is impossible if we assume that space and time are
infinitely subdivisible.

(Another version of this paradox is given in the reading from Goodstein, 1951,
in Chapter 2.B.)

2. But here is an argument from Zeno that shows that space and time cannot
terminate in indivisibles. It is a paraphrase of the *Stade* (Stadium) due to Boyer
(*A History of Mathematics*, p. 83).

Let's assume that we have three tapes divided into squares of the same size:

| A_1 | A_2 | A_3 |

| B_1 | B_2 | B_3 |

| C_1 | C_2 | C_3 |

B_1, B_2, B_3 move right so that each B_i passes one A_j in the smallest possible
instant of time. Simultaneously, C_1, C_2, C_3 move left so that each C_i passes one A_j

in an instant of time. Thus after one instant of time we have:

A_1	A_2	A_3

B_1	B_2	B_3

C_1	C_2	C_3

But C_1 will have passed two B_i's. Therefore, the instant cannot be the minimum time interval, for it must take less time for C_1 to pass one of the B_i.

The usual resolution of the paradox of Achilles and the tortoise has recourse to the calculus in terms of limits (Exercise 6). But that solution depends on many assumptions about the nature of the infinite, and it is precisely the infinite that is the problem here (in Chapter 6 we'll see more reasons to be uneasy about any solution that depends on limits and infinity). In the next chapter we present a resolution of this paradox that does not use infinities.

Exercises

1. Why is the paradox ascribed to Epimenides not a paradox as stated? How can you make it paradoxical?
2. Resolve the paradox of the barber. Can you resolve Russell's set theory paradox in the same way?
3. There is a fallacy in the purported letter from Fermat to Descartes, which surely Descartes would have spotted. What is it? (*Hint*: What other conclusion could Fermat have drawn?) Is this the same as the paradox of the barber?
4. Does Section A.3.c prove the existence of God?
5. The paradoxes and puzzles of Section A seem to be very much the same, but by working through these exercises you should begin to see differences. Try to contrast and classify the paradoxes according to principles which can be used to resolve them or principles they call into question.
6. Give a resolution of the paradox of Achilles and the tortoise in terms of limiting procedures from the calculus. Can the same sort of resolution be applied to the *Stade*?

Further Reading

Mates in his *Skeptical Essays* discusses the liar paradox and its history. He makes the important distinction, which we have glossed over, between an *antinomy*, which leads to a contradiction from plausible assumptions, and a *paradox*, which need only give rise to something odd, surprising, or wildly implausible.

Patrick Hughes and George Brecht have written an amusing and stimulating anthology of paradoxes called *Vicious Circles and Infinity*.

2 What Do the Paradoxes Mean?

(Optional)

The paradoxes of Chapter 1 raise questions about the foundations of mathematics and logic: What is the infinite and how are we to use it in mathematics? What is the right way to reason? Any resolution to those paradoxes is, at least implicitly, based on assumptions about the nature of mathematics.

In Section A of this chapter we consider the relation of philosophy to mathematics and, as an example, present Plato's influential view of mathematics as inhabiting a world of abstractions. In Section B we consider a resolution of the paradox of Achilles and the tortoise based on an understanding of the nature of numbers quite different from Plato's.

A. Philosophy and Mathematics

From *Philosophy and Mathematics* by Robert J. Baum

From the earliest times, man has searched for the answers to a multitude of questions. Some are quite specific and concrete: When will the next flooding of the Nile occur? What was the cause of this child's death? Why did the sky suddenly go black? Others are more general and abstract: What is justice? Is there life after death? What are the ultimate constituents of the universe? Although the particular concrete questions are of more immediate concern in the everyday contexts in which they normally arise, the more general abstract questions have been considered by many to be ultimately of more importance and greater interest. An adequate answer to the question about the sky going black requires reference to more abstract notions such as those of the eclipse of the sun, and the general principles of planetary motion. The real utility of abstract general knowledge is that just a few general principles are sufficient for answering innumerable questions.

But abstract general principles alone are not sufficient for providing adequate answers to our questions. Answers have always been available— too many answers. The Greek Sophists went so far as to claim that equally convincing arguments can be given in support of every logically possible answer to any question. The question thus arises: Which answer, if any, is the *true* answer? Traditionally the demand was often for not merely the most probable answer, but rather for that answer which is absolutely certain. The evidence was required not merely to remove any reasonable doubts, but to establish the truth of the statement beyond the shadow of any doubt. René Descartes echoed this ancient demand, in his *Meditations* :

> "I shall continue ... until I have found something certain, or at least, if I can do nothing else, until I have learned with certainty that there is nothing certain in this world. Archimedes, to move the earth from its orbit and place it in a new position, demanded nothing more than a fixed and immovable fulcrum; in a similar manner I shall have the right to entertain high hopes if I am fortunate enough to find a single truth which is certain and indubitable." ...

With few exceptions the authors ... before 1900 studied the nature of mathematical knowledge not for its own sake, but rather for the insights that such a study might provide into the nature of knowledge in general. Their concern was with *general* questions such as "Is certain knowledge possible?" and "What makes knowledge certain?" (It should be noted that many philosophers, particularly those before 1900, would have considered this wording redundant; for them "knowledge" meant "*certain* knowledge," and "uncertain knowledge" or "probable knowledge" involved an internal inconsistency as in "square circle." In present-day discussions the concepts of knowledge and certainty are usually defined independently.) Despite possible differences in motivation and perspective, the "traditional" philosophers arrived at conclusions which provide the foundations of and starting points for much of the work of today's philosophers of mathematics.

Baum, pp. 2–3

One of the earliest and still most influential views of mathematical knowledge was that of Socrates and Plato. To them, what is real resides in "the heaven above the heavens" and is imperceptible to our senses. Only our minds can perceive the true form of a circle, a square, a horse, a chair. What we call the world and life are only pale imitations. Mathematical knowledge is knowledge of these pure, eternal, imperishable forms.

From Jowett, *The Dialogues of Plato*

But of the heaven which is above the heavens, what earthly poet ever did or ever will sing worthily? It is such as I will describe; for I must dare to speak the truth, when truth is my theme. There abides the very being with which true knowledge is concerned; the colourless, formless, intangible essence, visible only to mind, the pilot of the soul. The divine intelligence, being nurtured upon mind and pure knowledge, and the intelligence of every soul which is capable of receiving the food proper to it, rejoices at beholding reality, and once more gazing upon truth, is replenished and made glad, until the revolution of the worlds brings her round again to the same place. In the revolution she beholds justice, and temperance, and knowledge absolute, not in the form of generation or relation, which men call existence, but knowledge absolute in existence absolute; and beholding the other true existences in like manner, and feasting upon them, she passes down into the interior of the heavens and returns home.

Phaedrus 247

[Socrates said:] Arithmetic has a very great and elevating effect, compelling the soul to reason about abstract number, and rebelling against the introduction of visible or tangible objects into the argument. You know how steadily the masters of the art repel and ridicule any one who attempts to divide absolute unity when he is calculating, and if you divide, they multiply, taking care that one shall continue one and not become lost in fractions.

That is very true.

Now suppose a person were to say to them: O my friends, what are these wonderful numbers about which you are reasoning, in which, as you say, there is a unity such as you demand, and each unit is equal, invariable, indivisible—what would they answer?

They would answer, as I should conceive, that they were speaking of those numbers which can only be realized in thought.

Then you see that this knowledge may truly be called necessary, necessitating as it clearly does the use of the pure intelligence in the attainment of pure truth?

Yes: that is a marked characteristic of it.

And you have further observed, that those who have a natural talent for calculation are generally quick at every other kind of knowledge; and even the dull, if they have had an arithmetical training, although they may derive no other advantage from it, always become much quicker than they would otherwise have been.

Very true, he said.

And indeed, you will not easily find a more difficult study, and not many as difficult.

You will not.

And for all these reasons, arithmetic is a kind of knowledge in which the best natures should be trained, and which must not be given up.

I agree.

Let this then be made one of our subjects of education. And next, shall we enquire whether the kindred science also concerns us?

You mean geometry.

Exactly so.

... the greater and more advanced part of geometry—whether that tends in any degree to make more easy the vision of the idea of good; and thither, as I was saying, all things tend which compel the soul to turn her gaze towards that

place, where is the full perfection of being, which she ought, by all means,
to behold.

True, he said.

Then if geometry compels us to view being, it concerns us; if becoming
only, it does not concern us?

Yes, that is what we assert.

Yet anybody who has the least acquaintance with geometry will not deny
that such a conception of the science is in flat contradiction to the ordinary
language of geometricians.

How so?

They have in view practice only, and are always speaking, in a narrow and
ridiculous manner, of squaring and extending and applying and the like—they
confuse the necessities of geometry with those of daily life; whereas knowledge
is the real object of the whole science.

Certainly, he said.

Then must not a further admission be made?

What admission?

That the knowledge at which geometry aims is knowledge of the eternal,
and not of aught perishing and transient.

That, he replied, may be readily allowed, and is true.

Then, my noble friend, geometry will draw the soul towards truth, and
create the spirit of philosophy, and raise up that which is now unhappily allowed
to fall down.

Nothing will be more likely to have such an effect.

Republic, VII 525–527

Plato's belief that abstract objects (such as numbers, rectangles, fields) exist
independently of us allows him to explain why mathematics is objective: theorems in
mathematics express truths about real objects and their properties. But since these
objects are not perceptible to our senses, how can we know anything about them?
Plato says that we know mathematical objects through the perception of our intellect,
which is analogous to but distinct from sense perception.

We shall call anyone who ascribes to these two beliefs a *platonist*, though
the term *realist* is often used (abstract objects are real). Actually, most platonists
believe something stronger: What we experience through our senses is the
perishable, changeable world, which is *less* real than the forms we perceive
through our intellect.

Plato's views seem to countenance and perhaps encourage the greatest forms of
abstraction in mathematics. But until very recently, and very much in the time of the
Greeks, mathematics has been tied to human experience. Greek mathematics was, in
the main, geometry. Euclid's *Elements*, still used in courses today, was concerned
with constructions—granted, a kind of abstract, theoretical construction—using
various tools such as straightedge and compass. His axioms were self-evident
because they corresponded to (were abstractions of) actual constructions: Every line

can be extended, through any two points one may draw a line, and so on. All his axioms, that is, except one:

> Given two lines *m* and *n* which intersect a third line *l*,
> if the interior angles α and β sum to less than two right angles,
> then the lines, if extended indefinitely, meet on that side of *l* :

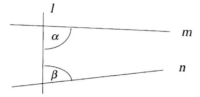

Given *l*, *m*, and *n*, how far do we have to go before we find that they intersect? How far is indefinitely? For large angles the distances would be enormous. With the other axioms, once we are given the points and lines of the hypothesis we can make a construction which convinces us that there are indeed the points and lines asserted to exist in the conclusion. Only for this one axiom is that not possible. It is equivalent to what is known as *Euclid's parallel postulate*:

> Given a line *l* and a point *P* not on *l* there is one and only one line
> *m* through *P* such that *m* is parallel to *l* .

For millenia mathematicians were intent on showing that this postulate could be proved from the other axioms in order to eliminate any reliance on the unintuitively abstract. In Chapter 7 we'll discuss the nineteenth century resolution of those attempts.

But a modern exponent of platonism, Gödel, argues that it is essential and unavoidable to rely on abstract objects in modern mathematics.

> Classes and concepts may, however, also be conceived of as real objects, namely as "pluralities of things" or as structures consisting of a plurality of things and concepts as the properties and relations of things existing independently of our definitions and constructions.
>
> It seems to me that the assumption of such objects is quite as legitimate as the assumption of physical bodies and there is quite as much reason to believe in their existence. They are in the same sense necessary to obtain a satisfactory system of mathematics as physical bodies are necessary for a satisfactory theory of our sense perceptions and in both cases it is impossible to interpret the propositions one wants to assert about these entities as propositions about the "data", i.e. in the latter case the actually occurring sense perceptions.
>
> Gödel, 1944, pp. 456–457

In the end, though, does it really matter whether we believe mathematical objects "exist" ?

From G. Kreisel's "Obituary of K. Gödel"

The second principal aim of this memoir is to substantiate Gödel's own view of the essential ingredient in his early successes, which solved problems directly relevant to principal interests of some of the most eminent mathematicians of this century. ... His view differs sharply from the impressions of many mathematical logicians who, over more than forty years, have looked in Gödel's work for the germs of some exceptionally novel mathematical constructions or for previously unheard-of subtle distinctions, but not very convincingly. Without losing sight of the permanent interest of his work, Gödel repeatedly stressed ... how little novel mathematics was needed; only attention to some quite commonplace (philosophical) distinctions; in the case of his most famous result: between arithmetical truth on the one hand and derivability by (any given) formal rules on the other. Far from being uncomfortable about so to speak getting something for nothing, he saw his early successes as special cases of a fruitful general, but neglected scheme:

By attention or, equivalently, analysis of suitable traditional philosophical notions and issues, adding possibly a touch of precision, one arrives painlessly at appropriate concepts, correct conjectures, and generally easy proofs—to be compared to the use of physical reasoning for developing mathematics or on a smaller scale, the use of geometry in algebra.

Kreisel, 1980, p. 150

B. Achilles and the Tortoise Revisited

Here is a resolution of a version of the paradox of Achilles and the tortoise which does not depend on the machinery of limits and possible infinities and which gives us a picture of mathematics quite different from Plato's.

The Introduction to *Constructive Formalism* by R. L. Goodstein

The great discoveries in mathematics are not in the nature of uncovered secrets, pre-existing timeless truths, but are rather constructions: and that which is constructed is a symbolism, not a proposition. The power of a living symbolism is the source of that insight into mathematics which is termed mathematical intuition.

In the foundations of mathematics a *formal calculus* plays the part which is taken by symbolism in the informal development. A symbolism leads on, a formal calculus leads back, and just as a formal calculus, rightly is felt by creative mathematicians as a barrier to the free expression of ideas, so in the critical study of the foundations, symbolism is a source of error and misconception.

The foremost question of the foundations of mathematics for the last twenty-five years concerns the legitimacy of certain methods of proof in mathematics. What makes this question so difficult is the absence of any absolute standard, outside mathematics, with which mathematics can be compared.

Philosophers have held that such a standard is to be found in a study of the Mind; that just as the laws of Nature are discovered by observation of, and experiment in, natural phenomena, so too the laws of mathematics are to be found as laws of thought, by a study of the thinking processes. Yet, if we consider, we find that the 'Laws of Nature' are but empirical hypotheses, subject to limitation and modification, admitting exceptions related to time and describing the world as it is, whereas the rules of mathematics *are* mathematics, timeless because they are outside time, independent of all observation and experiment and accordingly neither true nor false, expressing no property of the world, neither validating, nor validated by, any fact. The 'laws of thought', if by the term we mean laws formulated by experimental psychologists, no more form a standard by which the rules of mathematics can be tested, than the deductions of a Martian, from *observation* of the game, test the validity of the rules of chess.

What then is the meaning of the controversy between formalists and constructivists? The formalists say that the criteria by which formal systems are tested are the criteria of freedom-from-contradiction and completeness, and all their efforts in the past twenty-five years have been directed towards proving that a formal calculus, like *Principia Mathematica*, a calculus of implication, disjunction, and quantification, contains no insoluble problems, and in particular towards the construction of a proof of the non-contradictoriness of this calculus. This preoccupation with *contradiction* springs from two widely different sources. From the time when language first ceased to be only a *vehicle* of communication and became itself an *object* of discourse, men have invented paradoxes. Already in the oldest paradoxes of which we have written record, the paradox of the "Liar" and the infinity paradox of Zeno we find the prototypes of the paradoxes of the present day. The construction of formal systems, the very object of which was the resolution of these paradoxes, has accomplished only their multiplication. It seems as if the elimination of a paradox can, so to speak, be achieved on only one plane at a time and at the cost of fresh paradoxes on higher planes. Rather, this is the impression which the logistic technique of paradox resolution has produced, for in fact the roots of the paradoxes lie in this very technique.

The second source from which the fear of hidden, yet to be discovered, contradictions springs is the uncertainty which every thinker has felt, particularly in recent years, regarding the significance of postulational methods in mathematical philosophy, a feeling that the postulation of the existence of even a mathematical entity is entirely specious, metaphysical, and in no way comparable to the invention of a physical entity to serve as a medium of expression or a physical model.

Existence in mathematics. Problems regarding the existence of mathematical entities are of many different kinds. Contrast the questions. Are there numbers, do numbers exist? Does the real number "e" exist as something apart from the sequence $1,\ 1+1,\ 1+1+1/2!,\ 1+1+1/2!+1/3!\ \cdots\ ?$; Is there a prime number greater than 10^{10} ?; Is there a prime pair less than 10^{10} ? ;

greater than 10^{10} ? To the first question one may answer: Amongst the *signs* of our language we distinguish the *numerals*, or number-signs, which are constructed from the number sign "0" by the operation of placing a vertical stroke after a number sign; the term 'number' is thus a classification index of signs. The sense in which we can say that numbers exist is that number signs are used in our language. Such questions as "have numbers an objective reality", "are numbers subjects or objects of thought" are disguised questions concerning the grammar of the word "number" and ask whether or not we formulate such sentences as: That which you see, hear, taste, touch, etc. are numbers.

The second question is concerned with the meaning of limit-processes in mathematics and with the concept of an infinite set. To say that the real number "*e*" has an existence independent of the convergent sequence $1, \ 1 + 1, \ 1 + 1 + 1/2!, \ 1 + 1 + 1/2! + 1/3! \ \cdots$ is equivalent to saying that some infinite process is *completed*, for instance that the process of writing down *all* the digits in the decimal expansion of *e* has been carried through. In what sense can an infinite process be thought of as completed? An infinite process is, by definition, a process in which each stage of the process is followed by another stage just as each numeral is followed by another, formed by adding a vertical stroke to the end of the numeral. An infinite process is therefore an *unfinishable* process, a process which does not contain the possibility of being completed. A completed infinite process is a contradiction in terms.

The relation of Zeno's paradox to the formalist-finitist controversy.
It is, however, commonly argued that we *can* conceive of a completed infinite process; that in fact, were it not so, Zeno's famous argument would force us to deny the possibility of motion. For in passing from one position A, to another B, a body must pass through the mid-point A_1 of AB and then through the mid-point A_2 of A_1B, and then through the mid-point A_3 of A_2B, and so on. Thus the motion from A to B may be considered to consist in an unlimited (infinite) number of stages, viz., the stage of reaching A_1, the stage of reaching A_2, the stage of reaching A_3, and so on. After any stage A_n follows the stage A_{n+1} and no matter how many of the stages we have passed through we have not reached B, and so we *never* reach B. But if motion from a point A to any point B is not possible, then no motion is possible. Thus Zeno argues; and by *reductio ad absurdum* (for motion is certainly possible) it follows that the motion from A to B must be regarded as a completed infinite process. The fallacy in this discussion is by no means easy to detect and seems to have escaped the notice of many competent thinkers.

If we say that motion is possible we are appealing to our familiar experience of physical bodies changing their positions. Let us imagine a man running along a race track across which tapes are strung a few feet from the ground. We may suppose the track is 100 yards long and that we commence to string the tapes at the 50 yard mark. If we call the ends of the track A, B and the 50 yard mark A_1, then A_2 is the mid-point of A_1B and so on as above. At each of the points A_1, A_2, A_3, ... a tape is strung across the track. As a man runs from A to B he will break each of the tapes we set up, and if we suppose that a tape has been set up at each of the points A_1, A_2, A_3, ... then the runner will have broken an infinite number of tapes. In putting the argument in this form we have only placed the difficulty in a more obvious light, for we are now confronted with the task of setting up an unlimited number of tapes, or, looking at it from another point of view, of isolating an unlimited number of points. On the one hand we have the possibility of passing from A to B and the unlimited possibility of specifying points between A and B (an unlimited number of fractions between 0 and 100) and on the other hand the impossibility of isolating these points on the track. How is this apparent incompatibility resolved?

Think of a man *counting* from 0 to 100. He may say all the natural numbers from 0 to 100, or he may say only the "tens" or just "fifty", "hundred", or he may say "half, one, one and a half, two", and so on, by halves, up to a hundred. If he counts by tens can we say he has passed through all the integers between one and a hundred (or passed over them)? And if he counts by units, that he has passed through *all* the fractions between these units? One would not hesitate to answer that the man has counted, or passed through, only those numbers which in fact he counted, whatever they were, and that numbers which he did not count, even though such numbers could be inserted between the numbers which he counted, were *not* passed through by him in his counting. Correspondingly when a man runs from A to B he passes those points (or breaks those tapes) which we isolate, which we name, and these points only and what we name will be a finite number of points, however great. The Zeno argument achieves its end by confusing the physical possibility of motion with the logical possibility of naming as many points as we please.

It is sometimes maintained that the resolution of Zeno's paradox lies in the fact that a steadily increasing infinite sequence of numbers may be bounded; e.g., the sequence whose n^{th} term is $n/(n+1)$ is steadily increasing, because $n + 1/(n+2)$ exceeds $n/(n+1)$ by $1/(n+1)(n+2)$ and is bounded above by unity since $n/(n+1)$ is $1/(n+1)$ less than 1 . The fact is applied to Zeno's argument in the following way: Suppose the tape at the point A_1 is fixed in $1/2$ minute, the tape at the point A_2 in $(1/2)^2$ minutes, the tape at the point A_3 in $(1/2)^3$ minutes, and so on, then the first n tapes are fixed in $1 - (1/2)^n$ minutes, so that within one minute *all* the tapes are fixed, and an infinite operation has been completed. Thus although there always remains a tape to be fixed no matter how many have been set up, yet within a minute of starting there is no tape which has not yet been set up. This argument does not however resolve the paradox but merely restates it in a fresh plane, for the conclusion seems now to be that measurement of time is impossible, and this in its turn is bound up with the possibility of motion (for example, time may be measured by the motion of the hand of a clock, or the sun across the sky). The resolution of the paradox in this form is the same as the resolution of the motion paradox. If our criterion for the number of tapes fixed in a minute is the criterion of experiment, then no matter how rapidly the experiment is carried out the *unfinishable* task of setting up an unlimited number of tapes will not be finished. And if our criterion is just that $1 - (1/2)^n$ is less than unity, then this criterion tells us nothing about an actual experiment and we cannot appeal to the reality of the passage of time to generate the paradox.

Consider an analogous example. A line is drawn from the point 0 to the point 1. In what sense can we say that the line passes through infinitely many points, that *drawing* the line completes an infinity of operations, say the operations of joining 0, $1/2$ then $1/2$, $2/3$ then $2/3$, $3/4$ and so on ? Let us describe two operations. (1) Drawing the line from the point 0 to the point 1, and (2) drawing a line from 0 to $1/2$, a line from $1/2$ to $2/3$, a line from $2/3$ to $3/4$, and so on. The first operation has but a single stage, the second is an unfinishable operation by definition, since no *last* stage is defined. What have these operations in common and in what way do they differ? Zeno would persuade us that the first operation is indistinguishable from the second, thereby generating the paradox of a finished operation being identical with an unfinishable one. In drawing a line from 0 to 1 we have certainly drawn a line from 0 to $1/2$, and a line from $1/2$ to $2/3$ and a line from $2/3$ to $3/4$, and so on, so that by carrying out the first operation, there is no stage of the second operation that is unfinished. The fallacy in this argument is concealed beneath a dual usage of the expression "a line is drawn from a point A to a point B ". In describing the first operation, and in describing each of the stages of the second operation, the expression "a line is drawn from a point A to a point B " means a line whose endpoints are A and B, i.e., a physical mark, a stroke, terminating at A and B. The first operation consists in drawing a stroke from 0 to 1. The second operation consists in drawing successively strokes from 0 to $1/2$, from $1/2$ to $2/3$, from $2/3$ to $3/4$ and so on. Yet when we say that the stroke from 0 to 1 is also a stroke from 0 to $1/2$ (or $1/2$ to $2/3$, etc.) we have now changed the meaning of the expression "a stroke from A to B" for the stroke from 0 to 1 cannot be said to consist in strokes from 0 to $1/2$, from $1/2$ to $2/3$, etc. What constitutes a *stage* of the second operation, the termination of a stroke at one of the points $1/2$, $2/3$, $3/4$ $\cdots$ is precisely what is lacking in the first operation.

The resolution of Zeno's paradox may be expressed by saying that Zeno confuses a literal and metaphorical use of the expression "moving from one point to another". In the literal sense of this expression motion is change of the relative positions of physical objects, and a 'point' is a physical object; in this sense motion from one point to another passes through but a finite number of 'points', physical objects isolated and specified on the route. We may specify as many such objects as we please, but what we specify will have a number. The metaphorical use of the expression "moving from one point to another" gives this expression the sense of "a variable increasing from one value to another". As the variable x increases from 0 to 1 it passes through the values $1/2$, $2/3$, $3/4$ $\cdots$ and so on, and therefore, seemingly an endless succession of events is completed. But the expression "as x increases from 0 to 1 it passes through the values $1/2$, $2/3$, $3/4$ $\cdots$ and so on", says only that the function $m/(m+1)$ is one which increases with m, all its values lying in $(0,1)$. And the proof that the function is increasing and that its values lie in $(0,1)$ does not involve the possibility of completing an endless process, for what is proved is just that $m + 1/m + 2$ exceeds $m/(m+1)$ by $1/(m+1)(m+2)$ and that unity exceeds $m/(m+1)$ by $1/(m+1)$, i.e., that $(m+1)^2 = m(m+2) + 1$ and $(m+1) - 1 = m$.

[The remainder of the Introduction to *Constructive Formalism* by Goodstein appears in Chapter 5.G.2.]

Exercises

1. What is Plato's conception of mathematical objects? According to him, what does it mean to say that a circle exists? That a number exists? Give a concise explanation of why you agree or disagree with Plato's views. If Plato is right, how is it that we can use mathematics to build bridges?

2. In what ways does Goodstein view mathematics differently from Plato? Why would Goodstein's resolution of Zeno's paradoxes be unacceptable to a platonist? Is Goodstein's description of the resolution by means of the calculus apt?

3. Mathematicians often view their work as abstractions from experience. Is that viewpoint compatible with Plato's? With Goodstein's? Why?

Further Reading
For more of Plato as well as Aristotle on mathematics see Baum's *Philosophy and Mathematics*. Particularly illuminating are the metaphor of the cave and shadows in *Republic*, VII 514–517, and Socrates' quizzing of the slave boy in *Meno* 82–86, the latter of which is used to demonstrate that we remember mathematical truths rather than invent or discover them.

Abelson, in his article on definitions in *The Encyclopedia of Philosophy*, has a succinct summary of Plato's views as they pertain to the material in this book.

3 Whole Numbers

The whole numbers 1, 2, 3, . . . seem to be fundamental to all mathematics, and so it is with them that we begin.

A. Counting (Ordinal) vs. Quantity (Cardinal)

Counting: How many objects are there?

We point at one figure and say "1", at another and say "2", and at the last and say "3". We count with the words "1", "2", "3"; but then we use "3", the last, as a quantity and say that there are 3 objects.

We can compare cardinal numbers without counting, for example, are there the same number of chairs as people in this room? Pair off the people and chairs, and see if there are any chairs without people or people without chairs.

> It is true that the acts of matching off must be carried out one by one in temporal succession, even if it is simply the process of looking to see that each chair is occupied. Nonetheless, the ordinal *numbers* are not involved because we do not have to keep track of the relative order of which two chairs was first checked: we need only distinguish checked from non-checked. The concept of more or equal is prior to both cardinals and ordinals.
>
> Wang, pp. 59–60

One of the fundamental assumptions of this course will be that we understand how to count and that each of us can continue the sequence 1, 2, 3, ... We know what it means to add 1 and can continue to do so indefinitely. We understand what it means to say that there are 3 objects and that the rectangle in the diagram above is the second one from the left.

We will take the whole numbers 1, 2, 3, ... and 0 as primitive, undefined in terms of any other concepts. We refer to them as whole numbers, counting numbers, or more commonly when we include 0, as the *natural numbers.*

What exactly is this process of adding 1? We can describe it (not define it) in *unary notation* as: the whole number sequence begins with I, the next number is represented as I I, the next as I I I, and for any representation of a number in the sequence, the next number is represented by putting one more stroke on the right-hand side of the previous one.

In Chapter 26 we will reconsider whether we are justified in assuming that the natural number sequence is clear, unequivocal, and understood by us all.

B. Number Is All: √2

So fundamental did the counting numbers appear to Pythagoras and his followers that he declared, "All is number." And we have from Philolaus the dictum, "All things which can be known have number; for it is not possible that without number anything can be either conceived or known."

> The so-called Pythagoreans, having applied themselves to mathematics, first advanced that study; and having been trained in it they thought that the principles of mathematics were the principles of all things. Since of these principles numbers are by nature first, they thought they saw many similarities to things which exist and come into being in numbers ... justice being such and such a modification of numbers, soul and reason another, opportunity still another, and so with the rest, each being expressible numerically. Seeing, further, that the properties and ratios of the musical consonances were expressible in numbers, and that indeed all other things seemed to be wholly modelled in their nature upon numbers, they took numbers to be the whole of reality, the elements of numbers to be the elements of all existing things, and the whole heaven to be a musical scale and a number.
>
> Aristotle, *Metaphysics* i.5.985 b 23 in John M. Robinson, 1968, p. 69

> It had been a fundamental tenet of Pythagoreanism that the essence of all things, in geometry as well as in the practical and theoretical affairs of many, are explainable in terms of *arithmos*, or intrinsic properties of whole numbers or their ratios. The dialogues of Plato show, however, that the Greek mathematical community had been stunned by a disclosure that virtually demolished the basis for the Pythagorean faith in whole numbers. This was the discovery that within geometry itself the whole numbers and their ratios are inadequate to account for even simple fundamental properties.
>
> Boyer, *A History of Mathematics*, p. 79

The diagonal of a square with sides of length 1 is not a ratio of whole numbers. For suppose $\sqrt{2} = p/q$. Suppose also that p/q is in lowest terms; that is, no number divides both p and q. Then $p = \sqrt{2} \cdot q$ and $p^2 = 2 \cdot q^2$. Hence p^2 is even, so p also must be even. Therefore, since p/q is in lowest terms, q is odd. But if $p = 2\,r$, then $(2\,r)^2 = 2 \cdot q^2$ so that $4 \cdot r^2 = 2 \cdot q^2$. Thus $2 \cdot r^2 = q^2$, which means that q^2 is even and hence q is even. This is a contradiction. So $\sqrt{2} = p/q$ must be a false assumption.

> At first it must have appeared natural to the Greeks to assume that all magnitudes of the same kind are commensurable, and that, for example, any two lengths are multiples of the same unit. It follows from this assumption that all points on a line can be represented by rational numbers. The discovery that the square root of 2 is not rational, or in geometric terms, that the diagonal of a square is not commensurable with its sides, made it clear that there are points on a line which are not represented by rational numbers. It remains an unsettled historical question whether the irrationality of $\sqrt{2}$ was discovered by Pythagoras or his immediate pupils, or not long before 400 B.C. In any case, consequences of this discovery were drawn only at the beginning of the fourth century B.C. Eudoxus constructed a general theory of proportion which was adopted by Euclid and further developed by Archimedes. The theory of Eudoxus may be regarded as the beginning of a rigorous theory of irrational numbers. It leads to the question of determining all irrational numbers or all ratios of line segments which are not represented by fractions. Hardy considers the proof of the irrationality of $\sqrt{2}$ as one of two examples of beautiful and significant mathematics. Indeed, the proof is so simple and pure, and the theorem is so full of deep consequences, that it cannot fail to satisfy the desire to find simple keys to bodies of science.
>
> Wang, p. 72

Exercises

1. Describe the process of adding 1 in decimal notation.
2. a. To be sure you understand the proof that $\sqrt{2}$ is not rational, prove that $\sqrt{7}$ is not rational. Then show that there are arbitrarily many irrational numbers.
   b. Construct line segments corresponding to $\sqrt{2}, \sqrt{3}, \sqrt{4}, \sqrt{5}$ using straightedge and compass? (*Hint*: Use Pythagoras' theorem.)
   c. How do we represent irrational numbers in our system of decimal notation?
3. Show that we cannot get $\sqrt{7}$ from the rationals and $\sqrt{2}$ by using addition, subtraction, multiplication, and division. That is, show that there are no rational numbers a, b, c, and d such that $\sqrt{7} = a + b\sqrt{2} / c + d\sqrt{2}$.

4 Functions

Numbers, geometric figures, their properties and relations: all these are static. In this chapter we'll look at how we deal with processes in mathematics.

A. What Is a Function?

1. Black boxes

Here is an explanation of functions which we give to our first-year calculus students. A function is something calculated by a black box.

You put in an object, usually a number, crank the handle, and get out something. What you put in is called the *input*; what you get out is called the *output*. For example, the black box that adds 3 to a number:

We call this a "black box" because we don't care what actually goes on inside; only the inputs and outputs are important. If another black box gives us the same outputs for exactly the same inputs as the "+ 3" black box we'll say it's the same function, even though internally it might be first adding 4 then subtracting 1.

What distinguishes a function from just any black box is that a function *cannot equivocate*. For example, a black box that adds 2 and adds 3 and then tells us to take our choice for the output doesn't calculate a function:

Not a function

For every input there must be exactly one output.

Similarly, a black box that gives square roots isn't a function:

We can, however, convert the square root box into a function by agreeing always to take the nonnegative root as the output, ignoring the negative root. Then for every input there will be exactly one output. From now on that's what we'll mean: for example, $\sqrt{4}$ is the nonnegative number which when squared gives 4.

2. Domains and ranges

We said that a function gives to each input an output. For instance, to every positive real number the function $\sqrt{}$ gives as output the positive square root of that number. If we put in a negative number and crank the handle, nothing happens: we crank and crank fruitlessly.

Depending on what numbers we choose for our source of inputs, the $\sqrt{}$ function will give one output for every input, or will give exactly one output for certain specified inputs, and won't work for others. Sometimes we say that $\sqrt{}$ is a function *on* the real numbers which is *defined* only on the nonnegative real numbers.

We can give a name to all the numbers that are suitable as inputs within the given collection. We call these the *domain*. And we can collect all the outputs and call them collectively the *range*.

Beware: Terminology varies from author to author and even within the same textbook. Sometimes the *domain* is construed as the collection of numbers on which the function is given; for example the real numbers for the $\sqrt{\ }$ function. The words *the range* nearly always mean the collection of all outputs, but some authors inadvertently use them to mean the *codomain*, which is a collection of numbers within which the range can be found. For example, it's sometimes said that the $\sqrt{\ }$ function is a function from the real numbers to the real numbers that is defined for only the nonnegative real numbers.

We indicate the domain, X, and codomain, Y, of a function, f, by writing $f: X \rightarrow Y$. We read this as "f is a function from X to Y". When $X = Y$, we say that "f is a function *on X*".

3. Functions as rules, functions as collections of ordered pairs

a. Here is a definition of "function" taken from an elementary text:

> Let X and Y be two nonempty sets. Then a *function* from X to Y
> is a *rule* that assigns to each element $x \in X$ a unique $y \in Y$.

From this point of view, functions are processes which we can describe. This accords well with much of our experience, since the processes we may wish to model are those we can talk about. A function is not simply an assignment but a method of assignment.

If we take this view seriously then the rule assigning $+3$ to each number is a different function than the rule "$+4$, then -1". We have two functions, not one. We may, if we wish, call them *equivalent* because they match the same inputs to the same outputs. But most people who talk of functions as *assignments* or *rules* are speaking suggestively and don't wish to be taken too seriously, for they would say that the two rules give the same function.

b. In that case we are viewing a function not as an assignment or a rule but simply as a pairing. Actually we don't have a good word that describes the situation which doesn't indicate a process. But we may look at a function as just inputs and outputs so that, as with our black-box description, for any particular matching of inputs to outputs there is exactly one function. This is how functions are usually given in set theory. An example of such a definition is:

> Let X and Y be two nonempty sets. Then a *function* from X to Y
> is a collection of ordered pairs (x, y) where the first element of the pair

is from X, the second is from Y, and if (x,y) and (x,z) belong to the
collection then $y = z$.

The last clause just says that to every element of X there can be at most one element
of Y to which it is paired. That is, for each input there can be at most one output. In
this way functions have become objects; we have replaced the dynamic by the static.

c. The view of functions as collections of ordered pairs is an *extensionalist* view.
We may give lots of different names to a function, but the properties of the function
do not depend on our doing so. Usually this view is also platonist, in that the set of
ordered pairs is understood to exist regardless of whether we ever describe it or not.

The view of functions as rules is *nonextensional* in that it takes the name of
the function, that is, how we describe it, as being an essential property of the
function. Usually this is part of a stronger view, called *nominalism*, that a *name*,
a word or a description, is *all* that a function or any abstract object is.

The same distinctions can be applied to any mathematical "objects":
The properties of the object depend on the name (description) we give it, or
the properties are independent of our naming.

Now we, the authors, believe that viewing functions as ordered pairs is just a
further abstraction from our daily experience with processes. So we think it's all
right to begin our study of functions by viewing them as sets of ordered pairs, which
is how they are commonly viewed in modern mathematics. We can later return to
the less abstract view that the description of a function is an essential part of it.

B. Terminology and Notation
1. The λ-notation

We use variables such as x and y to represent inputs. Then one way to describe a
function is to set out the process explicitly:

$$x \mapsto 3x + 7$$

Here the symbol "$\mapsto$" indicates the assignment. It's a nice notation because it
suggests the dynamic aspect of functions.

If there isn't a standard symbol such as $\sqrt{}$ for the function, we usually name
the function either with a Roman letter, such as f or g, or with a Greek letter,
such as φ or ψ. So we might write $x \mapsto f(x)$.

If we write just $f(x)$ or just $f(x) = 3x + 7$, however, the notation is ambiguous. It's not clear whether we mean the function or whether we are saying: choose
some arbitrary input x and then look at the output for that particular x, which has
the form $3x + 7$. That is, it can mean two things: (1) the function itself (i.e., the
rule), or (2) x represents some particular number and $f(x)$ represents the value of
f applied to that number. The latter is called *the ambiguous value of f*. Compare:
"$3x + 7$ is differentiable" and "$3x + 7$ is less than 2." Similarly, when we write
$f(x) = 7$ do we mean that for some particular value of x, $f(x) = 7$? Or do we
mean the function with constant output 7?

Another context in which it is difficult to distinguish the reading occurs when we have a function of two variables. For instance, we can view the process of adding as requiring two inputs:

input 2, 3

output 5

Here the order of the inputs isn't important. We can first put in 2, then 3, or first 3, then 2, since $3 + 2 = 2 + 3$. But usually the order is important, and part of the rule is the order of inputs: for example, on the real numbers the function $(x, y) \mapsto x - y$. So we will always take the input of a function of several variables to be an ordered collection of numbers. Thus for $(x, y) \mapsto x + y$ we have $(2, 3) \mapsto 5$ and $(3, 2) \mapsto 5$.

We would like to distinguish between

$f(2, 3) = 2 + 3$, a function of two variables

and

$g(2) = 2 + 3$, a function of one variable

We will write $\lambda x(x + 3)$ to indicate we are viewing the function of two variables $(x, y) \mapsto x + y$ as a function of only the first variable, with the second variable held fixed as 3. Similarly, we will write $\lambda x(x + y)$ to mean that we are viewing the function of two variables $(x, y) \mapsto x + y$ as a function of only the first variable with the second variable held fixed. Here y is a *parameter*; that is, it is viewed as held fixed throughout the discussion, although we don't specify what particular value is being used. Depending on what we choose for y, we get a different function. For instance, if $y = 7$ we get the function $\lambda x(x + 7)$. We'll write $\lambda x \lambda y(x + y)$, or simply $\lambda x y (x + y)$, to mean we are viewing addition as a function of two variables. We'll use this λ-notation whenever the context might not make the meaning clear; for example, we can now write $\lambda x (7)$ for the function with constant output 7.

2. One–one and onto functions

Recall that for an assignment to be a function it must assign at most one output to each input. Symbolically, if $f(x) = y$ and $f(x) = z$, then $y = z$. For example, if $f(x) = x^2$ then if both $f(3) = y$ and $f(3) = z$, we must have $y = z = 9$.

For some functions the correspondence works in the other direction, too. Given some output we can find the input it came from because different inputs always give different outputs. That's not true for $\lambda x (x^2)$ on the real numbers because, for example, both 3 and –3 are assigned 9. Given a number in the range, here 9, we can't retrace our steps. We say a function is *1-1* or *one–one* (read "one to one") or *injective* if, symbolically, given $f(x) = z$ and $f(y) = z$, then $x = y$. For example, $\lambda x (x + 3)$ is a 1-1 function on the natural numbers.

Some functions use up all the numbers in the codomain. That is to say, every number in the codomain is the output of some number in the domain. An example is the function $\lambda x (\sqrt{x})$ from the nonnegative real numbers to the nonnegative real numbers: every nonnegative real number is the square root of some nonnegative real number, namely, its own square. Functions for which the range equals the codomain are called *onto* or *surjective*, and if $f: X \to Y$ is onto we say "f is a function from X onto Y". A function that is 1-1 and onto is called a *bijection*.

Here are the archetypal pictures:

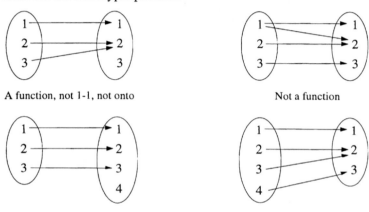

A function, not 1-1, not onto   Not a function

A 1-1 function, not onto   An onto function, not 1-1

3. Composition of functions

If $f: X \to Y$ and g is defined on the range of f, then we can compose g with f. The *composition* is $(g \circ f)(x) \equiv_{\text{Def}} g(f(x))$, as pictured below.

Thus, if $f(x) = 3x + 7$ and $g(x) = 2x^2$, then $g \circ f(x) = 18x^2 + 84x + 98$.

Beware: Some authors write $f \circ g$ instead of $g \circ f$.

Exercises

1. Rewrite each of the following functions on the natural numbers in each of the three types of functional notation described in this chapter.
   a. To each number assign its cube.
   b. To each number assign the number 47.
   c. To each number < 16 assign its square, to each number > 16 assign the number cubed minus 2, and assign 407 to 16 .

2. a. Prove that on the natural numbers $f(x) = 3x + 7$ is 1-1 .
   b. Prove that $\lambda x(x^4 + 2)$ is not 1-1 on the real numbers.

3. Which of the following functions on the real numbers are equal?
   a. $f(x) = x^2 + 2$
   b. $g(x) = 3x^3 - 2$
   c. $x \mapsto 3x^3 + x^2$
   d. $f \circ g$
   e. $g \circ f$
   f. $\lambda x\,(9x^6 - 12x^3 + 6)$
   g. $\lambda x\,(f(x) + g(x))$

5 Proofs

A. What Is a Proof?

What is a proof? How do we recognize when a mathematical statement has been proved? What are the criteria? In what ways is a proof in mathematics different from a proof in a court of law? In "The nature of mathematical proof", R. L. Wilder states his view:

> What is the role of proof? It seems to be only a testing process that we apply to these suggestions of our intuition.
>
> Obviously we don't possess, and probably will never possess, any standard of proof that is independent of time, the thing to be proved, or the person or school of thought using it.

Julia Robinson, a well-known logician, explained to her logic class in 1969:

> A proof is a demonstration that will be accepted by any reasonable person acquainted with the facts.

Most mathematicians do not concern themselves with clarifying precisely the notion of proof, for they do not need to. They know intuitively what is a correct proof and what is not. But where does this intuition come from? It is a result of imitation and correction as we learn mathematics, each generation passing on to the next a way of speaking mathematics, a culture of mathematics. Cultures change, however, and the standards used in proofs now are very different from those used in Euclid's time or in the seventeenth century when the calculus was developed by Newton and Leibniz. Most mathematicians believe the standard is higher now, that we do mathematics better than any previous generation. Certainly we make many distinctions in our work that were never made before. But are our proofs better? To think so places mathematical proofs outside the realm of culture and into a realm of absolute standards. One of the reasons for thinking that there is an absolute standard of proof is that we believe that proofs give us absolute, certain knowledge, a belief we will examine in Section G.1 below.

One thing, though, is essential to keep in mind: a proof is a form of communication. When you write a proof you are trying to convince someone else (or possibly yourself) that one statement follows from some others: if those others are true, then that statement must also be true. This is the case even if you think that proofs must adhere to an absolute platonic ideal, for in that case what you are supposedly communicating is your vision of the platonic ideal proof of the statement.

If proofs are forms of communication, then they are highly specialized forms. A proof in mathematics is different from one in a law court not by virtue of having specialized terms or rigid forms of communication, for law has those too, but by the particular forms of proof which are deemed acceptable.

In order that we can agree on some basic methods, we present a few forms of proof which are fundamental to mathematics. Then in Section G we will return to the question of what a proof is.

B. Induction

1. An example: prove that $1 + 2 + \cdots + n = 1/2 \, n \cdot (n + 1)$.

Proof: $1 = 1/2 \, 1 \cdot (1 + 1)$. This is called the *basis of the induction.*

Suppose that $1 + 2 + \cdots + n = 1/2 \, n \cdot (n + 1)$. This is called the *induction hypothesis.* Then

$$1 + 2 + \cdots + n + (n + 1) = \left[1/2 \, n \cdot (n + 1) \right] + (n + 1).$$
$$\text{So } 1 + 2 + \cdots + n + (n + 1) = 1/2 \, (n^2 + n) + 1/2 \, (2n + 2),$$
$$\text{so } 1 + 2 + \cdots + n + (n + 1) = 1/2 \, (n^2 + 3n + 2),$$
$$\text{so } 1 + 2 + \cdots + n + (n + 1) = 1/2 \, (n + 1) \cdot (n + 2).$$

That is, $1 + 2 + \cdots + n + (n + 1) = 1/2 \, (n + 1) \cdot ((n + 1) + 1)$,
which was to be proved.   ∎

The method is this: we show the statement is true for 1. Next we *assume* it is true for n, an arbitrary but fixed number, and show therefore that it's true for $n + 1$. Then we conclude that it's true for all numbers. Why? It's true for 1; so it's true for 2; since it's true for 2, it's therefore true for 3, and so on.

Those little words "and so on" carry a lot of weight. We believe that the natural numbers are completely specified by their method of generation: add 1, starting at 0 :

0   1   2   3   4   5   6   7 ...

To prove a statement A by induction, we first prove it for some starting point in this list of numbers, usually 1, but just as well 0 or 47. We then establish that we have a method of generating proofs which is exactly analogous to the method of generating natural numbers: if $A(n)$ is true, then $A(n + 1)$ is true. We have the list:

$A(0)$,	if $A(0)$, then $A(1)$,	if $A(1)$, then $A(2)$,	if $A(2)$, then $A(3)$, ...
	so $A(1)$	so $A(2)$	so $A(3)$

Then the statement is true for all natural numbers equal to or larger than our initial point, whether the statement be for all numbers larger than 1 or all numbers larger than 47 .

In essence we have only one idea: a process for generating objects one after another without end, in one case numerals or numbers, and in the other, proofs. We believe induction is a correct form of proof because the two applications of the single idea are matched up. To deny proof by induction amounts to denying that the natural numbers are completely determined by the process of adding 1 or that we can deduce a proposition C from the propositions *if B, then C* and *B*.

2.   Here is an example in which the basis of the induction is neither 0 nor 1 : prove that $1 + 2^n < 3^n$ for $n \geq 2$.

Proof: Note that the statement is false for n $=1$.

    Basis: $1 + 2^2 = 5 < 3^2 = 9$.

    Induction step: Assume for a fixed $n \geq 2$ that $1 + 2^n < 3^n$. Then

$$
\begin{aligned}
1 + 2^{n+1} &= 1 + (2 \cdot 2^n) \\
&= (1 + 2^n) + 2^n \\
&< (1 + 2^n) + (1 + 2^n) \\
&< 3^n + 3^n \quad \text{by the induction hypothesis} \\
&< 3^n + 3^n + 3^n \\
&= 3^{n+1}
\end{aligned}
$$

That is, $1 + 2^{n+1} < 3^{n+1}$, which was to be proved.   ∎

3. We can use induction for any objects that we can number. Here is an example in which we apply induction to a collection of objects, in this case finite sets of points in the plane, where not only one but many different objects can be associated with each natural number.

> Given any collection of n points in the plane no three of which lie on a line, there are exactly $1/_2 n \cdot (n - 1)$ line segments connecting them.

Proof: The smallest number n to which the theorem could apply is 2. Given 2 distinct points on the plane, there is exactly 1 line segment joining them, and $1/_2 \, 2 \cdot (2 - 1) = 1$. So the theorem is true for $n = 2$.

Now suppose it is true for n ; we will show it for $n + 1$. Suppose we have a collection of $n + 1$ points in the plane. Call one of them P. Then the entire collection except for P contains n points and so we can apply the induction hypothesis: there are $1/_2 \, n \cdot (n - 1)$ line segments connecting these. The only other line segments that can be drawn in this collection are those joining P to one of the other n points. There are n such segments. So in total there are

$\left(\frac{1}{2} n \cdot (n-1)\right) + n$ line segments joining points in the entire collection. And

$$
\begin{aligned}
\left[\tfrac{1}{2} n \cdot (n-1)\right] + n &= \left(\tfrac{1}{2} n \cdot (n-1)\right) + \tfrac{1}{2} 2 n \\
&= \tfrac{1}{2}(n^2 + n) \\
&= \tfrac{1}{2}(n+1) \cdot n
\end{aligned}
$$

which was to be proved. ∎

4.   The label 'mathematical induction' though well established, is misleading. Outside of mathematics, induction (here called 'ordinary induction' to avoid confusion) means a process of generalization on the basis of properties of a random or specially selected sample. ... [the] conclusion, though reasonable, is precarious and must be regarded as no more than probably true. It can be refuted by a single counterexample. By contrast, the conclusion of a mathematical induction is quite certain, if no mistake has been made in the reasoning: there can be no question of its being only probably true, or of the possibility of failure in exceptional cases. Mathematical induction is not a special form of ordinary induction; it is a variety of strict proof: a demonstration or deduction.

Max Black, in the *Encyclopaedia Americana*, vol. 15, 1971, p. 100

C. Proof by Contradiction (*Reductio ad Absurdum*)

The method of proof by contradiction is to assume that the conclusion is false and hence that its negation is true. From that we derive a contradiction. Therefore, the conclusion must be true. An example is the proof in Chapter 3.B that $\sqrt{2}$ is not rational.

This method is based on two assumptions: (1) If a statement implies something false, it must be false, and (2) For every statement, either it or its negation is true. The latter is called *the law of excluded middle* or *tertium non datur* (no third way is given) because it asserts that there is no third choice between true and false.

D. Proof by Construction

To show something exists we construct it. For example, in euclidean plane geometry without the parallel postulate we can prove that given any line *l* and a point *P* not on *l* there is *at least* one line *m* through *P* which is parallel to *l*:

Given *l* and *P* not on *l*, we construct the perpendicular to *l* from *P*, meeting *l* at point *Q*. Then we construct a perpendicular to the line *PQ* at *P*, calling it *m*. Both *m* and *l* are perpendicular to *PQ* and so they are parallel. Thus we have the desired parallel line.

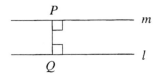

The construction is described anthropomorphically and can be fleshed out by describing precisely how to construct perpendiculars. But lines are not something that can be drawn: lines have no breadth nor width. Constructions in mathematics, no matter how anthropomorphic they may sound, are theoretical constructions. We do not exhibit a parallel line in the same manner as we exhibit a woman over 2 meters tall.

E. Proof by Counterexample

Proof by counterexample is related to proof by construction. To show that a proposition about some class of objects is not true, just "exhibit" one that fails to have the property. For example, to prove the proposition "All cats can swim" is false, we only need to find a cat that doesn't swim and throw it in a lake. To show that "All primes are odd" is false, we merely need to exhibit the number 2. The answer to Exercise 2.b of Chapter 4 should be a proof by counterexample.

F. On Existence Proofs

We can show that something exists by a mathematical construction. But may we not also use *reductio ad absurdum?*

For a finite collection a proof by contradiction can be transformed into a proof by construction. Since we have convinced ourselves that there must be some object satisfying the condition, we can "look" through the finite collection, testing each object in turn until we find the one we want.

When we want to show something exists in some potentially infinite collection, however, the situation is quite different. A proof by contradiction may give us no information about how to actually produce the object we are looking for. Consider the following proof that there are irrational numbers a, b such that a^b is rational:

Consider $\sqrt{2}^{\sqrt{2}}$. Either it is rational or not. If it is, then we are done. If not, take $a = \sqrt{2}^{\sqrt{2}}$ and $b = \sqrt{2}$. Then $a^b = (\sqrt{2}^{\sqrt{2}})^{\sqrt{2}} = 2$.

In Section G.2 below Goodstein discusses whether this is a legitimate proof.

G. The Nature of Proof: Certainty and Existence
(Optional)

1. From "Mathematical proofs: the genesis of reasonable doubt" by Gina Bari Kolata

Do proofs in mathematics guarantee certain knowledge? In the following extract Gina Bari Kolata reviews some recent work that makes us doubt that.

Investigators are finding that even theoretically decidable questions may have proofs so long that they can never be written down, either by humans or by computers.

To circumvent the problem of impossibly long proofs, Michael Rabin of the Hebrew University in Jerusalem proposes that mathematicians relax their definition of proof. In many cases it may be possible to "prove" statements with the aid of a computer if the computer is allowed to err with a predetermined low probability. Rabin demonstrated the feasibility of this idea with a new way to quickly determine, with one chance in a billion of being wrong, whether or not an arbitrarily chosen large number is a prime. Because Rabin's method of proof goes against deeply ingrained notions of truth and beauty in mathematics, it is setting off a sometimes heated controversy among investigators.

Rabin became convinced of the utility of a new definition of proof when he considered the history of attempts to prove theorems with computers. About 5 years ago, there was a great deal of interest in this way of proving theorems. This interest arose in connection with research in artificial intelligence and, specifically, in connection with such problems as designing automatic de-bugging procedures to find errors in computer programs. Researchers soon found, however, that proofs of even the simplest statements tend to require unacceptable amounts of computer time. Rabin believes that this failure at automatic theorem proving may be due to the inevitably great lengths of proofs of many decidable statements rather than to a lack of ingenuity in the design of the computer algorithms.

About 4 years ago, Albert Meyer of the Massachusetts Institute of Technology demonstrated that computer proofs of some arbitrarily chosen statements in a very simple logical system will necessarily be unfeasibly long. The system consists of sets of integers and one arithmetic operation—the addition of the number 1 to integers. It had long been known that any statement in this logical system can be proved true or false with a finite number of steps, but Meyer showed that this number of steps can be an iterated exponential, that is, an exponential of an exponential of an exponential, and so on. A statement of length n can require [a proof that uses]

$$2^{2^{2^{2^{\cdot^{\cdot^{\cdot}}}}}}$$

steps, in which the number of powers is proportional to n. ...

[Meyer and Stockmeyer] defined "completely impossible" as requiring a computer network of 10^{123} components which, according to Meyer, is an estimate of the number of proton-sized objects that would densely fill the known universe. Then, they showed that in order to [be able to] prove an[y] arbitrary statement consisting of 617 or fewer symbols, a computer would require 10^{123} components.

The problem with proofs, Rabin decided, is the demand that they be correct. Yet humans constantly make errors in mathematics and all other endeavors. Perhaps because of this, humans who solve problems tend to finish their tasks, whereas computers often stop for lack of time. ...

Rabin found that if n is not a prime, at least half the integers between 1 and n will fail [a particular] test.* Thus if some number between 1 and n is chosen at random and tested, there is at least a $\frac{1}{2}$ chance it will fail the test if n is not a prime. If two numbers are chosen at random and tested, there is at least a $\frac{3}{4}$ chance that one of them will fail if n is not a prime. If 30 numbers are chosen at random, there is at least a $1 - (\frac{1}{2})^{30}$ chance that one will fail the test if n is not a prime. The chance that 30 randomly chosen numbers between 1 and n all pass the test and that n is not a prime, then, is only $(\frac{1}{2})^{30}$ or 1 in 1 billion. This probabilistic method involves the testing of relatively few integers. The number of integers tested is independent of the size of n, but does depend upon what chance of being wrong is risked.

Rabin's probabilistic test is far more rapid than exact tests. Exact tests take so long that the only numbers larger than about 10^{60} that have been tested are of special forms.** Rabin can test numbers of that size in about 1 second of computer time. As an example, he and Vaughn Pratt of the Massachusetts Institute of Technology showed that $2^{400} - 593$ passes his test and thus is a prime "for all practical purposes." ...

Typical of the reactions of many mathematicians is that of one who said he does not accept a probabilistic method of proof because the "glory of mathematics is that existing methods of proof are essentially error-free." Ronald Graham of Bell Laboratories in Murray Hill and others reply that they have more confidence in results that could be obtained by probabilistic methods such as Rabin's prime test than in many 400-page mathematical proofs. ...

Graham is concerned that long and involved proofs are becoming the norm rather than the exception in mathematics, at least in certain fields such as

* João Meidanis has pointed out to us that R. Solovay and V. Strassen ("A fast Monte-Carlo test for primality," *SIAM J. Comput.* 6, 1977, pp. 84–85; erratum, 7, 1978, p. 118) showed that when n is composite at least half of the integers between 1 and n will fail the test. Rabin ("Probabilistic algorithm for testing primality," *J. Number Theory*, 12, 1980, pp. 128–138) showed that the test can be strengthened so that only one-fourth of the integers between 1 and n will pass it when n is composite. The method described here as Rabin's method is known as the Solovay–Strassen algorithm. Rabin's algorithm would require half of the tests for the same chance of error. —Ed.

** Nowadays there are exact primality tests that can handle numbers of 60 decimal digits in an average time of about 10 seconds and of 200 decimal digits in about 10 minutes (H. Cohen and H.W. Lenstra, Jr. "Primality tests and Jacobi sums," *Math. Comput.* 42, 1984, pp. 297–330). —Ed.

group theory. ... He and Paul Erdös believe that already some of the long proofs being published are at the limit of the amount of information the human mind can handle. Thus Graham and others stress that verification of theorems by computers may necessarily be part of the future of mathematics. And mathematicians may have to revise their notions of what constitutes strong enough evidence to believe a statement is true.

2. The Introduction to *Constructive Formalism* by R. L. Goodstein (Concluded)

Here Goodstein argues that a proof by contradiction can never justify the existence of anything: the only legitimate proofs of existence are those that exhibit the object.

The infinitude of primes. We come now to the third question "Is there a prime number greater than 10^{10}"? Consider first the question: "Is there a prime number between 10^{10} and $10^{10} + 10$"? The nine numbers $10^{10} + 1$, $10^{10} + 2$, $10^{10} + 3$, $10^{10} + 4$, $10^{10} + 5$, $10^{10} + 6$, $10^{10} + 7$, $10^{10} + 8$, $10^{10} + 9$, can be tested to find whether or not they are prime, that is to say, each of the numbers may be divided in turn by the numbers 2, 3, 4, 5, up to 10^5 and if one of the nine numbers leaves a remainder not less than unity for each of the divisions then that number is prime; if however each of the nine numbers leaves a zero remainder for some division then none of the nine numbers is prime. In the same way we can test whether any of the numbers between $10^{10} + 10$ and $10^{10} + 20$ is prime, and, of course, the test is applicable to any finite series (i.e. a series in which the *last* member is given). Thus the question "is there a prime number between *a* and *b*" may be decided one way or the other in a specifiable number of steps, depending only upon *a* and *b*, whatever numbers *a* and *b* may be. When, however, we ask whether there is a prime number greater than 10^{10} the test is no longer applicable since we have placed no bound on the number of experiments to be carried out. However many numbers greater than 10^{10} we tested, we might not find a prime number and yet should remain *always* unable to say that there was no prime greater than 10^{10}. We might, in the course of the experiment, chance upon a prime number, but unless this happened the test would be inconclusive. To show that the test can really be decisive it is necessary that we should be able in some way to limit the number of experiments required, and this was achieved by Euclid when he proved that, for each value of *n*, the chain of numbers from *n* to $n! + 1$ inclusive, contains at least one prime number. [$n!$ is the product of the whole numbers from 1 to *n* inclusive.] The underlying ideas of this proof are just that $n! + 1$ leaves the remainder unity when divided by any of the numbers from 2 to *n*, and that the *least* number, above unity, which divides any number is necessarily prime (every number has a divisor greater than unity, namely, the number itself, and the least divisor is prime since its factors will also divide the number and so must be unity or the least divisor itself); thus the *least* divisor (greater than unity) of $n! + 1$ is prime and greater than *n*. What Euclid's proof accomplished is not the discovery or specification of a prime number but the construction of a function whose values are prime numbers. We shall have further occasion to observe how often mathematics answers the question "is there a number with such and such properties" by *constructing* a function; the manner and kind of such constructions will form the subject of later considerations.

When we turn to the question concerning the existence of a prime pair greater than 10^{10} we are faced with the *endless* task of testing, one after the other, the primes great than 10^{10}, of which, as we have seen, we can determine as many as we please, to find whether there are two primes which differ by 2. In this instance no function has been constructed whose values form prime pairs, and there is no way of deciding the question negatively. We have asked a question—if *question* it be—to which there is no possibility of answering *no* and to which the answer *yes* could be given only if we *chanced* to find, in the course of the endless task of seeking through a succession of primes, a pair of primes which differed by 2. The formalists maintain that we can conceive of this endless task as completed and accordingly can say that the sentence "there is a prime pair greater than 10^{10}" must be either true or false; to this constructivists reply that a "completed endless task" is a self-contradictory concept and that the sentence "there is a prime pair greater than 10^{10}" may be true but could never be shown to be false, so that if it be a defining characteristic of sentences that they be either true or false (the principle of the excluded middle) then "there is a prime pair greater than 10^{10}" is no sentence. This dilemma has led some constructivists to deny the principle of the excluded middle, which means they have changed the definition of "sentence", others to retain the principle, and, albeit unwillingly, reject the unlimited existential proposition, whilst the formalist retains both the principle of excluded middle and the unlimited existential proposition together with an uneasy preoccupation with the problem of freedom-from-contradiction. The real dispute between formalists and constructivists is not a dispute concerning the legitimacy of certain methods of proof in mathematics; the constructivists deny and the formalists affirm the possibility of completing an endless process.

Exercises

1. Prove by induction: $1^2 + 2^2 + \cdots + n^2 = \frac{1}{6} n \cdot (n + 1) \cdot (2n + 1)$.

2. Prove by induction: Given any collection of *n* points in the plane no three of which lie on a line, there are exactly $\frac{1}{6} n \cdot (n - 1) \cdot (n - 2)$ triangles that can be formed by the line segments joining the points.

3. We will prove that in every finite collection of natural numbers all of the numbers are equal, using induction on the number of natural numbers in a collection:

    The statement is true for any collection with just one natural number, a, for $a = a$.

    Now suppose it is true for any collection of *n* natural numbers.

    Let $a_1, a_2, \ldots, a_n, a_{n+1}$ be any collection of $n + 1$ natural numbers.

By induction hypothesis, $a_1 = a_2 = \cdots = a_n$. But also we have $a_2 = \cdots = a_n = a_{n+1}$, because here, too, there are only n numbers. And so $a_1 = a_2 = \cdots = a_n = a_{n+1}$.

What is wrong with this "proof"?

4. Prove by induction the *Fundamental Theorem of Arithmetic*:
   Any natural number ≥ 2 can be expressed as a product of primes
   in one and only one way, except for the order of the primes.

5. The following questions refer to the article by Gina Bari Kolata in SectionG.1.
   a. What is the difference between an error in a proof done by a computer and an
      error in a proof written by a mathematician? Why do we sometimes accept
      proofs in mathematics even though they may have "trivial" errors?
   b. What is wrong with the following statement as a description of the method of
      testing for primes: "In many cases it may be possible to "prove" statements
      with the aid of a computer if the computer is allowed to err with a predeter-
      mined low probability"?
   c. Is the following assertion accurate: "The glory of mathematics is that existing
      methods of proof are essentially error-free"?

6. a. In his "Introduction", Goodstein sketches a proof of Euclid's Theorem that
      given any prime p we can find another prime between p and $p! + 1$. Write
      up that proof in mathematical notation filling in all the details.
   b. Prove that there is not an unlimited number of primes separated by 3: that is,
      show that there is some number n such that there are no numbers p and
      $p + 3$ greater than n that are prime. Does this mean that we have completed
      an endless task of checking all primes? Why?

Further Reading

Another discussion of the use of computers in proofs is "The philosophical implications of
the four-color problem" by E.R. Swart.

R.C. Buck in his paper "Mathematical induction and recursive definitions" has many
interesting examples of induction; they are also of interest for Chapter 11.

6 Infinite Collections?

In the latter part of the nineteenth century Georg Cantor devised a theory of sets in which he introduced infinite collections into mathematics. He was originally motivated by problems about trigonometric series. The subject, however, soon took on a life of its own, and in the hands of Cantor and others it was used to provide foundations for the calculus. Integers were defined as equivalence classes of ordered pairs of natural numbers, rationals as equivalence classes of ordered pairs of integers, and real numbers as *infinite collections* of rationals. Roughly, a real number is defined as a set of rationals which is bounded below but has no least element. In another version, a real number is said to be an infinite equivalence class of all sequences of rationals which "have the same limit".

Since that time infinite collections have become standard in modern mathematics, some even say essential. We don't intend to go into why that's the case or explain the details of the "construction" of the real numbers from the natural numbers. All we're going to do here is shake hands with some infinite collections so we'll know what people are talking about.

A. How Big Is Infinite?

Let's assume for now that we can collect into one set all the natural numbers. We'll call that set $\mathbf{N} = \{\ 0, 1, 2, 3, 4, \dots\ \}$.

Let's go further and say that we understand what it means to collect all the integers together, $\mathbf{Z} = \{\ \dots, -3, -2, -1, 0, 1, 2, 3, \dots\ \}$. And we'll also collect all the rationals together, naming that collection $\mathbf{Q}$. It's a little hard to suggest the list of elements in $\mathbf{Q}$, but you know what we mean. Don't you?

Finally, we'll call the collection of all real numbers $\mathbf{R}$. And from here on we'll use the words "collection" and "set" interchangeably.

The most fundamental question that occurs (at least to us) is: Just how big are these sets? In particular, is there one level of infinity or more?

To answer that question we must decide what it means to say that two infinite sets have the same number of elements (a platonist would say "discover" instead of "decide"). Let's recall how we can determine if two finite sets have the same number of elements. We can count both; if we get the same number each time then, assuming we've made no mistake, they have the same number of elements. And we can even say what that number is.

Or we can proceed in what Wang called a more fundamental manner and match off the elements of one set against those of the other; for instance, matching chairs to people in a room. If after having paired off as many as we can there are some unpaired elements in one collection and none unpaired in the other, then the one with unpaired elements is larger than the other. This pairing is a process, so we can use the language of functions to describe it:

> For *finite* sets: two sets have the same number of elements
> if and only if there is a 1-1, onto function from one to the other.

Depending on whether you take matching or counting as fundamental, either this is a fundamental, irreducible fact about finite sets, or it is the most basic theorem we can prove about them.

Since we have no intuition about infinite collections to guide us in determining whether two infinite sets have the same number of elements, we will extrapolate from our experience with finite sets and make the following *definition*:

> For *infinite* sets: two sets have the same number of elements
> if there is a 1-1, onto function from one to the other.

We write $A \approx B$ to mean that there is a one-one, onto function from A to B, and in that case we say that A is *equivalent* (or *equipollent*) to B.

Do all infinite sets have the same number of elements?

B. Enumerability: The Rationals Are Countable

The simplest infinite collection is **N**. We say that a collection of objects A is *countable* (or *enumerable* or *denumerable*) if it is a finite (possibly empty) set or is equivalent to **N**. That is, for some n there is a bijection between the natural numbers less than n and A or a bijection between all of **N** and A. The bijection is called an *enumeration*. If a collection is countable and not finite, we sometimes stress that by saying that it is *countably infinite*.

1. The collection of all even numbers is enumerable:

0	1	2	3	4	$\cdots$	$n \cdots$
↓	↓	↓	↓	↓	$\cdots$	↓
0	2	4	6	8	$\cdots$	$2n \cdots$

It is characteristic of an infinite collection that it can be put into 1-1 correspondence with a part of itself. Indeed, some mathematicians use this as a definition of what it means to be infinite.

2. Since the rationals are dense in the real line, it seems there should be more of them than natural numbers. But that's just an illusion: We can enumerate **Q**. Here is (a variation on) *Cantor's tour of the rationals.*

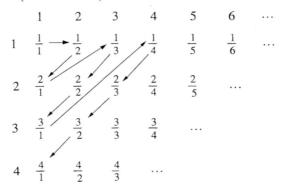

From the picture it seems clear that we can follow the path, skipping any fraction we've come to before, thus obtaining an enumeration of the positive rationals.
If the positive rationals are enumerable, then so are all the rationals (Exercise 3).

For a more formal proof, we first show that the set **P** of all ordered pairs of natural numbers is countable. We define

$$J(m, n) = \tfrac{1}{2}[(m + n)(m + n + 1)] + m$$

This is the ordering in the picture except that we also allow pairs with first or second component 0. The order is according to increasing $m + n$, where if the sum $m + n$ is the same for two pairs then they are arranged with first components in increasing order. In Exercise 4 you're asked to show that J is 1-1 and onto.

Now the set of positive rationals, **Q$^+$**, is equivalent to a subset of **P** via $a \mapsto (p, q)$, where $a = \dfrac{p}{q}$ in lowest terms (and $q = 1$ is allowed). Finally, we leave to you Exercise 5.d that every subset of a countable set is countable.

C. The Reals Are Not Countable

The rationals are countable and are dense on the real line. So surely the reals should be countable, too? Don't count on it.

Give the name $[0,1)$ to the interval of reals ≥ 0 and < 1. Then $[0,1)$ is equivalent to the reals ≥ 0 via the bijection $g(x) = \dfrac{x}{1-x}$, so you can show that $[0,1) \approx \mathbf{R}$ (Exercise 8).

Now we'll show that $[0,1)$ isn't countable. We'll represent the numbers in $[0,1)$ by decimal expansions, $x = .x_0 x_1 \cdots x_n \cdots$. Because, for example, $.1 = .099\ldots9\ldots$ we ensure the uniqueness of each representation by requiring that none of our decimals ends in a tail of 9's.

Now suppose by way of contradiction that $N \approx [0,1)$. So there is an enumeration of $[0,1)$:

$$
\begin{aligned}
a_0 &= .a_{00}\ a_{01}\ a_{02}\ \cdots \\
a_1 &= .a_{10}\ a_{11}\ a_{12}\ \cdots \\
a_2 &= .a_{20}\ a_{21}\ a_{22}\ \cdots \\
&\ \vdots \\
a_n &= .a_{n0}\ a_{n1}\ a_{n2}\ \cdots\ a_{nn}\ \cdots
\end{aligned}
$$

Define $b = .b_0\ b_1\ \cdots\ b_n\ \cdots$ where $b_n = \begin{cases} a_{nn} + 1 & \text{if } a_{nn} < 8 \\ a_{nn} - 1 & \text{if } a_{nn} \geq 8 \end{cases}$

Then $0 \leq b < 1$ and b does not end in a tail of 9s. But it can't be on our list since it disagrees with each a_n on the diagonal! So there is no enumeration of $[0,1)$.

Thus, there are at least two levels of infinity.

D. Power Sets and the Set of All Sets

But there are more levels of infinity still, an infinity of them.

Given a set A, we define *the power set of A*, written $\mathcal{P}(A)$, to be the set of all subsets of A. We claim that $A \neq \mathcal{P}(A)$.

Suppose by way of contradiction that $A \approx \mathcal{P}(A)$. Let $f: A \to \mathcal{P}(A)$ be onto. Denote $f(a) = A_a$. Let B be all those elements x of A such that $x \notin A_x$. Then B is a subset of A, and hence $B = A_b$ for some b. But then we have:

If $b \in B$, then by definition $b \notin A_b$, so $b \notin B$.
If $b \notin B$, then by definition $b \in A_b$, so $b \in B$.

Therefore, we have a contradiction. So $A \neq \mathcal{P}(A)$.

Now let's consider the set of *all* sets, which we'll call S. We've just shown that the power set of S is "bigger" than S. But that's paradoxical since by definition S is the most inclusive of all sets. This is called *Cantor's antinomy* and was known to Cantor as early as 1899.

If set theory and infinite sets are to be the underpinnings for a solid foundation of the calculus, what are we to make of Cantor's antinomy and Russell's paradox (Chapter 1.A.5)? Can we really trust our work with infinite sets when our intuition is likely to lead us astray, as with $Q \approx N$, and when paradoxes, contradictions, are lurking in the corners?

Exercises

1. Show that $\approx$ is an equivalence relation. That is, show for all sets A, B, C
   a. $A \approx A$.
   b. If $A \approx B$, then $B \approx A$.
   c. If $A \approx B$ and $B \approx C$, then $A \approx C$.

2. Show that the following are countable:
   a. The collection of all odd natural numbers
   b. The integers
   c. The collection of all primes

3. Suppose that $f : \mathbf{N} \to \mathbf{Q}^+$ is 1-1 and onto. Using f exhibit $g : \mathbf{N} \to \mathbf{Q}$ that is 1-1 and onto.

4. Prove that J is 1-1 and onto by showing that

$$J(m,n) = \begin{cases} \text{the number of pairs } (x \, , y \,) \text{ such that} \\ x + y < m + n \quad \text{or} \quad (x + y = m + n \quad \text{and } x < m) \end{cases}$$

   (*Hint:* See Chapter 5.B.1.)

5. Show that the following are denumerable:
   a. All triples of natural numbers.
   b. All n-tuples of natural numbers for a fixed n.
   † c. The collection of all n-tuples of natural numbers for all n.
      (*Hint*: Divide $\mathbf{N}$ into countably many countably infinite sets by taking the nth set for $n \geq 1$ to be all those numbers divisible by 2^{n-1} and no higher power of 2.)
   d. Every subset of a denumerable set. (*Hint:* It's immediate for finite subsets. Otherwise pick it out from an enumeration of the whole set.)

6. Let A be any countable collection, which we shall call the *alphabet*; for example, A could be the letters of the Roman alphabet or it could be the rationals. Define a *word* to be any concatenation (finite sequence) of the objects in A, for example, *abaabx* or $\frac{1}{2} \, 0 \, \frac{1}{2} \, \frac{1}{2}$. Show that the collection of words over A is countable.

7. We say that a real number is *algebraic* if it satisfies an equation
   $a_n x^n + a_{n-1} x^{n-1} + \cdots + a_1 x + a_0 = 0$ where $n \geq 1$, $a_n \neq 0$, and all the coefficients are rational. Otherwise we call it transcendental. Show that the set of algebraic numbers is countable. Conclude that there are transcendental numbers. Evaluate this existence proof by Goodstein's standards (Chapter 5.G.2).

8. a. Show that the function $g : [0,1) \to$ the nonnegative real numbers given by $g(x) = \frac{x}{1-x}$ is 1-1 and onto.
   b. Prove that $[0,1) \approx \mathbf{R}$.

9. Show that there are the same number of points on the real line as there are on the plane. (*Hint:* Show that $[0,1) \approx$ the unit square less the top and right-hand edges by interweaving the binary expansions of x and y for the point (x,y).)
   What happened to our concept of dimension?

10. Compare the proof that $A \neq \mathcal{P}(A)$ with the Liar Paradox and Russell's set theory paradox.

Further Reading

Cantor's original writings on this subject, *Transfinite Numbers*, are still some of the clearest and most interesting.

A very good simple introduction to set theory is *Set Theory* by Seymour Lipschutz. For a fuller presentation see Hausdorff's *Set Theory*. For discussions about the paradoxes of set theory we recommend *Foundations of Set Theory* by Fraenkel, Bar-Hillel, and Levy.

7   Hilbert "On the Infinite"

(Optional)

The strangeness of the results about sizes of infinities and the confusion engendered by Russell's paradox led many mathematicians at the beginning of the twentieth century to question the legitimacy of using infinite collections in mathematics. A similar situation had occurred in the previous century when Bolyai and Lobachevsky developed non-euclidean geometries. They showed that one could add the denial of Euclid's axiom on parallels to his other axioms and have a geometry that, although it seemed odd and contradicted the experience of euclidean geometry, nonetheless appeared to have an internal consistency. Eventually it was shown by Beltrami, Klein, and Poincaré that if euclidean geometry was free from contradiction, then so, too, were the geometries of Bolyai and Lobachevsky. Their method was to exhibit a model of the new geometries within euclidean geometry. Thus, the new geometries were as secure as euclidean geometry, whose consistency was not in doubt.

Hilbert played a major role in the formalization of these geometries. In 1898–1899 in his book *Foundations of Geometry* he presented an axiomatization of plane geometry which started with a core of axioms to which could be added either Euclid's parallel axiom or a form of its denial due to Riemann. He then proved sufficient theorems in his formal system to show that the two sorts of geometry could each be fully characterized by his axiomatizations.

Later, Hilbert wished to carry over this approach to justify the use of the infinite in mathematics. Various axiomatizations of set theory were available by the 1920s. The difficulty was to show that at least one of those axiomatizations was free from contradiction. In the paper presented here, Hilbert reviews physical theories of the world and concludes that we have no reason to believe there is anything in the world which corresponds to an infinite collection. Thus we cannot justify the axiomatizations by a model.

How then did Hilbert hope to justify the infinite in mathematics?

Warning: The paper is long and it's difficult to grasp it all on the first reading. You'll want to come back to it after we've studied how Hilbert's program was formalized and whether it was successful. Nonetheless, you should be able to answer the exercises at the end of the chapter.

From "On the Infinite" by David Hilbert *

As a result of his penetrating critique, Weierstrass has provided a solid foundation for mathematical analysis. By elucidating many notions, in particular those of minimum, function, and differential quotient, he removed the defects which were still found in the infinitesimal calculus, rid it of all confused notions about the infinitesimal, and thereby completely resolved the difficulties which stem from that concept. If in analysis today there is complete agreement and certitude in employing the deductive methods which are based on the concepts of irrational number and limit, and if in even the most complex questions of the theory of differential and integral equations, notwithstanding the use of the most ingenious and varied combinations of the different kinds of limits, there nevertheless is unanimity with respect to the results obtained, then this happy state of affairs is due primarily to Weierstrass's scientific work.

And yet in spite of the foundation Weierstrass has provided for the infinitesimal calculus, disputes about the foundations of analysis still go on.

These disputes have not terminated because the meaning of the *infinite*, as that concept is used in mathematics, has never been completely clarified. Weierstrass's analysis did indeed eliminate the infinitely large and the infinitely small by reducing statements about them to [statements about] relations between finite magnitudes. Nevertheless the infinite still appears in the infinite numerical series which defines the real numbers and in the concept of the real number system which is thought of as a completed totality existing all at once.

In his foundation for analysis, Weierstrass accepted unreservedly and used repeatedly those forms of logical deduction in which the concept of the infinite comes into play, as when one treats of *all* real numbers with a certain property or when one argues that *there exist* real numbers with a certain property.

Hence the infinite can reappear in another guise in Weierstrass's theory and thus escape the precision imposed by his critique. It is, therefore, *the problem of the infinite* in the sense just indicated which we need to resolve once and for all. Just as in the limit processes of the infinitesimal calculus, the infinite in the sense of the infinitely large and the infinitely small proved to be merely a figure of speech, so too we must realize that the infinite in the sense of an infinite totality, where we still find it used in deductive methods, is an illusion. Just as operations with the infinitely small were replaced by operations with the finite which yielded exactly the same results and led to exactly the same elegant formal relationships, so in general must deductive methods based on the infinite be replaced by finite procedures which yield exactly the same results;

* Text of an address delivered June 4, 1925, before a congress of the Westphalian Mathematical Society in Münster, in honor of Karl Weierstrass. Translated by Erna Putnam and Gerald Massey from *Mathematische Annalen* (Berlin) no. 95, pp. 161-190.

i.e., which make possible the same chains of proofs and the same methods of getting formulas and theorems.

The goal of my theory is to establish once and for all the certitude of mathematical methods. This is a task which was not accomplished even during the critical period of the infinitesimal calculus. This theory should thus complete what Weierstrass hoped to achieve by his foundation for analysis and toward the accomplishment of which he has taken a necessary and important step.

But a still more general perspective is relevant for clarifying the concept of the infinite. A careful reader will find that the literature of mathematics is glutted with inanities and absurdities which have had their source in the infinite. For example, we find writers insisting, as though it were a restrictive condition, that in rigorous mathematics only a *finite* number of deductions are admissible in a proof—as if someone had succeeded in making an infinite number of them.

Also old objections which we supposed long abandoned still reappear in different forms. For example, the following recently appeared: Although it may be possible to introduce a concept without risk, i.e., without getting contradictions, and even though one can prove that its introduction causes no contradictions to arise, still the introduction of the concept is not thereby justified. Is not this exactly the same objection which was once brought against complex-imaginary numbers when it was said: "True, their use doesn't lead to any contradictions. Nevertheless their introduction is unwarranted, for imaginary magnitudes do not exist"? If, apart from proving consistency, the question of the justification of a measure is to have any meaning, it can consist only in ascertaining whether the measure is accompanied by commensurate success. Such success is in fact essential, for in mathematics as elsewhere success is the supreme court to whose decisions everyone submits.

As some people see ghosts, another writer seems to see contradictions even where no statements whatsoever have been made, viz., in the concrete world of sensation, the "consistent functioning" of which he takes as special assumption. I myself have always supposed that only statements, and hypotheses insofar as they lead through deductions to statements, could contradict one another. The view that facts and events could themselves be in contradiction seems to me to be a prime example of careless thinking.

The foregoing remarks are intended only to establish the fact that the definitive clarification of *the nature of the infinite*, instead of pertaining just to the sphere of specialized scientific interests, is needed for *the dignity of the human intellect* itself.

From time immemorial, the infinite has stirred men's *emotions* more than any other question. Hardly any other *idea* has stimulated the mind so fruitfully. Yet no other *concept* needs *clarification* more than it does.

Before turning to the task of clarifying the nature of the infinite, we should first note briefly what meaning is actually given to the infinite. First let us see what we can learn from physics. One's first naive impression of natural events and of matter is one of permanency, of continuity. When we consider a piece of metal or a volume of liquid, we get the impression that they are unlimitedly divisible, that their smallest parts exhibit the same properties that the whole does. But wherever the methods of investigating the physics of matter have been sufficiently refined, scientists have met divisibility boundaries which do not result from the shortcomings of their efforts but from the very nature of things. Consequently we could even interpret the tendency of modern science as emancipation from the infinitely small. Instead of the old principle *natura non facit saltus*, we might even assert the opposite, viz., "nature makes jumps."

It is common knowledge that all matter is composed of tiny building blocks called "atoms," the combinations and connections of which produce all the variety of macroscopic objects. Still physics did not stop at the atomism of matter. At the end of the last century there appeared the atomism of electricity which seems much more bizarre at first sight. Electricity, which until then had been thought of as a fluid and was considered the model of a continuously active agent, was then shown to be built up of positive and negative *electrons*.

In addition to matter and electricity, there is one other entity in physics for which the law of conservation holds, viz., energy. But it has been established that even energy does not unconditionally admit of infinite divisibility. Planck has discovered *quanta of energy*.

Hence, a homogeneous continuum which admits of the sort of divisibility needed to realize the infinitely small is nowhere to be found in reality. The infinite divisibility of a continuum is an operation which exists only in thought. It is merely an idea which is in fact impugned by the results of our observations of nature and of our physical and chemical experiments.

The second place where we encounter the question of whether the infinite is found in nature is in the consideration of the universe as a whole. Here we must consider the expanse of the universe to determine whether it embraces anything infinitely large. But here again modern science, in particular astronomy, has reopened the question and is endeavoring to solve it, not by the defective means of metaphysical speculation, but by reasons which are based on experiment and on the application of the laws of nature. Here, too, serious objections against infinity have been found. *Euclidean* geometry necessarily leads to the postulate that space is infinite. Although euclidean geometry is indeed a consistent conceptual system, it does not thereby follow that euclidean geometry actually holds in reality. Whether or not real space is euclidean can be determined only through observation and experiment. The attempt to prove the infinity of space by pure speculation contains gross errors. From the fact that outside a certain portion of space there is always more space, it follows only that space is unbounded, not that it is infinite. Unboundedness and finiteness are compatible. In so-called *elliptical* geometry, mathematical investigation furnishes the natural model of a finite universe. Today the abandonment of euclidean geometry is no longer merely a mathematical or philosophical speculation but is suggested by considerations which originally had nothing to do with the question of the finiteness of the universe. Einstein has shown that euclidean geometry must be abandoned. On the basis of his gravitational theory, he deals with cosmological questions and shows that a finite universe is possible.

Moreover, all the results of astronomy are perfectly compatible with the postulate that the universe is elliptical.

We have established that the universe is finite in two respects, i.e., as regards the infinitely small and the infinitely large. But it may still be the case that the infinite occupies a justified place *in our thinking*, that it plays the role of an indispensable concept. Let us see what the situation is in mathematics. Let us first interrogate that purest and simplest offspring of the human mind, viz., number theory. Consider one formula out of the rich variety of elementary formulas of number theory, e.g., the formula

$$1^2 + 2^2 + \ldots + n^2 = \tfrac{1}{6} n \cdot (n+1) \cdot (2n+1)$$

Since we may substitute any integer whatsoever for n, for example $n = 2$ or $n = 5$, this formula implicitly contains *infinitely many* propositions. This characteristic is essential to a formula. It enables the formula to represent the solution of an arithmetical problem and necessitates a special idea for its proof. On the other hand, the individual numerical equations

$$1^2 + 2^2 = \tfrac{1}{6} \cdot 2 \cdot 3 \cdot 5$$
$$1^2 + 2^2 + 3^2 + 4^2 + 5^2 = \tfrac{1}{6} \cdot 5 \cdot 6 \cdot 11$$

can be verified simply by calculation and hence individually are of no especial interest.

We encounter a completely different and quite unique conception of the notion of infinity in the important and fruitful method of *ideal elements*. The method of ideal elements is used even in elementary plane geometry. The points and straight lines of the plane originally are real, actually existent objects. One of the axioms that hold for them is the axiom of connection: one and only one straight line passes through two points. It follows from this axiom that two straight lines intersect at most at one point. There is no theorem that two straight lines always intersect at some point, however, for the two straight lines might well be parallel. Still we know that by introducing ideal elements, viz., infinitely long lines and points at infinity, we can make the theorem that two straight lines always intersect at one and only one point come out universally true. These ideal "infinite" elements have the advantage of making the system of connection laws as simple and perspicuous as possible. Moreover, because of the symmetry between a point and a straight line, there results the very fruitful principle of duality for geometry.

Another example of the use of ideal elements are the familiar *complex-imaginary* magnitudes of algebra which serve to simplify theorems about the existence and number of roots of an equation.

[Here Hilbert talks about ideal elements in algebra.]

We now come to the most aesthetic and delicately erected structure of mathematics, viz., analysis. You already know that infinity plays the leading role in analysis. In a certain sense, mathematical analysis is a symphony of the infinite.

The tremendous progress made in the infinitesimal calculus results mainly from operating with mathematical systems of infinitely many elements. But, as it seemed very plausible to identify the infinite with the "very large", there soon arose inconsistencies which were known in part to the ancient sophists, viz., the so-called paradoxes of the infinitesimal calculus. But the recognition that many theorems which hold for the finite (for example, the part is smaller than the whole, the existence of a minimum and a maximum, the interchangeability of the order of the terms of a sum or a product) cannot be immediately and unrestrictedly extended to the infinite, marked fundamental progress. I said at the beginning of this paper that these questions have been completely clarified, notably through Weierstrass's acuity. Today, analysis is not only infallible within its domain but has become a practical instrument for using the infinite.

But analysis alone does not provide us with the deepest insight into the nature of the infinite. This insight is procured for us by a discipline which comes closer to a general philosophical way of thinking and which was designed to cast new light on the whole complex of questions about the infinite. This discipline, created by Georg Cantor, is set theory. In this paper we are interested only in that unique and original part of set theory which forms the central core of Cantor's doctrine, viz., the theory of *transfinite* numbers. This theory is, I think, the finest product of mathematical genius and one of the supreme achievements of purely intellectual human activity. What, then, is this theory?

Someone who wished to characterize briefly the new conception of the infinite which Cantor introduced might say that in analysis we deal with the infinitely large and the infinitely small only as limiting concepts, as something becoming, happening, i.e., with the *potential infinite*. But this is not the true infinite. We meet the true infinite when we regard the totality of number 1, 2, 3, 4, ... itself as a completed unity, or when we regard the points of an interval as a totality of things which exist all at once. This kind of infinity is known as *actual infinity*.

Frege and Dedekind, the two mathematicians most celebrated for their work in the foundations of mathematics, independently of each other used the actual infinite to provide a foundation for arithmetic which was independent of both intuition and experience. This foundation was based solely on pure logic and made use only of deductions that were purely logical. Dedekind even went so far as not to take the notion of finite number from intuition but to derive it logically by employing the concept of an infinite set. But it was Cantor who systematically developed the concept of the actual infinite. Consider the two examples of the infinite already mentioned

1. 1, 2, 3, 4, ...
2. The points of the interval 0 to 1 or, what comes to the same thing, the totality of real numbers between 0 and 1.

It is quite natural to treat these examples from the point of view of their size. But such a treatment reveals amazing results with which every mathematician today is familiar. For when we consider the set of all rational numbers, i.e., the fractions $\frac{1}{2}$, $\frac{1}{3}$, $\frac{2}{3}$, $\frac{1}{4}$, ... , $\frac{3}{7}$, ... , we notice that—from the sole standpoint of its size—this set is no larger than the set of integers. Hence we say that the rational numbers can be counted in the usual way; i.e., that they are enumerable.

The same holds for the set of all roots of numbers, indeed even for the set of all algebraic numbers. The second example is analogous to the first. Surprisingly enough, the set of all points of a square or cube is no larger than the set of points of the interval 0 to 1. Similarly for the set of all continuous functions. On learning these facts for the first time, you might think that from the point of view of size there is only one unique infinite. No, indeed! The sets in examples (1) and (2) are not, as we say, "equivalent". Rather, the set (2) cannot be enumerated, for it is larger than the set (1). [For a detailed exposition of the material above see Chapter 6.] We meet what is new and characteristic in Cantor's theory at this point. The points of an interval cannot be counted in the usual way, i.e., by counting 1, 2, 3, ... But since we admit the actual infinite, we are not obliged to stop here. When we have counted 1, 2, 3, ... , we can regard the objects thus enumerated as an infinite set existing all at once in a particular order. If, following Cantor, we call the type of this order ω , then counting continues naturally with $\omega + 1$, $\omega + 2$, ... up to $\omega + \omega$ or $\omega \cdot 2$, ... [Here he discusses counting further in the infinities.]

On the basis of these concepts, Cantor developed the theory of transfinite numbers quite successfully and invented a full calculus for them. Thus thanks to the Herculean collaboration of Frege, Dedekind, and Cantor, the infinite was made king and enjoyed a reign of great triumph. In daring flight, the infinite had reached a dizzy pinnacle of success.

But reaction was not lacking. It took in fact a very dramatic form. It set in perfectly analogously to the way reaction had set in against the development of the infinitesimal calculus. In the joy of discovering new and important results, mathematicians paid too little attention to the validity of their deductive methods. For, simply as a result of employing definitions and deductive methods which had become customary, contradictions began gradually to appear. These contradictions, the so-called paradoxes of set theory, though at first scattered, became progressively more acute and more serious. In particular, a contradiction discovered by Zermelo and Russell [see p.4 above] had a downright catastrophic effect when it became known throughout the world of mathematics. Confronted by these paradoxes, Dedekind and Frege completely abandoned their point of view and retreated. Dedekind hesitated a long time before permitting a new edition of his epoch-making treatise *Was sind und was sollen die Zahlen* to be published. In an epilogue, Frege too had to acknowledge that the direction of his book *Grundgesetze der Mathematik* was wrong. Cantor's doctrine, too, was attacked on all sides. So violent was this reaction that even the most ordinary and fruitful concepts and the simplest and most important deductive methods of mathematics were threatened and their employment was on the verge of being declared illicit. The old order had its defenders, of course. Their defensive tactics, however, were too fainthearted and they never formed a united front at the vital spots. Too many different remedies for the paradoxes were offered, and the methods proposed to clarify them were too variegated.

Admittedly, the present state of affairs where we run up against the paradoxes is intolerable. Just think, the definitions and deductive methods which everyone learns, teaches, and uses in mathematics, the paragon of truth and certitude, lead to absurdities! If mathematical thinking is defective, where are we to find truth and certitude?

There is, however, a completely satisfactory way of avoiding the paradoxes without betraying our science. The desires and attitudes which will help us find this way and show us what direction to take are these:

1. Wherever there is any hope of salvage, we will carefully investigate fruitful definitions and deductive methods. We will nurse them, strengthen them, and make them useful. No one shall drive us out of the paradise that Cantor has created for us.
2. We must establish throughout mathematics the same certitude for our deductions as exists in elementary number theory, which no one doubts and where contradictions and paradoxes arise only through our own carelessness.

Obviously these goals can be attained only after we have fully elucidated *the nature of the infinite.*

We have already seen that the infinite is nowhere to be found in reality, no matter what experiences, observations, and knowledge are appealed to. Can thought about things be so much different from things? Can thinking processes be so unlike the actual processes of things? In short, can thought be so far removed from reality? Rather is it not clear that, when we think that we have encountered the infinite in some real sense, we have merely been seduced into thinking so by the fact that we often encounter extremely large and extremely small dimensions in reality?

Does material logical deduction somehow deceive us or leave us in the lurch when we apply it to real things or events?* No! Material logical deduction is indispensable. It deceives us only when we form arbitrary abstract definitions, especially those which involve infinitely many objects. In such cases we have illegitimately used material logical deduction; i.e., we have not paid sufficient attention to the preconditions necessary for its valid use. In recognizing that there are such preconditions that must be taken into account, we find ourselves in agreement with the philosophers, notably with Kant. Kant taught—and it is an integral part of his doctrine—that mathematics treats a subject matter which is given independently of logic. Mathematics, therefore, can never be grounded solely on logic. Consequently, Frege's and Dedekind's attempts to so ground it were doomed to failure.

As a further precondition for using logical deduction and carrying out logical operations, something must be given in conception, viz., certain extra-logical concrete objects which are intuited as directly experienced prior to all thinking. For logical deduction to be certain, we must be able to see every aspect of these objects, and their properties, differences, sequences, and

* Throughout this paper the German word "inhaltlich" has been translated by the words "material" or "materially" which are reserved for this purpose and which are used to refer to matter in the sense of the traditional distinction between matter and content and logical form.—Translator

contiguities must be given, together with the objects themselves, as something which cannot be reduced to something else and which requires no reduction. This is the basic philosophy which I find necessary not just for mathematics, but for all scientific thinking, understanding, and communicating. The subject matter of mathematics is, in accordance with this theory, the concrete symbols themselves whose structure is immediately clear and recognizable.

Consider the nature and methods of ordinary finitary number theory. It can certainly be constructed from numerical structures through intuitive material considerations. But mathematics surely does not consist solely of numerical equations and surely cannot be reduced to them alone. Still one could argue that mathematics is an apparatus which, when applied to integers, always yields correct numerical equations. But in that event we still need to investigate the structure of this apparatus thoroughly enough to make sure that it in fact always yields correct equations. To carry out such an investigation, we have available only the same concrete material finitary methods as were used to derive numerical equations in the construction of number theory. This scientific requirement can in fact be met, i.e., it is possible to obtain in a purely intuitive and finitary way—the way we attain the truths of number theory—the insights which guarantee the validity of the mathematical apparatus.

Let us consider number theory more closely. In number theory we have the numerical symbols

1, 11, 111, 11111

where each numerical symbol is intuitively recognizable by the fact it contains only 1's. These numerical symbols which are themselves our subject matter have no significance in themselves. But we require in addition to these symbols, even in elementary number theory, other symbols which have meaning and which serve to facilitate communication, for example the symbol 2 is used as an abbreviation for the numerical symbol 11, and the numerical symbol 3 as an abbreviation for the numerical symbol 111. Moreover, we use symbols like $+$, $=$, and $>$ to communicate statements. $2 + 3 = 3 + 2$ is intended to communicate the fact that $2 + 3$ and $3 + 2$, when abbreviations are taken into account, are the self-same numerical symbol, viz., the numerical symbol 11111. Similarly, 3 > 2 serves to communicate the fact that the symbol 3, i.e., 111, is longer than the symbol 2, i.e., 11; or, in other words, that the latter symbol is a proper part of the former.

We also use the letters **a**, **b**, **c** for communication.* Thus **b** $>$ **a** communicates the fact that the numerical symbol **b** is longer than the numerical symbol **a**. From this point of view, $\mathbf{a} + \mathbf{b} = \mathbf{b} + \mathbf{a}$ communicates only the fact that the numerical symbol $\mathbf{a} + \mathbf{b}$ is the same as $\mathbf{b} + \mathbf{a}$. The content of this communication can also be proved through material deduction. Indeed, this kind of intuitive material treatment can take us quite far.

But let me give you an example where this intuitive method is outstripped. The largest known prime number is (39 digits)

---

* We use boldface letters where Hilbert used German script.—Ed.

$$\mathbf{p} = 170\ 141\ 183\ 460\ 469\ 231\ 731\ 687\ 303\ 715\ 884\ 105\ 727$$

By a well-known method due to Euclid we can give a proof, one which remains entirely within our finitary framework, of the statement that between $\mathbf{p} + 1$ and $\mathbf{p}! + 1$ there exists at least one new prime number. The statement itself conforms perfectly to our finitary approach, for the expression "there exists" serves only to abbreviate the expression: it is certain that $\mathbf{p} + 1$ or $\mathbf{p} + 2$ or $\mathbf{p} + 3$ ... or $\mathbf{p}! + 1$ is a prime number. Furthermore, since it obviously comes down to the same thing to say: there exists a prime number which

1. $> \mathbf{p}$, and at the same time is
2. $\leq \mathbf{p}! + 1$,

we are led to formulate a theorem which expresses only a part of what the euclidean theorem expresses; viz., the theorem that there exists a prime number $> \mathbf{p}$. Although this theorem is a much weaker statement in terms of content— it asserts only part of what the euclidean theorem asserts—and although the passage from the euclidean theorem to this one seems quite harmless, that passage nonetheless involves a leap into the transfinite when the partial statement is taken out of context and regarded as an independent statement.

How can this be? Because we have an existential statement, "there exists"! True, we had a similar expression in the euclidean theorem, but there the "there exists" was, as I already mentioned, an abbreviation for: $\mathbf{p} + 1$ or $\mathbf{p} + 2$ or $\mathbf{p} + 3$ ... or $\mathbf{p}! + 1$ is a prime number—just as when, instead of saying "either this piece of chalk or this piece or this piece ... or this piece is red" we say briefly "there exists a red piece of chalk among these pieces." A statement such as "there exists" an object with a certain property in a finite totality conforms perfectly to our finitary approach. But a statement like "either $\mathbf{p} + 1$ or $\mathbf{p} + 2$ or $\mathbf{p} + 3$ ... or (ad infinitum) ... has a certain property" is itself an infinite logical product. Such an extension into the infinite is, unless further explanation and precautions are forthcoming, no more permissible than the extension from finite to infinite products in calculus. Such extensions, accordingly, usually lapse into meaninglessness.

From our finitary point of view, an existential statement of the form "there exists a number with a certain property" has in general only the significance of a partial statement; i.e., it is regarded as part of a more determinate statement. The more precise formulation may, however, be unnecessary for many purposes.

In analyzing an existential statement whose content cannot be expressed by a finite disjunction, we encounter the infinite. Similarly, by negating a general statement, i.e., one which refers to arbitrary numerical symbols, we obtain a transfinite statement. For example, the statement that if $\mathbf{a}$ is a numerical symbol, then $\mathbf{a} + 1 = 1 + \mathbf{a}$ is universally true, is from our finitary perspective *incapable of negation*. We will see this better if we consider that this statement cannot be interpreted as a conjunction of infinitely many numerical equations by means of "and" but only as a hypothetical judgment which asserts something for the case when a numerical symbol is given.

From our finitary viewpoint, therefore, we cannot argue that an equation like the one just given, where an arbitrary numerical symbol occurs, either holds for every symbol or is disproved by a counter example. Such an argument, being an application of the law of excluded middle, rests on the presupposition that the statement of the universal validity of such an equation is capable of negation.

At any rate, we note the following: if we remain within the domain of finitary statements, as indeed we must, we have as a rule very complicated logical laws. Their complexity becomes unmanageable when the expressions "all" and "there exists" are combined and when they occur in expressions nested within other expressions. In short, the logical laws which Aristotle taught and which men have used ever since they began to think, do not hold. We could, of course, develop logical laws which do hold for the domain of finitary statements. But it would do us no good to develop such a logic, for we do not want to give up the use of the simple laws of Aristotelian logic. Furthermore, no one, though he speak with the tongues of angels, could keep people from negating general statements, or from forming partial judgments, or from using *tertium non datur*. What, then, are we to do?

Let us remember that *we are mathematicians* and that as mathematicians we have often been in precarious situations from which we have been rescued by the ingenious method of ideal elements. I showed you some illustrious examples of the use of this method at the beginning of this paper. Just as $i = \sqrt{-1}$ was introduced to preserve in simplest form the laws of algebra (for example, the laws about the existence and number of the roots of an equation); just as ideal factors were introduced to preserve the simple laws of divisibility for algebraic whole numbers (for example, a common ideal divisor for the numbers 2 and $1 + \sqrt{-5}$ was introduced, though no such divisor really exists); similarly, to preserve the simple formal rules of ordinary Aristotelian logic, we must *supplement the finitary statements with ideal statements*. It is quite ironic that the deductive methods which Kronecker so vehemently attacked are the exact counterpart of what Kronecker admired so enthusiastically in Kummer's work on number theory which Kronecker extolled as the highest achievement of mathematics.

How do we obtain *ideal statements*? It is a remarkable as well as a favorable and promising fact that to obtain ideal statements, we need only continue in a natural and obvious fashion the development which the theory of the foundations of mathematics has already undergone. Indeed, we should realize that even elementary mathematics goes beyond the standpoint of intuitive number theory. Intuitive, material number theory, as we have been construing it, does not include the method of algebraic computation with letters. Formulas were always used exclusively for communication in intuitive number theory. The letters stood for numerical symbols and an equation communicated the fact that the two symbols coincided. In algebra, on the other hand, we regard expressions containing letters as independent structures which formalize the material theorems of number theory. In place of statements about numerical symbols, we have formulas which are themselves the concrete objects of intuitive study. In place of number-theoretic material proof, we have the derivation of a formula from another formula according to determinate rules.

Hence, as we see even in algebra, a proliferation of finitary objects takes place. Up to now the only objects were numerical symbols like 1, 11, ... , 11111. These alone were the objects of material treatment. But mathematical practice goes further, even in algebra. Indeed, even when from our finitary viewpoint a formula is valid with respect to what it signifies as, for example, the theorem that always

a + b = b + a

where **a** and **b** stand for particular numerical symbols, nevertheless we prefer not to use this form of communication but to replace it instead by the formula

$a + b = b + a.$

This latter formula is in no wise an immediate communication of something signified but is rather a certain formal structure whose relation to the old finitary statements,

$$2 + 3 = 3 + 2,$$
$$5 + 7 = 7 + 5,$$

consists in the fact that, when a and b are replaced in the formula by the numerical symbols 2, 3, 5, 7, the individual finitary statements are thereby obtained, i.e., by a proof procedure, albeit a very simple one. We therefore conclude that a, b, $=$, $+$, as well as the whole formula $a + b = b + a$ mean nothing in themselves, no more than the numerical symbols meant anything. Still we can derive from that formula other formulas to which we do ascribe meaning, viz., by interpreting them as communications of finitary statements. Generalizing this conclusion, we conceive mathematics to be a stock of two kinds of formulas: first, those to which the meaningful communications of finitary statements correspond; and, secondly, other formulas which signify nothing and which are the *ideal structures of our theory*.

Now what was our goal? In mathematics, on the one hand, we found finitary statements which contained only numerical symbols, for example,

$$3 > 2, \quad 2 + 3 = 3 + 2, \quad 2 = 3, \quad 1 \neq 1,$$

which from our finitary standpoint are immediately intuitable and understandable without recourse to anything else. These statements can be negated, truly or falsely. One can apply Aristotelian logic unrestrictedly to them without taking special precautions. The principle of non-contradiction holds for them; i.e., the negation of one of these statements and the statement itself cannot both be true. *Tertium non datur* holds for them; i.e., either a statement or its negation is true. To say that a statement is false is equivalent to saying that its negation is true. On the other hand, in addition to these elementary statements which present no problems, we also found more problematic finitary statements; e.g., we found finitary statements that could not be split up into partial statements. Finally, we introduced ideal statements in order that the ordinary laws of logic would hold universally. But since these ideal statements, viz., the formulas, do not mean anything insofar as they do not express finitary statements, logical operations cannot be materially applied to them as they can be to finitary statements. It is, therefore, necessary to formalize the logical operations and the mathematical proofs themselves. This formalization necessitates translating logical relations into formulas. Hence, in addition to mathematical symbols, we must also introduce logical symbols such as

&,   ∨,   →,   –
(and)   (or) (implies) (not)

and in addition to the mathematical variables a, b, c, ... we must also employ logical variables, viz., the propositional variables A, B, C... .

How can this be done? Fortunately that same preestablished harmony which we have so often observed operative in the history of the development of science, that same preestablished harmony which aided Einstein by giving him the general invariant calculus already fully developed for his gravitational theory, comes also to our aid: we find the logical calculus already worked out in advance. To be sure, the logical calculus was originally developed from an altogether different point of view. The symbols of the logical calculus originally were introduced only in order to communicate. Still it is consistent with our finitary viewpoint to deny any meaning to logical symbols, just as we denied meaning to mathematical symbols, and to declare that the formulas of the logical calculus are ideal statements which mean nothing in themselves. We possess in the logical calculus a symbolic language which can transform mathematical statements into formulas and express logical deduction by means of formal procedures. In exact analogy to the transition from material number theory to formal algebra, we now treat the signs and operation symbols of the logical calculus in abstraction from their meaning. Thus we finally obtain, instead of material mathematical knowledge which is communicated in ordinary language, just a set of formulas containing mathematical and logical symbols which are generated successively, according to determinate rules. Certain of the formulas correspond to mathematical axioms. The rules whereby the formulas are derived from one another correspond to material deduction. Material deduction is thus replaced by a formal procedure governed by rules. The rigorous transition from a naive to a formal treatment is effected, therefore, both for the axioms (which, though originally viewed naively as basic truths, have been long treated in modern axiomatics as mere relations between concepts) and for the logical calculus (which originally was supposed to be merely a different language).

We will now explain briefly how *mathematical proofs* are formalized.
[Here Hilbert discusses a formalization of logical deduction, a version of which is presented in Chapters 19 and 21 of this text.]

Thus we are now in a position to carry out our theory of proof and to construct the system of provable formulas, i.e., mathematics. But in our general joy over this achievement and in our particular joy over finding that indispensable tool, the logical calculus, already developed without any effort on our part, we must not forget the essential condition of our work. There is just one condition, albeit an absolutely necessary one, connected with the method of ideal elements. That condition is a *proof of consistency*, for the extension of a domain by the addition of ideal elements is legitimate only if the extension does not cause contradictions to appear in the old, narrower domain, or, in other words, only if the relations that obtain among the old structures when the ideal structures are deleted are always valid in the old domain.

The problem of consistency is easily handled in the present circumstances. It reduces obviously to proving that from our axioms and according to the rules we laid down we cannot get "$1 \neq 1$" as the last formula of a proof, or, in other words, that "$1 \neq 1$" is not a provable formula. This task belongs just as much to the domain of intuitive treatment as does, for example, the task of finding a proof of the irrationality of $\sqrt{2}$ in materially constructed number theory—i.e., a proof that it is impossible to find two numerical symbols **a** and **b** which stand in the relation $\mathbf{a}^2 = 2\mathbf{b}^2$, or in other words, that one cannot produce two numerical symbols with a certain property. Similarly, it is incumbent on us to show that one cannot produce a certain kind of proof. A formalized proof, like a numerical symbol, is a concrete and visible object. We can describe it completely. Further, the requisite property of the last formula; viz., that it read "$1 \neq 1$", is a concretely ascertainable property of the proof. And since we can, as a matter of fact, prove that it is impossible to get a proof which has that formula as its last formula, we thereby justify our introduction of ideal statements.

It is also a pleasant surprise to discover that, at the very same time, we have resolved a problem which has plagued mathematicians for a long time, viz., the problem of proving *the consistency of the axioms of arithmetic*. For, wherever the axiomatic method is used, the problem of proving consistency arises. Surely in choosing, understanding, and using rules and axioms we do not want to rely solely on blind faith. In geometry and physical theory, proof of consistency is effected by reducing their consistency to that of the axioms of arithmetic. But obviously we cannot use this method to prove the consistency of arithmetic itself. Since our theory of proof, based on the method of ideal elements, enables us to take this last important step, it forms the necessary keystone of the doctrinal arch of axiomatics. What we have twice experienced, once with the paradoxes of the infinitesimal calculus and once with the paradoxes of set theory, will not be experienced a third time, nor ever again.

The theory of proof which we have sketched not only is capable of providing a solid basis for the foundations of mathematics but also, I believe, supplies a general method for treating fundamental mathematical questions which mathematicians heretofore have been unable to handle.

In a sense, mathematics has become a court of arbitration, a supreme tribunal to decide fundamental questions—on a concrete basis on which everyone can agree and where every statement can be controlled.

The assertions of the new so-called "intuitionism" [see Chapter 26]—modest though they may be—must in my opinion first receive their certificate of validity from this tribunal.

An example of the kind of fundamental questions which can be so handled is the thesis that every mathematical problem is solvable. We are all convinced that it really is so. In fact one of the principal attractions of tackling a mathematical problem is that we always hear this cry within us: There is the problem, find the answer; you can find it just by thinking, for there is no *ignorabimus* in mathematics. Now my theory of proof cannot supply a general method for solving every mathematical problem—there just is no such method. Still the proof (that the assumption that every mathematical problem is solvable is a consistent assumption) falls completely within the scope of our theory.

I will now play my last trump. The acid test of a new theory is its ability to solve problems which, though known for a long time, the theory was not expressly designed to solve. The maxim "By their fruits ye shall know them" applies also to theories.

[Here he claims to be able to resolve the continuum problem: are any infinities bigger than **N** and smaller than **R**?]

In summary, let us return to our main theme and draw some conclusions from all our thinking about the infinite. Our principal result is that the infinite is nowhere to be found in reality. It neither exists in nature nor provides a legitimate basis for rational thought—a remarkable harmony between being and thought. In contrast to the earlier efforts of Frege and Dedekind, we are convinced that certain intuitive concepts and insights are necessary conditions of scientific knowledge, that logic alone is not sufficient. Operating with the infinite can be made certain only by the finitary.

The role that remains for the infinite to play is solely that of an idea—if one means by an idea, in Kant's terminology, a concept of reason which transcends all experience and which completes the concrete as a totality—that of an idea which we may unhesitatingly trust with the framework erected by our theory.

Lastly, I wish to thank P. Bernays for his intelligent collaboration and valuable help, both technical and editorial, especially with the proof of the continuum theorem.

Exercises

1. Why would a model (in the world) of a collection of axioms justify that the axioms are free from contradiction?

2. a. What was the motive for Hilbert's paper?
   b. What did Weierstrass do that Hilbert so admired?
   c. Do you agree with Hilbert when he says, "in mathematics as elsewhere success is the supreme court to whose decisions everyone submits"? Explain.
   d. What was the paradise that Cantor created?
   e. Why does Hilbert say that the logical laws of Aristotle do not hold? How does he plan to deal with that?
   f. What are ideal statements in arithmetic?

g. When are we justified in using ideal statements?

h. Why was Hilbert especially concerned about proving the consistency of arithmetic?

i. According to Hilbert, what is the subject matter, i.e., the objects, which mathematics studies?

j. Hilbert's view of mathematics as presented here is called *formalism*. Is that name apt?

k. What was the role of logic in Hilbert's program? How did that differ from the role of logic in Frege's program?

3. A platonist would disagree with Hilbert on many points, but most fundamentally on what justifies our use of infinities in mathematics. Explain.

4. Goodstein is a constructivist. How would he object to Hilbert's use of ideal statements in mathematics?

Further Reading
Michael Hand clarifies this reading a lot in his "Hilbert's iterativistic tendencies".
Constance Reid's biography of Hilbert is excellent reading.

II

COMPUTABLE FUNCTIONS

8 Computability

Hilbert's approach is now called *formalism*. He wished to ground the infinite in a contradiction-free formal system the validity of which he could prove by finitary means. But what do "finitary means" encompass? If Hilbert could have given a constructive proof of the consistency of his system, then the question of delimiting the finitary from the infinitary would not need to be answered. But Hilbert and his school failed time and again to give such a proof. Finally Gödel showed that no finitistic method of a certain limited sort could be used to analyze the problem of consistency. Was his proof definitive or did he only show that *some* finitistic means won't work? An analysis of the notion of computability is essential to be able to answer that question.

Do not read any further before doing the following:
Write up what *you* believe are reasonable criteria for a procedure, limited to manipulation of natural numbers, to be finitistic, mechanical, and/or computable. List conditions which are necessary and/or sufficient.

A. Algorithms

We know many algorithms. Perhaps the earliest one you learned in mathematics was adding, which you did by reference to a fixed table. Similarly subtracting and multiplying were originally presented to you as algorithms.

One of the first general algorithms you may have learned was how to solve quadratic equations with real number coefficients. Given

$$ax^2 + bx^2 + c = 0$$

where a, b, c are real numbers, then if $b^2 - 4ac \geq 0$ the solutions to the equation are

$$x = \frac{-b \pm \sqrt{b^2 - 4ac}}{2a}$$

The solution of the quadratic led in the sixteenth century to solutions of the general third- and fourth-degree equations by similar algorithms. But for three centuries no one was able to give an algorithm for the solutions for the general fifth-degree equation

$$a_5 x^5 + a_4 x^4 + a_3 x^3 + a_2 x^2 + a_1 x + a_0 = 0$$

There was finally general agreement that what was required was a solution using only the coefficients and the operations $+, \cdot, -, \div$, and nth roots for any n. By clearly defining the class of algorithms that were deemed acceptable, it was possible for Abel to prove in 1824 that no such general solution existed.

Most of us are also familiar with algorithms for constructing geometric figures using straightedge and compass, such as those for constructing a parallel (Chapter 5.D), or a right angle, or an equilateral triangle. A long-standing problem about constructions was: given an arbitrary angle, is it possible to trisect it using only straightedge and compass? Methods similar to those which established that the quintic and higher degree equations can't be solved by radicals show there can be no such algorithm.

B. General Criteria for Algorithms

We need a general notion of algorithm if we wish to show that for some problems there is *no* algorithmic solution. By this point you should have tried to give your own criteria for what a computable procedure is. Let's look at some criteria that others have proposed.

1. Mal'cev's criteria, from *Algorithms and Recursive Functions*

   a. An algorithm is a process for the successive construction of quantities. It proceeds in discrete time so that at the beginning there is an initial finite system of quantities given and at every succeeding moment the system of quantities is obtained by means of a definite law (program) from the system of quantities at hand at the preceding moment of time (*discreteness of the algorithm*).

   b. The system of quantities obtained at some (not the initial) moment of time is uniquely determined by the system of quantities obtained in the preceding moments of time (*determinacy of the algorithm*).

   c. The law for obtaining the succeeding system of quantities from the preceding must be simple and local (*elementarity of the steps of the algorithm*).

   d. If the method of obtaining the succeeding quantity from any given quantity does not give a result, then it must be pointed out what must be considered to be the result of the algorithm (*direction of the algorithm*).

   e. The initial system of quantities can be chosen from some potentially infinite set (*massivity of the algorithm*). ...

   The intuitive concept of an algorithm, although it is nonrigorous, is clear to the extent that in practice there are no serious cases when mathematicians disagree in their opinion about whether some concretely given process is an algorithm or not.

Mal'cev, pp. 18–19

Are Mal'cev's criteria complete? Consider the following function: We construct a machine that can pick up a pair of dice, shake them, roll them out, read the top two faces, and print out the sum of the dots on them; to the number n we associate the nth result of this process. Since this is done by a machine it is surely *mechanical*. But is it an algorithm for computing a function? We believe not, since it is not *duplicable*. Duplicability is an essential characteristic of any computable procedure. Mal'cev's criteria do not rule out this function as computable unless we are willing to argue that the entire system of machine and dice is not physically determined (see Exercise 2).

2. Hermes, from *Enumerability, Decidability, Computability*

Introductory Reflections on Algorithms

The concept of algorithm

The concept of algorithm, i.e. of a "general procedure", is more or less known to all mathematicians. In this introductory paragraph we want to make this concept more precise. In doing this we want to stress what is to be considered essential.

Algorithms as general procedures. The specific way of mathematicians to draw up and to enlarge theories has various aspects. Here we want to single out and discuss more precisely an aspect characteristic of many developments. Whenever mathematicians are occupied with a group of problems it is at first mostly isolated facts that captivate their interests. Soon however they will proceed to finding a connection between these facts. They will try to systema-tize the research more and more with the aim of attaining a comprehensive view and an eventual complete mastery of the field in question. Frequently the method of attaining such mastery consists in separating special classes of questions such that each class can be dealt with by the help of an algorithm. An algorithm is a general procedure such that for any appropriate question the answer can be obtained by the use of a simple computation according to a specified method.

Examples of general procedures can be found in every mathematical discipline. We only need to think of the division procedure for the natural numbers given in the decimal notation, of the algorithm for the computation of approximating decimal expressions of the square root of a natural number, or of the method of the decomposition into partial fractions for the computation of integrals with rational functions as integrands.

In this book we shall understand by a *general procedure* a process the execution of which is clearly specified to the smallest details. Among other things this means that we must be able to express the instructions for the execution of the process in a *finitely long* text.*

There is no room left for the practice of the creative imagination of the executor. He has to work slavishly according to the instructions given to him, which determine everything to the smallest detail.**

The requirements for a process to be a general procedure are very strict. It must be clear that the ways and means which a mathematician is used to of describing a general procedure are in general too vague to come up really to the required standard of exactness. This applies for instance to the usual description of methods for the solution of a linear equation system. Among other things it is left open in this description in which way the necessary additions and multiplications should be executed. It is however clear to every mathematician that in this case and in cases of the same sort the instruction can be supplemented to make a complete instruction which does not leave anything open.—The instructions according to which the not specially mathematically trained assistants work in a calculating pool come relatively near to the ideal we have fixed our eyes upon.

There is a case, which we feel is worth mentioning here, in which a mathematician is used to speaking of a general procedure by which he does not intend to characterize an unambiguous way of proceeding. We are thinking of calculi with several rules such that it is not determined in which sequence the rules should be applied. But these calculi are closely connected with the completely unambiguously described procedures. ... In this book we want to adopt the convention of *calling procedures general procedures only if the way of proceeding is completely unambiguous.*

There are *terminating algorithms,* whereas other algorithms can be continued as long as we like. The Euclidean algorithm for the determination of the greatest common divisor of two numbers terminates; after a finite number of steps in the computation we obtain an answer, and the procedure is at an end. The well-known algorithm of the computation of the square root of a natural number given in decimal notation does *not,* in general, terminate. We can continue with the algorithm as long as we like, and we obtain further and further decimal fractions as closer approximations to the root.

Realization of algorithms. A general procedure, as it is meant here, means in any case primarily an operation (action) with concrete things. The separation of these things from each other must be sufficiently clear. They can be pebbles

* One cannot produce an infinitely long instruction. We can however imagine the construction of one which is potentially infinitely long. This can be obtained by first giving a finite beginning of the instruction, and then giving a finitely long set of rules which determines exactly how in every case the already existing part of our instruction is to be extended. But then we can say that the finite beginning together with the finitely long set of rules is the actual (finite) instruction.

** Obviously the schematical execution of a given general procedure is (after a few tries) of no special interest to a mathematician. Thus we can state the remarkable fact that by the specifically mathematical achievement of developing a general method a creative mathematician, so to speak, mathematically depreciates the field he becomes master of by this very method.

(counters, small wood beads) as e.g. on the classical *abacus* or on the Japanese *soroban*, they can be symbols as in mathematical usage (e.g. 2, x, +, (, ʃ), but they can also be the cogwheels of a small calculating machine, or electrical impulses as it is usual in big computers. The operation consists in bringing spatially and temporally ordered things into new configurations.

For the practice of applied mathematics it is absolutely essential which *material* is used to execute a procedure. However, we want to deal with the algorithms from the theoretical point of view. In this case the material is irrelevant. If a procedure is known to work with a certain material then this procedure can also be transferred (more or less successfully) to another material. Thus the addition in the domain of natural numbers can be realized by the attachment of strokes to a line of strokes, by the adding or taking away of beads on an abacus, or by the turning of wheels in a calculating-machine.

Since we are only interested in such questions in the domain of general procedures which are independent of the material realization of these procedures, we can take as a basis of our considerations a realization which is mathematically especially easy to deal with. It is therefore preferred in the mathematical theory of algorithms to consider such algorithms which take effect in altering a *line of signs*. A line of signs is a finite linear sequence of *symbols* (*single signs, letters*). It will be taken for granted that for each algorithm there is a finite number of letters (at least one) the collection of which forms the *alphabet* which is the basis of the algorithm. The finite lines of signs, which can be composed from the alphabet, are called *words*. It is sometimes convenient to allow the *empty word* □ , which contains no letters.— If **A** is an alphabet and W a word which is composed only of letters of **A**, we call W a *word over* **A** .

The letters of an alphabet **A** which is the basis of an algorithm are in a certain sense non-essential. Namely, if we alter the letters of **A** and so obtain a corresponding new alphabet **A´**, then we can, without difficulty, give an account of an algorithm for **A´** which is "isomorphic" to the original algorithm, and which functions, fundamentally, in the same way.

Gödel numbering.* We can, in principle, make do with an alphabet which contains only a single letter, e.g. the letter I . The words of this alphabet are (apart from the empty word): I, II, III, etc. These words can in a trivial way be identified with the natural numbers 0, 1, 2, ... Such an extreme standardization of the "material" is advisable for some considerations. On the other hand it is often convenient to have at our disposal the diversity of an alphabet consisting of several elements. ...

The use of an alphabet consisting of *one element* only does not imply an essential limitation. We can, as a matter of fact, associate the words W over an alphabet **A** consisting of N elements with natural numbers $G(W)$ (in such a way that each natural number is associated with at most one word), i.e. with words of an alphabet consisting of *one* element. Such a representation of G is called a *Gödel numbering*, and $G(W)$ the *Gödel number* (with respect to G)

---

* Also often called *arithmetization*.

of the word W. Gödel ... was the first to use such a representation. The following are the requirements for an arithmetization G.

1. If $W_1 \neq W_2$ then $G(W_1) \neq G(W_2)$.
2. There exists an algorithm such that for any given word W the corresponding natural number $G(W)$ can be computed in a finite number of steps by the help of this algorithm.
3. For any natural number n it can be decided in a finite number of steps, whether n is the Gödel number of a word W over $\mathbf{A}$.
4. There exists an algorithm such that if n is the Gödel number of a word W over $\mathbf{A}$, then this word W (which according to 1 must be unique) can be constructed in a finite number of steps by the help of this algorithm.

Hermes, pp. 1–4

C. Numbering

Here is an example of a Gödel numbering as described by Hermes. Take as alphabet the letters a, b, c, and say that a *word* is any finite *concatenation* of these— that is, a placement of these side by side in a line. For example, *abacca* is a word. We can then number the words as follows:

Given a word $x_1 x_2 \cdots x_n$ where each x_i is a, b, or c, we assign to it the number $2^{d_0} \cdot 3^{d_1} \cdot \ldots \cdot p_n^{d_n}$, where p_i is the ith prime (2 is the 0th prime) and

$$d_i = \begin{cases} 1 & \text{if } x_i \text{ is } a \\ 2 & \text{if } x_i \text{ is } b \\ 3 & \text{if } x_i \text{ is } c \end{cases}$$

The empty word is given number 0.

For example, the word *abac* has number $2^1 \cdot 3^2 \cdot 5^1 \cdot 7^3 = 30{,}870$. The word *bbc* has the number $2^2 \cdot 3^2 \cdot 5^3 = 4500$. The number 360 represents *cba* because $360 = 2^3 \cdot 3^2 \cdot 5^1$.

To show that this numbering satisfies Hermes criteria we need to invoke the *Fundamental Theorem of Arithmetic* (Exercise 5 of Chapter 5):

Any natural number ≥ 2 can be represented as a product of primes, and that product is, except for the order of the primes, unique.

We may number all kinds of objects, not just alphabets. In general, the criteria for a numbering to be useful are:

1. No two objects have the same number.
2. Given any object, we can "effectively" find the number for it.
3. Given any number, we can "effectively" find if it is assigned to an object and, if so, to which object.

D. Algorithm vs. Algorithmic Function

Numbering is one way of giving names to objects. And as long as we're on the subject of names, let's recall that we agreed to treat objects extensionally: They have properties independent of how we name them (see Chapter 4.A.3.c). Does the distinction really matter? Consider the following function from natural numbers to natural numbers:

$$f(x) = \begin{cases} 1 & \text{if a consecutive run of } exactly\, x\; 5\text{s in a row} \\ & \text{occurs in the decimal expansion of } \pi \\ 0 & \text{otherwise} \end{cases}$$

No algorithm is known for computing f. It may be that none exists, but to be able to claim there is none we would need a precise definition of "algorithm".

Now consider the function

$$g(x) = \begin{cases} 1 & \text{if a consecutive run of } at\,least\, x\; 5\text{s in a row} \\ & \text{occurs in the decimal expansion of } \pi \\ 0 & \text{otherwise} \end{cases}$$

We claim that g is computable; that is, there is an algorithm for computing g. Consider the following list of functions:

$$
\begin{aligned}
h(x) &= 1 \quad \text{for all } x \\
h_0(x) &= 1 \quad \text{if } x = 0; \text{ for all other } x,\ h_0(x) = 0 \\
h_1(x) &= 1 \quad \text{if } x = 0 \text{ or } 1; \text{ for all other } x,\ h_1(x) = 0 \\
h_2(x) &= 1 \quad \text{if } x = 0, 1, \text{ or } 2; \text{ for all other } x,\ h_2(x) = 0 \\
&\ \vdots \\
h_k(x) &= 1 \quad \text{if } x = 0, 1, 2, \ldots, \text{ or } k; \text{ for all other } x,\ h_k(x) = 0 \\
&\ \vdots
\end{aligned}
$$

Each of these functions is computable. And g must be on the list: if there are no 5's in the expansion of π, then g is h_0; if there is a longest run of 5's in π, say n, then g is h_n; if there are arbitrarily long runs of 5's in π then g is h.

Of course, we can't tell *which* of these descriptions is correct for g, but we've shown that one of them must match the same inputs to the same outputs as the previous description of g. We must not confuse the fact that we have chosen a bad name for g (from the extensionalist viewpoint)—that is, one which won't allow us to distinguish which h_k or h it is—with the fact that whichever function on the list g *is* is computable. Properties of functions are independent of how we describe the functions.

An algorithm is a *description*, a name for a function. From the extensionalist point of view, *algorithm ≠ algorithmic function.*

E. Approaches to Formalizing Computability

We now know why we'd like to formalize the notion of computability, we have some guidelines, and we know that if we can formalize the notion for some kind of objects (words, numbers), we will have accomplished it for other kinds via translations by numbering. Before we look at any particular formalization let's survey the different approaches.

1. *Representability in a formal system* (Church, 1933, Gödel [and Herbrand], 1934). In this approach a formal system of arithmetic (axioms and rules of proof) is taken and a function is declared computable if for every m and n for which $f(m) = n$ we can prove $f(m) = n$ in the system. We will see this approach in Part III when we formalize arithmetic.

2. *The λ-calculus* (Church, 1936). This is closely related to representability in a formal system. Church takes a simple formal alphabet and language together with a notion of derivability that mocks the idea of proof yet seems simpler. Everything is reduced to manipulation of the symbols, and $f(m) = n$ if we can derive this in the system. See, for example, Rosser, 1984.

3. *Arithmetical descriptions* (Kleene, 1936). This approach is based on generalizing the notion of definition by induction. A class of functions that includes $+$ and $\cdot$ is "closed under" some simple rules, like definition by induction. This yields the class of (μ-) *recursive functions* and is the easiest system to work with mathematically. We will develop this one fully in Chapters 11 and 14–16.

4. *Machine-like descriptions.* There have been several attempts (mostly before the advent of working computers) to give a mathematical model of *machine*. Each tries to formalize the intuitive notion by giving a description of every possible machine.
   a. *Turing machines* (1936). We'll study these in Chapter 9.
   b. *Markov algorithms.* See Markov, 1954.
   c. *Unlimited register machines* (Shepherdson and Sturgis, 1963). This model is an idealization of computers with unlimited time, unlimited memory, and no errors.

What all these formalizations have in common is that they are all purely syntactical, despite the often anthropomorphic descriptions. They are methods for pushing symbols around. Here is what Mostowski has to say in his excellent survey.

> However much we would like to "mathematize" the definition of computability, we can never get completely rid of the semantic aspect of this concept. The process of computation is a linguistic notion (presupposing that our notion of language is sufficiently general); what we have to do is to delimit a class of those functions (considered as abstract mathematical objects) for which there exists a corresponding linguistic object (a process of computation).
>
> Mostowski, p. 35

computability — a semantic, intuitive concept

computation — a syntactic, purely formal concept

Exercises

1. Give an algorithm for finding the largest natural number that divides two given natural numbers. Does your algorithm satisfy Mal'cev's criteria? (Describe your algorithm in English, not "computerese".)

2. Do you agree that the dice-rolling procedure of Section B.1 is mechanical but not computable? How could you use Mal'cev's criterion (b) to rule it out? (Be careful not to implicitly use the notion of computability in explaining what you mean by "determined".)

3. Hermes says we can identify the words □, I, I I, I I I, ... with the natural numbers 0, 1, 2, 3, ... in a trivial way. (Does he mean *numbers* or *numerals*?) Give the identification explicitly (cf. Exercise 3.1).

4. Give two examples from your daily life in which numbers are assigned by criteria (1), (2), and (3) of Section C.

5. Consider the alphabet $(\,,\,)$, $\rightarrow$, $\neg$, p_0, p_1, p_2, ... We define a word as follows:
   i. (p_i) is a word for $i = 0, 1, 2, ...$.
   ii. If **A** and **B** are words, so too are $(\mathbf{A} \rightarrow \mathbf{B})$ and $(\neg \mathbf{B})$.
   iii. A string of symbols is a word if and only if it arises via applications of (i) and (ii).

   Number all words effectively. (*Hint:* Consider the numbering in Section C.)

Further Reading
Odifreddi in *Classical Recursion Theory,* Chapter I.8, discusses the issue of a process being mechanical vs. being computable.

9 Turing Machines

A. Turing on Computability
(Optional)

One of the first analyses of the notion of computability, and certainly the most influential, is due to Turing.

Alan M. Turing, from "On Computable Numbers, with an Application to the Entscheidungsproblem, " 1936

The "computable" numbers may be described briefly as the real numbers whose expressions as a decimal are calculable by finite means. . . . According to my definition, a number is computable if its decimal can be written down by a machine. p. 116

[Turing then gives his formal definitions and in particular says that for a real number or function on the natural numbers to be computable it must be computable by a machine that gives an output for every input.]

No attempt has yet been made to show that the "computable" numbers include all numbers which would naturally be regarded as computable. All arguments which can be given are bound to be, fundamentally, appeals to intuition, and for this reason rather unsatisfactory mathematically. The real question at issue is "What are the possible processes which can be carried out in computing a number?"

The arguments which I shall use are of three kinds.

a. A direct appeal to intuition.

b. A proof of the equivalence of two definitions (in case the new definition has a greater intuitive appeal). [In an appendix to the paper Turing proves that a function is calculable by his definition if and only if it is one of Church's effectively calculable functions.]

c. Giving examples of large classes of numbers which are computable. ...

[I.] Computing is normally done by writing certain symbols on paper. We may suppose this paper is divided into squares like a child's arithmetic book. In elementary arithmetic the two-dimensional character of the paper is sometimes used. But such a use is always avoidable, and I think that it will be agreed that the two-dimensional character of paper is no essential of computation. I assume then that the computation is carried out on one-dimensional paper, *i.e.* on a tape divided into squares. I shall also suppose that the number of symbols which may be printed is finite. If we were to allow an infinity of symbols, then there would be symbols differing to an arbitrarily small extent. The effect of this restriction of the number of symbols is not very serious. It is always possible to use sequences of symbols in place of single symbols. Thus an Arabic numeral such as 17 or 999999999999999 is normally treated as a single symbol. Similarly in any European language words are treated as single symbols (Chinese, however, attempts to have an enumerable infinity of symbols). The differences from our point of view between the single and compound symbols is that the compound symbols, if they are too lengthy, cannot be observed at one glance. This is in accordance with experience. We cannot tell at a glance whether 9999999999999999 and 999999999999999 are the same.

The behaviour of the computer at any moment is determined by the symbols which he is observing, and his "state of mind" at that moment. We may suppose that there is a bound B to the number of symbols or squares which the computer can observe at one moment. If he wishes to observe more, he must use successive observations. We will also suppose that the number of states of mind which need be taken into account is finite. The reasons for this are of the same character as those which restrict the number of symbols. If we admitted an infinity of states of mind, some of them will be "arbitrarily close" and will be confused. Again, the restriction is not one which seriously affects computation, since the use of more complicated states of mind can be avoided by writing more symbols on the tape.

Let us imagine the operations performed by the computer to be split up into "simple operations" which are so elementary that it is not easy to imagine them further divided. Every such operation consists of some change of the physical system if we know the sequence of symbols on the tape, which of these are observed by the computer (possibly with a special order), and the state of mind of the computer. We may suppose that in a simple operation not more than one symbol is altered. Any other changes can be split up into simple changes of this kind. The situation in regard to the squares whose symbols may be altered in this way is the same as in regard to the observed squares. We may therefore, without loss of generality, assume that the squares whose symbols are changed are always "observed" squares.

Besides these changes of symbols, the simple operations must include changes of distribution of observed squares. The new observed squares must be immediately recognisable by the computer. I think it is reasonable to suppose that they can only be squares whose distance from the closest of the immediately previously observed squares does not exceed a certain fixed amount. Let us say that each of the new observed squares is within L squares of an immediately previously observed square.

In connection with "immediate recognisability", it may be thought that there are other kinds of square which are immediately recognisable. In particular, squares marked by special symbols might be taken as immediately recognisable. Now if these squares are marked only by single symbols there can be only a finite number of them, and we should not upset our theory by adjoining these marked squares to the observed squares. If, on the other hand, they are marked by a sequence of symbols, we cannot regard the process of recognition as a simple process. This is a fundamental point and should be illustrated. In most mathematical papers the equations and theorems are numbered. Normally the numbers do not go beyond (say) 1000. It is, therefore, possible to recognise a theorem at a glance by its number. But if the paper was very long, we might reach Theorem 157767733443477; then, further on in the paper, we might find "... hence (applying Theorem 157767733443477) we have ...". In order to make sure which was the relevant theorem we should have to compare the two numbers figure by figure, possibly ticking the figures off in pencil to make sure of their not being counted twice. If in spite of this it is still thought that there are other "immediately recognisable" squares, it does not upset my contention so long as these squares can be found by some process of which my type of machine is capable. This idea is developed in [III] below.

The simple changes must therefore include:

a. Changes of the symbol on one of the observed squares.

b. Changes of one of the squares observed to another square within L squares of one of the previously observed squares.

It may be that some of these changes necessarily involve a change of state of mind. The most general single operation must therefore be taken to be one of the following:

A. A possible change (**a**) of symbol together with a possible change of state of mind.

B. A possible change (**b**) of observed squares, together with a possible change of state of mind.

The operation actually performed is determined, as has been suggested [above] by the state of mind of the computer and the observed symbols. In particular, they determine the state of mind of the computer after the operation is carried out.

We may now construct a machine to do the work of this computer. To each state of mind of the computer corresponds an "m-configuration" of the machine. The machine scans B squares corresponding to the B squares observed by the computer. In any move the machine can change a symbol on a scanned square or can change any one of the scanned squares to another square distant not more than L squares from one of the other scanned squares. The move which is done, and the succeeding configuration, are determined by the scanned symbol and the m-configuration. ...

[III] We suppose, as in [I], that the computation is carried out on a tape; but we avoid introducing the "state of mind" by considering a more physical and definite counterpart of it. It is always possible for the computer to break off from his work, to go away and forget all about it, and later to come back and go on with it. If he does this he must leave a note of instructions (written in some standard form) explaining how the work is to be continued. This note is the counterpart of the "state of mind". We will suppose that the computer works in such a desultory manner that he never does more than one step at a sitting. The note of instructions must enable him to carry out one step and write the next note. Thus the state of progress of the computation at any stage is completely determined by the note of instructions and the symbols on the tape. That is, the state of the system may be described by a single expression (sequence of symbols), consisting of the symbols on the tape followed by Δ (which we suppose not to appear elsewhere) and then by the note of instructions. This expression may be called the "state formula". We know that the state formula at any given stage is determined by the state formula before the last step was made, and we assume that the relation of these two formulae is expressible in the functional calculus [see Chapter 21 of this text]. In other words, we assume that there is an axiom **A** which expresses the rules governing the behaviour of the computer, in terms of the relation of the state formula at any stage to the state formula at the preceding stage. If this is so, we can construct a machine to write down the successive state formulae, and hence to compute the required number.

Turing, pp. 135–140

B. Descriptions and Examples of Turing Machines

We shall describe a machine according to the conditions prescribed by Turing in his article given above. We are going to assume that the computer can scan only one square at a time (Turing's bound B will be taken to be 1), and can move at most one square to the left or right (his bound L will be taken to be 1). The same arguments that convinced us that having any bound at all was no restriction should also convince us that we can simulate any higher bounds with these. We are also going to assume that the only symbol other than a blank square which the machine can recognize is 1. Since we can use unary notation to represent numbers, this will be no restriction on what we can compute. This version is due in essence to Kleene, 1952, Chapter XIII; see Odifreddi, 1989, for a detailed explanation of why the restrictions are inessential.

The machine is composed of the following parts:

1. A *tape* divided into squares; the tape is assumed to be finite but additional blank squares can be added to the right end or left end at any time; that is, the tape is "potentially infinite".
2. A device called a *head* which can do the following:
   i. Observe one square at one moment (the *scanned square*).
   ii. Read whether the square is blank or has a 1 written in it.
   iii. Write or delete a symbol 1.
   iv. Move to the square immediately left or right of the one it is observing.

We further assume that the machine is always in one of a finite number of *states* ("of mind", or, as Turing says, it works according to a finite supply of notes of instructions).

The operation of the machine is determined by the current state and the current symbol being observed (either a blank, which we notate by "0" below, or a "1"), which generate the following single operations and a new (or possibly the same) state:

    a. Write the symbol 1 on the observed square   1
    b. Delete whatever symbol appears on the observed square   0
    c. Move one square to the right of the observed square   R
    d. Move one square to the left of the observed square   L

Thus, a complete instruction to the machine consists of a *quadruple*

$$(q_i, S, Op, q_j)$$

where q_i is the current state, $S \in \{0,1\}$ is the current symbol (recall that "0" simply means the tape is blank), $Op \in \{ 1, 0, R, L \}$ is one of the operations above, and q_j is the new state. Note that we allow both $(q_i,0,0,q_j)$ and $(q_i,1,1,q_j)$, understanding by these that the machine only changes states.

To help you visualize the operation of a Turing machine, imagine the tape to be a railroad track that can be extended at will in either direction, with the squares being the spaces between the ties. On the track is a boxcar that has an opening in the bottom just big enough to see one square, and that has a lever which can move it one square in either direction. Think of a man or a woman in the machine (but not both together as that will complicate matters) and he or she has n different cards labeled $q_1, q_2, \dots , q_n$, which carry instructions. At every stage just one of these cards is on a board that he or she is looking at. The instruction has a conditional form: *if* you see this, then do that and pick up card number ...

We make some *conventions* about the machines:

1. A Turing machine (*TM*) always starts in its lowest numbered state, which for convenience we require to be q_1.
2. If there is no possible instruction to follow, the machine stops (*halts*).
3. The quadruples in a program for a TM never present conflict in instructions: in any one program there are no quadruples that have the same first two coordinates and disagree on one of the last two.

A *program* for a Turing machine is a finite collection of quadruples subject to these conventions.

There is no difference (in the theory) if we think of a unique TM running all different programs or if we have a TM dedicated to each program, and we will indifferently refer to a set of quadruples as a program or a machine.

We call a *configuration* of a TM an ordered triple:

(contents of the tape, scanned square, state)

The quadruples, then, can be seen as functions that transform configurations into new configurations. Parentheses and commas are omitted in the quadruples below. The sequence of 0's and 1's on the left represents the contents of the machine, and a number below that indicates the position of the head in that state of the machine.

Example 1   Write n 1's in a row on a blank tape of a TM and have the head return to the starting point.

The number of 1's that the machine writes will be controlled by the number of quadruples that write 1's. The figures below denote configurations, and numbers under the squares denote the states.

(a) $0\ \underset{1}{0}\ 0\ 0\ \cdots$ Recall that the machine starts in state q_1. We give instruction $q_1\ 0\ 1\ q_1$ to make the head write on the square it is now scanning. This produces (b).

(b) $0\ \underset{1}{1}\ 0\ 0\ \cdots$ We now have to move the head: $q_1\ 1\ R\ q_2$ produces (c).

(c) $0\ 1\ \underset{2}{0}\ 0\ \cdots$ Now we write the next $(n-1)$ 1's by repeating the instructions above using different states,

$$q_2\ 0\ 1\ q_2,\quad q_2\ 1\ R\ q_3,\quad \ldots,\quad q_n\ 0\ 1\ q_{n+1},$$

which produces (d).

(d) $0\ 1\ 1\cdots\ \underset{n+1}{1}\ 0\ \cdots$ To return to the starting position we don't have to count: we merely move left until we find a blank, and then move one square to the right.

(e) $0\ \underset{n+2}{1}\ 1\ 1\cdots 1\ 0\ \cdots$ $q_{n+1}\ 1\ L\ q_{n+1},\quad q_{n+1}\ 0\ R\ q_{n+2}$

The whole program then consists of $2n+1$ quadruples using $n+2$ states:

$$q_1\ 0\ 1\ q_1,\qquad q_1\ 1\ R\ q_2,\qquad q_2\ 0\ 1\ q_2,\qquad q_2\ 1\ R\ q_3,\quad \ldots,$$
$$q_n\ 0\ 1\ q_{n+1},\qquad q_{n+1}\ 1\ L\ q_{n+1},\qquad q_{n+1}\ 0\ R\ q_{n+2}\ .$$

Example 2   Write a TM which when started with its head at the leftmost 1 of a sequence of 1's on a tape that contains nothing else, duplicates the number of 1's and stops with its head at the leftmost 1.

We will make a machine that begins at the leftmost 1 and (i) erases the 1, (ii) moves right to the end of the string of 1's, (iii) skips over the first blank there, (iv) if it finds a blank, goes to the next step; if it finds a 1, goes to the rightmost 1 of that string, (v) writes two new 1's, (vi) returns to the leftmost 1 of the original string and repeats, until (vii) it finds a blank when looking for that 1, then moves to the right and stops on the leftmost 1 of the new string.

$q_1\ 1\ 0\ q_2$ Sees a 1 of the input and deletes it.

$q_2\ 0\ R\ q_3$

q_3 1 R q_3
q_3 0 R q_4 Moves one square beyond the blank to the right of the input string.

q_4 0 0 q_5
q_4 1 R q_4 Either that square is blank, or it moves right to the first blank.

q_5 0 1 q_5
q_5 1 R q_6 Writes two 1's to the right.
q_6 0 1 q_7

q_7 1 L q_7
q_7 0 L q_8 Moves to the leftmost 1 of the original input string.
q_8 1 L q_8
q_8 0 R q_9

q_9 1 1 q_1 Repeats the entire procedure if there are any 1's remaining from
q_9 0 R q_{10} the input; otherwise goes to the leftmost 1 of the new string of 1's
and stops.

Let's see how this TM works on a concrete example, duplicating three 1's. Successive configurations of the machine are

(1) 0 $\underset{1}{1}$ 1 1 0   (2) 0 0 $\underset{2}{1}$ 1 0   (3) 0 0 1 $\underset{3}{1}$ 0   (4) 0 0 1 $\underset{3}{1}$ 0

(5) 0 0 1 1 $\underset{3}{0}$ (6) 0 0 1 1 0 $\underset{4}{0}$ (7) 0 0 1 1 0 $\underset{5}{0}$ (8) 0 0 1 1 0 $\underset{5}{1}$

(9) 0 0 1 1 0 1 $\underset{6}{0}$ (10) 0 0 1 1 0 1 $\underset{7}{1}$

You should now write out for yourself all the remaining configurations of the machine until it halts.

Example 3   Write $2n$ 1's on a blank tape, stopping at the leftmost 1.

This problem can be solved in the same way as Example 1, using $2n + 2$ states. But there is a more economical way (in terms of the number of states): we can use the machine of Example 1 to write n 1's using $n + 2$ states, then connect the machine of Example 2 to duplicate that sequence. The only question is how do we "connect" two machines? Here we can rename the states of the second machine so that the first state is now q_{n+2}, the last is q_{n+11}, and the ith is $q_{(n+1)+i}$. We thus use $n + 11$ instead of $2n + 2$ states.

C. Turing Machines and Functions

We need to make some conventions in order to be able to interpret a TM as calculating a function. To represent numbers we will use strings of 1's, $111 \ldots 1$; for $n \geq 1$ we denote the string of n 1's by 1^n. We say that a tape is in *standard configuration* if it is either blank or contains only one string of the form 1^n.

We will say that a Turing machine M *calculates output m on input n* if:

i. M starts by scanning the leftmost 1 of a string 1^{n+1} on an otherwise blank tape,

ii. M begins in its lowest numbered state,

iii. M stops in its highest numbered state and

    a. $m = 0$ and the tape is blank

      or

    b. $m \neq 0$ and M is scanning the leftmost 1 of 1^m on an otherwise blank tape.

Note that we use 1^{n+1} to represent input n but 1^m to represent output m.

For functions of several variables, we say that M calculates output m on input $n_1, n_2, \ldots, n_k$ if

i. M starts by scanning the leftmost 1 of a string
$$1^{n_1+1} 0 1^{n_2+1} 0 \ldots 0 1^{n_k+1}$$

    on an otherwise blank tape,

ii. and **iii.** as before.

We say that a Turing machine M *calculates the function f* of k variables if for every $(n_1, n_2, \ldots, n_k)$, $f(n_1, n_2, \ldots, n_k) = m$ iff M calculates output m on input $(n_1, n_2, \ldots, n_k)$.

Finally, we say a function is *Turing machine computable* (*TM computable*) if there is a Turing machine that calculates it. This is an extensional definition. We ask you to show in Exercise 5 that if there is one machine that calculates a function f, then there are arbitrarily many others that also calculate f. Nonetheless, when we are discussing a particular machine that calculates f it is sometimes convenient to refer to it as T_f.

Here are some examples of TM computable functions.

The successor function $S(n) = n + 1$

All we need to do is start the machine on 1^{n+1} and stop. The machine $T_S = \{\, q_1\, 1\, 1\, q_2 \,\}$ does just that.

But if T_S is started on a tape in nonstandard configuration, it also does nothing. So functions of more than one variable are not computed by T_S.

The zero function $Z(n) = 0$

We need to erase all the 1's on the tape. We take T_Z to be:

1) $q_1\, 1\, 0\, q_2$ 2) $q_1\, 0\, R\, q_3$ 3) $q_2\, 0\, R\, q_1$ 4) $q_3\, 1\, 1\, q_1$

In this case, however, the machine also calculates the constant zero-valued function for any number of inputs.

Addition $Add(n, m) = n + m$

We define a machine that starting from $1^{n+1} 0 1^{m+1}$ fills the blank between the blocks, goes to the left and deletes the first three 1's, and then stops (in a standard configuration). This is done by the machine T_{add}:

q_1 1 R q_1	Fill the blank between the input strings.
q_1 0 1 q_2	
q_2 1 L q_2	Look for the leftmost 1.
q_2 0 R q_3	
q_3 1 0 q_4	Delete the leftmost 1 and go right.
q_4 0 R q_5	
q_5 1 0 q_6	Delete the next 1 and go right.
q_6 0 R q_7	
q_7 1 0 q_8	Delete the third 1, go right, and stop.
q_8 0 R q_9	

Multiplication $Mult(m,n) = m \cdot n$

Presented with a string $1^{m+1}01^{n+1}$, we use the first string as a counter device to control the number of repetitions of the second string: we delete a 1 from the string 1^{n+1} and repeat the resulting string m times. The difficulty here is that the value of m has to be read from the input, in contrast to being controlled by the number of states as in Example 3. Here is a description of how our machine T_{mult} will work.

1. Starting from $1^{m+1}01^{n+1}$, T_{mult} deletes the leftmost 1 from 1^{m+1};

   Subcase **i.** if there are no more 1's, it deletes the rest of the ones on the tape and stops (because $m = 0$)

   Subcase **ii.** if there are 1's remaining, it deletes the rightmost 1 from 1^{n+1}.

2. If there are no 1's remaining in the second string, T_{mult} deletes everything and stops (because in this case $n = 0$).

3. If there are 1's remaining in both strings, T_{mult} starts a shifting subroutine of moving the string 1^n exactly n squares to the right:

$$\cdots 0\ 0\ \underbrace{1\ 1 \cdots 1}_{m\ 1\text{'s}}\ 0\ \underbrace{1\ 1 \cdots 1}_{n\ 1\text{'s}} \cdots \quad \text{results in}$$

$$\cdots 0\ 0\ \underbrace{1\ 1 \cdots 1}_{m\ 1\text{'s}}\ 0\ \underbrace{0\ 0 \cdots 0}_{n\ \text{blanks}}\ \underbrace{1\ 1 \cdots 1}_{n\ 1\text{'s}} \cdots$$

   Then T_{mult} deletes the leftmost 1 from 1^m and repeats this process of shifting the block of n 1's so long as there are 1's to be deleted in 1^n.

4. When T_{mult} finds the last 1 in the counter block, it deletes it, moves two squares to the right, and changes all 0's to 1's (going right), until the first 1 is found. At that point we have m successive blocks of n 1's, and the machine goes to the leftmost 1 and stops.

The machine T_{mult}

q_1 1 0 q_1	Starting at the leftmost 1 of $1^{m+1}01^{n+1}$, delete the leftmost 1
q_1 0 R q_2	and go right.
q_2 0 R q_3	If $m = 0$, delete the rest of the tape and stop.

q_3 1 0 q_4
q_4 0 R q_3
q_3 0 0 q_{39}

q_2 1 1 q_5 If $m \neq 0$, go to the second 1 of 1^{n+1}.
q_5 1 R q_5
q_5 0 R q_6
q_6 1 R q_{17}

q_{17} 0 L q_{10} If this square has a 0, then delete everything and stop
q_{10} 1 0 q_{35} (the product is 0 because $n = 0$).
q_{35} 0 L q_{36}
q_{36} 0 L q_{36}
q_{36} 1 1 q_{37}
q_{37} 1 0 q_{38}
q_{38} 0 L q_{37}
q_{37} 0 L q_{39}

q_{17} 1 1 q_7 If this square has a 1, then $n \neq 0$, so delete the rightmost 1
q_7 1 R q_7 of 1^{n+1}.
q_7 0 L q_8
q_8 1 0 q_9

q_9 0 L q_{14} Return to the leftmost 1 of 1^m.
q_{14} 1 L q_{14}
q_{14} 0 L q_{15}
q_{15} 1 L q_{15}
q_{15} 0 R q_{16}

q_{16} 1 1 q_{18} Delete the leftmost 1 from 1^m and move right.
q_{18} 1 0 q_{18}
q_{18} 0 R q_{19} If there are no more 1's in 1^m, fill the blanks with 1's while
q_{19} 0 R q_{32} moving right and return to a standard configuration.
q_{32} 0 1 q_{33}
q_{33} 1 R q_{32}
q_{32} 1 L q_{34}
q_{34} 1 L q_{34}
q_{34} 0 R q_{39}

q_{19} 1 R q_{20} While there are 1's in 1^m, enter into a shifting subroutine:
q_{20} 1 R q_{20} go to the leftmost 1 of 1^n and delete it.
q_{20} 0 R q_{21} Then go right one square.
q_{21} 0 R q_{21}
q_{21} 1 0 q_{22}
q_{22} 0 R q_{23}

q_{23} 0 1 q_{29} If this square is empty, then the whole block 1^n has been shifted.
q_{29} 1 L q_{30} Go to the beginning of the counter block.
q_{30} 0 L q_{30}
q_{30} 1 L q_{31}
q_{31} 1 L q_{31}
q_{31} 0 R q_{18}

q_{23} 1 R q_{24} If this square is not empty, then keep shifting and deleting 1's
q_{24} 1 R q_{24} from 1^n.
q_{24} 0 R q_{25}
q_{25} 1 R q_{26}
q_{26} 1 R q_{26}
q_{26} 0 1 q_{27}
q_{25} 0 1 q_{27}
q_{27} 1 L q_{27}
q_{27} 0 L q_{28}
q_{28} 1 L q_{28}
q_{28} 0 R q_{21}

Composition of functions

If f and g are functions of one variable and T_f calculates f with highest state q_n, and T_g calculates g, then to produce a machine which calculates the composition $g \circ f$ we do the following:

i. Add quadruples to convert output to input by writing a 1 to the right and returning to the leftmost 1:

$$q_n 1 R q_n, \quad q_n 0 1 q_{n+1}, \quad q_{n+1} 1 L q_{n+1}, \quad q_{n+1} 0 R q_{n+2}$$

ii. Relabel the states of T_g from q_i to q_{i+n+1} so that it will start where we left off.

iii. Collect these quadruples together as $T_{g \circ f}$.

There are many other functions that are TM computable and many ways we can combine machines to form new functions from those. But if you haven't already noticed, Turing machines are an unwieldy way to calculate. It's difficult to show that even very simple functions are TM computable: we offer a prize of $4.17 to anyone who can provide us with a Turing machine which calculates the exponential function $f(x,y) = x^y$ along with a clear enough explanation we can understand.

The point of Turing machines, at least as far as we're concerned with them in this book, is to provide an analysis of computability by breaking that notion into its smallest components. Rather than dwell on Turing machines now, we are going to look at computability from the point of view of arithmetic descriptions of functions and then show in Chapter 18 that the two approaches are equivalent.

Exercises ─────────────────────────────────────

1. a. Does the machine of Example 2 calculate a function?

   b. Define a TM (i.e., give a collection of quadruples) that for every n duplicates a string of the form 1^n, creating $1^n 0 1^n$. Does this machine calculate any function?

2. a. Define a TM that calculates the *projection* function on the first coordinate, $P(m, n) = m$.

   †b. For every k and i such that $k \geq i \geq 1$ define a TM that calculates the *projection on the ith coordinate*, $P_k^i (n_1, n_2, \ldots, n_k) = n_i$.

3. Show that for every n the *constant* function $\lambda x(n)$ is TM computable. (*Hint*: Modify Example 1. Does your modification also calculate $\lambda x \lambda y(n)$?)

4. Prove that there are infinitely many distinct TM computable functions.

5. Prove that if a function is TM computable, then there are infinitely many different Turing machines that calculate it.

6. Show that the *equality* function

$$E(m,n) = \begin{cases} 1 & \text{if } m = n \\ 0 & \text{if } m \neq n \end{cases}$$

   is TM computable.

†7. Give an effective numbering of all Turing machines which satisfies criteria (1), (2), (3) of Chapter 8.C.

8. *The halting problem for Turing machines*

   From Exercise 7 we can list all Turing machines $M_0, M_1, \ldots, M_n, \ldots$ (let the nth one be the machine which has the nth smallest number assigned to it). Each calculates a function of one variable (although that may be undefined for every input). Show that the function

$$h(m,n) = \begin{cases} 1 & \text{if } M_m \text{ halts on input } n \\ 0 & \text{otherwise} \end{cases}$$

   known as the *halting problem for Turing machines*, is not Turing machine computable.

   (*Hint*: If there were a machine H that computed h, we could define another Turing machine that, given input n,

   a. writes $1^{n+1} 0 1^{n+1}$

   b. implements H on that input, and then

      if the result is the blank tape, it writes a 1 and halts

      if the result is a single 1 on the tape, it goes into a loop and never halts.

   What would the number of that new machine be?)

†9. *The Busy Beaver problem*

This problem was proposed by T. Rado in 1962 in order to give a concrete example of a function that is not TM computable.

Given a TM, define its *productivity* to be the number of 1's in the tape if it halts in a standard configuration starting from a blank tape and 0 otherwise. For each $n \geq 1$ we define $p(n)$ to be the *maximal* productivity of any machine with n states.

a. Show that $p(1) \geq 1$.

b. Show that $p(n + 11) \geq 2n$. (*Hint:* See Example 3.)

c. Show that $p(n + 1) > p(n)$.

Conclude that for all i, j, if $p(i) \geq p(j)$ then $i \geq j$.

d. Show that if there exists a TM P with k states that computes the function p, then $p(n + 2 + 2k) \geq p[p(n)]$.

(*Hint:* connect P twice with the machine T_n of Example 1, which writes n 1's on a blank tape (suggestively) $P[P(T_n)]$.)

e. Conclude that p is not TM computable.

(*Hint:* $p(n + 13 + 2k) \geq p[p(n + 11)]$ and then apply parts (c) and (b) to get a contradiction.)

Further Reading

For more about Turing machines consult Martin Davis' *Computability and Unsolvability*, Rózsa Péter's *Recursive Functions*, or Stephen Kleene's *Introduction to Metamathematics*.

10 The Most Amazing Fact and Church's Thesis

A. The Most Amazing Fact

We have studied one formalization of the notion of computability. In succeeding chapters we will study two more: recursive functions and functions representable in a formal system.

The Most Amazing Fact
All the attempts at formalizing the intuitive notion of computable function yield exactly the same class of functions.

So if a function is Turing machine computable, it can also be computed in any of the other systems described in Chapter 8.E. This is a mathematical fact which requires a proof. In Chapters 18 and 22 we do it for the two formalizations mentioned above; Odifreddi, 1989 establishes all the equivalences. Once you're quite familiar with one system it'll be easier to follow such a proof.

The Most Amazing Fact is stated about an extensional class of functions, but it can be stated constructively: Any computation procedure for any of the attempts at formalizing the intuitive notion of computable function can be translated into any other formalization in such a way that the two formalizations have the same outputs for the same inputs.

In 1936, even before these equivalences were established, Church said,

> We now define the notion, already discussed, of an *effectively calculable* function of positive integers by identifying it with the notion of a recursive function of positive integers (or of a λ-definable function of positive integers). This definition is thought to be justified by the considerations which follow, so far as positive justification can ever be obtained for the selection of a formal definition to correspond to an intuitive notion.
>
> Church, 1936, p. 100

[Note: Church's definition of "recursive function" is different from the one commonly used now.]

So we have

Church's Thesis: A function is computable iff it is λ-definable .

This is a nonmathematical thesis: it equates an intuitive notion (computability) with a precise, formal one (λ-definability). By our amazing fact this thesis is equivalent to

A function is computable iff it is Turing machine computable.

Turing devised his machines in a conscious attempt to capture in simplest terms what computability is. That his model turned out to give the same class of functions as Church's (as established by Turing in the paper cited above) was strong evidence that it was the "right" class. Later we will consider some criticisms of Church's Thesis in that the notion of computability should coincide with either a larger or a smaller class than the Turing machine computable ones.

Prior to that we are going to study this class from a more purely arithmetical point of view, not using a machine definition at all. Turing machines break up the notion of computability into its most basic parts, but at the cost of getting a definition that is very cumbersome to use. By turning to *recursive functions* we'll have an arithmetical system we can use more easily.

But first let's look at another formalization of the notion of computability given by Post, along with his comments on Church's Thesis.

B. Emil L. Post on Computability (Optional)

Post's analysis of computability was done independently of Turing, though not of Church. It is therefore surprising how very similar it is to Turing's analysis in his paper in Chapter 9 (similarities to our formalization of Turing's ideas are not so remarkable since we've been influenced by developments since then, including Post's paper). Post, too, attempts to justify his formulation in intuitive terms. Note that, unlike Church, he does not view Church's Thesis as a *definition* but claims that if, as it turned out, the Most Amazing Fact holds, then Church's Thesis amounts to a *natural law*.

"Finite Combinatory Processes —Formulation 1" *

The present formulation should prove significant in the development of symbolic logic along the lines of Gödel's theorem on the incompleteness of

* Received October 7, 1936. The reader should compare an article by A. M. Turing, "On computable numbers," shortly forthcoming in the *Proceedings of the London Mathematical Society*. The present article, however, although bearing a later date, was written entirely independently of Turing's. *Editor* [of *The Journal of Symbolic Logic*].

symbolic logics[1] and Church's results concerning absolutely unsolvable problems.[2]

We have in mind a *general problem* consisting of a class of *specific problems*. A solution of the general problem will then be one which furnishes an answer to each specific problem.

In the following formulation of such a solution two concepts are involved: that of a *symbol space* in which the work leading from problem to answer is to be carried out,[3] and a fixed unalterable *set of directions* which will both direct operations in the symbol space and determine the order in which those directions are to be applied.

In the present formulation the symbol space is to consist of a two way infinite sequence of spaces or boxes, i.e., ordinally similar to the series of integers ... , –3, –2, –1, 0, 1, 2, 3, ... The problem solver or worker is to move and work in this symbol space, being capable of being in, and operating in but one box at a time. And apart from the presence of the worker, a box is to admit of but two possible conditions, i.e., being empty or unmarked, and having a single mark in it, say a vertical stroke.

One box is to be singled out and called the starting point. We now further assume that a specific problem is to be given in symbolic form by a finite number of boxes being marked with a stroke. Likewise the answer is to be given in symbolic form by such a configuration of marked boxes. To be specific, the answer is to be the configuration of marked boxes left at the conclusion of the solving process.

The worker is assumed to be capable of performing the following primitive acts: [4]

    (a) *Marking the box he is in (assumed empty),*

    (b) *Erasing the mark in the box he is in (assumed marked),*

    (c) *Moving to the box on his right,*

    (d) *Moving to the box on his left,*

    (e) *Determining whether the box he is in, is or is not marked.*

The set of directions which, be it noted, is the same for all specific problems and thus corresponds to the general problem, is to be of the following form. It is to be headed:

Start at the starting point and follow direction 1.

It is then to consist of a finite number of directions to be numbered 1, 2, 3, ... n. The ith direction is then to have one of the following forms:

    (A) *Perform operation O_i [O_i = (a), (b), (c), or (d)] and then follow direction j_i,*

    (B) *Perform operation (e) and according as the answer is yes or no correspondingly follow direction $j_i{'}$ or $j_i{''}$,*

    (C) *Stop.*

Clearly but one direction need be of type C. Note also that the state of the

---

[1] Kurt Gödel, [1931].

[2] Alonzo Church, [1936].

[3] Symbol space, and time.

[4] As well as otherwise following the directions described below.

symbol space directly affects the process only through directions of type B.

A set of directions will be said to be *applicable* to a given general problem if in its application to each specific problem it never orders operation (a) when the box the worker is in is marked, or (b) when it is unmarked.[5] A set of directions applicable to a general problem sets up a deterministic process when applied to each specific problem. This process will terminate when and only when it comes to the direction of type (C). The set of directions will then be said to set up a *finite 1-process* in connection with the general problem if it is applicable to the problem and *if the process it determines terminates for each specific problem.* A finite 1-process associated with a general problem will be said to be a *1-solution* of the problem if the answer it thus yields for each specific problem is always correct.

We do not concern ourselves here with how the configuration of marked boxes corresponding to a specific problem, and that corresponding to its answer, symbolize the meaningful problem and answer. In fact the above assumes the specific problem to be given in symbolized form by an outside agency and, presumably, the symbolic answer likewise to be received. A more self-contained development ensues as follows. The general problem clearly consists of at most an enumerable infinity of specific problems. We need not consider the finite case. Imagine then a one-to-one correspondence set up between the class of positive integers and the class of specific problems. We can, rather arbitrarily, represent the positive integer *n* by marking the first *n* boxes to the right of the starting point. The general problem will then be said to be *1-given* if a finite 1-process is set up which, when applied to the class of positive integers as thus symbolized, yields in one-to-one fashion the class of specific problems constituting the general problem. It is convenient further to assume that when the general problem is thus 1-given each specific process at its termination leaves the worker at the starting point. If then a general problem is 1-given and 1-solved, with some obvious changes we can combine the two sets of directions to yield a finite 1-process which gives the answer to each specific problem when the latter is merely given by its number in symbolic form.

With some modification the above formulation is also applicable to symbolic logics. We do not now have a class of specific problems but a single initial finite marking of the symbol space to symbolize the primitive formal assertions of the logic. On the other hand, there will now be no direction of type (C). Consequently, assuming applicability, a deterministic process will be set up which is *unending*. We further assume that in the course of this process certain recognizable symbol groups, i.e., finite sequences of marked and unmarked boxes, will appear which are not further altered in the course of the process. These will be the derived assertions of the logic. Of course the set of directions corresponds to the deductive processes of the logic. The logic may then be said to be *1-generated*.

An alternative procedure, less in keeping, however, with the spirit of

[5] While our formulation of the set of directions could easily have been so framed that applicability would immediately be assured it seems undesirable to do so for a variety of reasons.

symbolic logic, would be to set up a finite 1-process which would yield the nth theorem or formal assertion of the logic given n, again symbolized as above.

Our initial concept of a given specific problem involves a difficulty which should be mentioned. To wit, if an outside agency gives the initial finite marking of the symbol space there is no way for us to determine, for example, which is the first and which the last marked box. This difficulty is completely avoided when the general problem is 1-given. It has also been successfully avoided whenever a finite 1-process has been set up. In practice the meaningful specific problems would be so symbolized that the bounds of such a symbolization would be recognizable by characteristic groups of marked and unmarked boxes.

The root of our difficulty however, probably lies in our assumption of an infinite symbol space. In the present formulation the boxes are, conceptually at least, physical entities, e.g., contiguous squares. Our outside agency could no more give us an infinite number of these boxes than he could mark an infinity of them assumed given. If then he presents us with the specific problem in a finite strip of such a symbol space the difficulty vanishes. Of course this would require an extension of the primitive operations to allow for the necessary extension of the given finite symbol space as the process proceeds. A final version of a formulation of the present type would therefore also set up directions for generating the symbol space.[6]

The writer expects the present formulation to turn out to be logically equivalent to recursiveness in the sense of the Gödel–Church development.[7] Its purpose, however, is not only to present a system of a certain logical potency but also, in its restricted field, of psychological fidelity. In the latter sense wider and wider formulations are contemplated. On the other hand, our aim will be to show that all such are logically reducible to formulation 1. We offer this conclusion at the present moment as a *working hypothesis*. And to our mind such is Church's identification of effective calculability with recursiveness.[8]

[6] The development of formulation 1 tends in its initial stages to be rather tricky. As this is not in keeping with the spirit of such a formulation the definitive form of this formulation may relinquish some of its present simplicity to achieve greater flexibility. Having more than one way of marking a box is one possibility. The desired naturalness of development may perhaps better be achieved by allowing a finite number, perhaps two, of physical objects to serve as pointers, which the worker can identify and move from box to box.

[7] The comparison can perhaps most easily be made by defining a 1-function and proving the definition equivalent to that of recursive function. (See Church, loc. cit., p. 350.) A 1-function $f(n)$ in the field of positive integers would be one for which a finite 1-process can be set up which for each positive integer n as problem would yield $f(n)$ as answer, n and $f(n)$ symbolized as above.

[8] Cf. Church, loc. cit., pp. 346, 356–58. Actually the work already done by Church and others carries this identification considerably beyond the working hypothesis stage. But to mask this identification under a definition hides the fact that a fundamental discovery in the limitations of the mathematicizing power of Homo Sapiens has been made and blinds us to the need of its continual verification.

Out of this hypothesis, and because of it apparent contradiction to all mathematical development starting with Cantor's proof of the non-enumerability of the points of a line, independently flows a Gödel-Church development. The success of the above program would, for us, change this hypothesis not so much to a definition or to an axiom but to a *natural law*. Only so, it seems to the writer, can Gödel's theorem concerning the incompleteness of symbolic logics of a certain general type and Church's results on the recursive unsolvability of certain problems be transformed into conclusions concerning all symbolic logics and all methods of solvability.

Post, 1936

11 Primitive Recursive Functions

While the most convincing definition of mechanical procedures is by means of Turing's concept of abstract machines, the equivalent concept of recursive functions first appeared historically as more or less a culmination of extensions of the simple recursive definitions of addition and multiplication.

Wang, p. 87

A. Definition by Induction

When you first learned about exponentiation you probably were told that

$$x^n = \underbrace{x \cdot x \cdot \cdots \cdot x}_{n \ \text{times}}$$

That was suggestive and probably convinced you that you could compute the function. Later you learned a proper definition by induction: $x^0 = 1$ and $x^{n+1} = x^n \cdot x$.

Similarly, the factorial function is usually introduced as $n! = n \cdot (n-1) \cdot \cdots \cdot 2 \cdot 1$. An inductive definition of it would be: $0! = 1$ and $(n+1)! = (n+1) \cdot (n!)$.

In its simplest general form, a definition of a function f by induction from another function g looks like $f(0) = m$ and $f(n+1) = g(f(n))$. We have confidence that this method of definition really gives us a function because we can convince ourselves that the generation of values of f can be matched to the generation of natural numbers and is completely determined at each stage:

To	0	1	2	3	
Assign	$f(0) = m$	$f(1) = g(m)$	$f(2) = g(f(1))$	$f(3) = g(f(2))$	...

This convincing ourselves cannot be reduced to a proof by induction, for to apply that method here we'd already need to have f in hand.

Moreover, since generating the natural number series is effective (computable), if g is computable then without doubt f will also be computable. So let's consider the functions we can obtain using induction and composition, starting with a few simple indisputably computable functions.

B. The Definition of the Primitive Recursive Functions

The class of functions we described in intuitive terms in Section A is composed entirely of computable functions. But for a function to be computable, there must be an algorithm or procedure for computing it. So in our formal definition of this class of functions we must replace the intuitive, semantic ideas of Section A with precise descriptions of the functions, exactly as we did in Chapter 9.

To begin, we take as variables the letters n, x_1, x_2, ... though we will continue to use x, y, and z informally. We'll write $\vec{x}$ for $(x_1, \ldots, x_k)$.

Next we list the basic, incontrovertibly computable functions that we will use as building blocks for all others.

1. Basic (initial) functions

zero $Z(n) = 0$ for all n

successor $S(n) = \begin{cases} \text{that number which follows } n \\ \text{in the natural number series} \end{cases}$

projections $P_k^i(x_1, \ldots, x_k) = x_i$ for $1 \leq i \leq k$

We sometimes call the projections the *pick-out* functions, and P_1^1 the *identity* function, written $id(x) = x$. We don't say that $S(x) = x + 1$ because addition is a more complicated function which we intend to define.

Next, we specify the ways we allow new functions to be defined from ones we already have.

2. Basic operations

Composition

If g is a function of m-variables and $h_1, \ldots, h_m$ are functions of k variables, which are already defined, then composition yields the function

$$f(\vec{x}) = g(h_1(\vec{x}), \ldots, h_m(\vec{x}))$$

Primitive recursion

For functions of one variable the schema is:

$$f(0) = d$$
$$f(n+1) = h(f(n), n)$$

where d is a number and h is a function already defined.

For functions of two or more variables, if g and h are already defined then f is given by *primitive recursion on h with basis g* as:

$$f(0, \vec{x}) = g(\vec{x})$$
$$f(n+1, \vec{x}) = h(f(n, \vec{x}), n, \vec{x})$$

[The reason we allow n and $\vec{x}$ as well as $f(n, \vec{x})$ to appear in h is that we may wish to keep track of both the stage we're at and the input, so that we can have, for example, $f(5,47) = f(10,47)$, but $f(6,47) \neq f(11,47)$.]

3. An inductive definition of the class of functions

Finally, we complete the definition by stipulating that the *primitive recursive functions* are exactly those which are either basic or can be obtained from the basic ones by a finite number of applications of the basic operations. This is an *inductive definition* of the class of functions. To see that, think of assigning:

0	to all the basic functions
1	to all those functions which can be obtained by one or no application of a basic operation to functions which have been assigned 0 (so the basic functions are also assigned 1)
$\vdots$	
$n + 1$	to all those functions which can be obtained by at most one application of a basic operation to functions which have been assigned a number less than $n + 1$.

Then a function is primitive recursive if and only if it is assigned some number n.

Another way the class of primitive recursive functions is sometimes described is by saying that it is the *smallest* class containing the basic functions and *closed under* the basic operations, where "smallest" is understood to mean the set-theoretic intersection and "closed under" means that whenever one of the operations is applied to elements of the set, the resulting object is also in the set. That way of talking presupposes that the entire completed infinity of the class of functions exists as an intersection of other infinite classes of functions, whereas the inductive definition is nothing more than a constructive way of handing out the label "primitive recursive" to various functions. Since we wish to avoid the use of infinities in our analysis of computability, when we speak sometimes of a class closed under an operation we will understand that as shorthand for an inductive definition.

Thus, to demonstrate that a function is primitive recursive we need to show that it has a description, a definition that precisely fits the criteria above. Though here, as for Turing machine computable functions, if a function has one definition then it will have arbitrarily many others (Exercise 9).

C. Examples

1. The constants

For any natural number n, the function $\lambda x \, f(x) = n$ can be defined as

$$\lambda x \underbrace{S\,(S\,(...S\,(Z\,(x))...))}_{n\ S's}$$

But the use of "..." is precisely what we are trying to avoid. We define inductively a sequence of functions: $C_0 = Z$; $C_{n+1} = S \circ C_n$, so that $\lambda x_1\, C_n(x_1) = n$.

Here again we have defined the natural numbers by a unary representation, reflecting that "zero and the idea of one more", rather than "whole number and zero", is our primitive concept.

2. Addition

We can define $x + n$ by viewing it as a function of one variable, n, with the other variable held fixed as parameter. That is, we define addition by x, $\lambda n\,(x + n)$, as:

$x + 0 = x$
$x + (n + 1) = (x + n) + 1$

But that's not a proper definition according to our description of primitive recursive functions. So let's try again:

$+ (0, x) = x$
$+ (n + 1, x) = S(+ (n, x))$

This seems like a careful formal definition, but it still doesn't have the required form. A definition that exactly fits the criteria given in Section B for a function to be classified as primitive recursive begins by first defining $S(P_3^1(x_1, x_2, x_3))$, which is primitive recursive since it's a composition of initial functions. Then,

$$+ (0, x_1) = P_1^1(x_1)$$
$$+ (n + 1, x_1) = S(P_3^1(+ (n, x_1), n, x_1))$$

3. Multiplication

Now that we have addition, we can give an inductive definition of multiplication. We use x as a parameter to define $x \cdot n$ as multiplication by x, so $x \cdot 0 = 0$ and $x \cdot (n + 1) = (x \cdot n) + x$. Or, to write it in functional notation,

$\cdot (0, x) = 0$ and $\cdot (n + 1, x) = + (x, \cdot (n, x))$

This definition looks formal enough, but again it isn't in a form specified in Section B necessary to justify that multiplication is primitive recursive. The first homework exercise below is for you to give such a definition.

4. Exponentiation

Formally in our system of primitive recursive functions we define:

$Exp(0, x_1) = 1$
$Exp(n + 1, x_1) = h(Exp(n, x_1), n, x_1)$

where
$$h(x_1, x_2, x_3) = \cdot (P_3^1(x_1, x_2, x_3), P_3^3(x_1, x_2, x_3))$$

5. Signature and zero test

The signature function is:

$sg(0) = 0$
$sg(n + 1) = 1$

The zero test function is:

$\overline{sg}(0) = 1$
$\overline{sg}(n + 1) = 0$

Exercise 2 asks for definitions of these functions that fulfill the criteria of Section B.

6. Half

We can't divide odd numbers by 2, but we can find the largest natural number less than or equal to one-half of n:

$$half(n) = \begin{cases} \frac{n}{2} & \text{if } n \text{ is even} \\ \frac{n-1}{2} & \text{if } n \text{ is odd} \end{cases}$$

To give a primitive recursive definition of this function, we first need to be able to separate out the case where n is odd:

$$Odd(n) = \begin{cases} 1 & \textit{if } n \text{ is odd} \\ 0 & \textit{if } n \text{ is even} \end{cases}$$

We ask you to show that *Odd* is primitive recursive in Exercise 3. Then

$half(0) = 0$ and $half(n + 1) = h(half(n), n)$

where
$$h(x_1, x_2) = + (P_2^1(x_1, x_2), Odd(P_2^2(x_1, x_2)))$$

7. Predecessor and limited subtraction

In order to define addition, we started with the successor function which adds 1. To define subtraction, we start with the predecessor function which subtracts 1, namely, $P(0) = 0$; $P(n + 1) = n$. Exercise 4 below asks you to show that this function is primitive recursive.

Since we can't define subtraction on the natural numbers, we define *limited subtraction*:

$$x \div n = \begin{cases} x - n & \textit{if } n \leq x \\ 0 & \textit{if } n > x \end{cases}$$

Keeping x fixed, as n increases the value of $x \div n$ goes down until 0 is reached. So we can define: $x \div 0 = x$; $x \div (n + 1) = P(x \div n)$, which you can convert into a correct formal definition (Exercise 4).

Exercises Part 1 ————————————————————————————————

1. Give a definition of multiplication as a primitive recursive function that precisely fits the specifications of Section B. Compare that definition to the Turing machine definition in Chapter 9.C.

2. Demonstrate that $\overline{sg}$ and sg are primitive recursive.

3. Show that *Odd* is primitive recursive.

4. Show that the predecessor and limited subtraction functions are primitive recursive.

5. Give a primitive recursive definition of the factorial function of Section A.

6. Demonstrate that the following functions are primitive recursive:
$$< (x,y) = \begin{cases} 1 & if\ x < y \\ 0 & if\ x \ge y \end{cases} \qquad E(x,y) = \begin{cases} 1 & if\ x = y \\ 0 & if\ x \ne y \end{cases}$$

7. Show that the function f "defined" by $f(n) = 0 + 1 + \cdots + n$ is primitive recursive.

8. Denote the *maximum* of $x_1, \ldots, x_n$ by $max(x_1, \ldots, x_n)$. Show that this is primitive recursive. (*Hint*: See Section C.1; there is one function for each n.)

9. Show that if f has a primitive recursive definition, then there are arbitrarily (countably) many other primitive recursive definitions that give rise to f.

†10. A famous function defined by induction is the Fibonacci series:
$$1, 1, 2, 3, 5, 8, 13, \ldots, \quad u_{n+2} = u_{n+1} + u_n$$
To calculate u_n we need to know what's been calculated in the previous two steps, which, backtracking, we can do once we get to the first two terms. Devise a definition of $f(n) = u_n$ as a primitive recursive function.

D. Other Operations That Are Primitive Recursive

We wouldn't be surprised if you had difficulty showing that the Fibonacci series (Exercise 10) is primitive recursive. It's clearly computable, but primitive recursion allows us to use only the last value of the function at the induction stage, not the previous two values. Rather than tackle that function, it would be much more useful to show that any definition that begins with primitive recursive functions and uses any of the previous values of the function at the induction step in a primitive recursive fashion always results in a primitive recursive function.

We call an operation *primitive recursive* if whenever it is applied to primitive recursive functions it yields a primitive recursive function. In that case it can be simulated by using composition, primitive recursion, and auxiliary primitive recursive functions. In this section we're going to show that the operation described above, and others, are legitimate ways to form primitive recursive functions from primitive recursive functions.

1. Addition and multiplication of functions

If f and g are primitive recursive, then $f + g$ is primitive recursive, where
$(f + g)(x) = f(x) + g(x)$ (composition of primitive recursive functions).
Similarly $(f \cdot g)(x) = f(x) \cdot g(x)$ is primitive recursive if f and g are.
Generally, we define

$$\sum_{i=1}^{n} f_i(\vec{x}) \equiv_{\text{Def}} f_1(\vec{x}) + \cdots + f_n(\vec{x})$$

$$\prod_{i=1}^{n} f_i(\vec{x}) \equiv_{\text{Def}} f_1(\vec{x}) \cdot \cdots \cdot f_n(\vec{x})$$

In Exercise 11 we ask you to give a correct inductive definition of these that does not
use "..." and to show that for each n, if $f_1, \ldots, f_n$ are primitive recursive, so are
$\sum_{i=1}^{n} f_i(\vec{x})$ and $\prod_{i=1}^{n} f_i(\vec{x})$.

2. Functions defined according to conditions

As an example consider

$$f(n) = \begin{cases} 2n & \text{if } n \text{ is even} \\ 3n & \text{if } n \text{ is odd} \end{cases}$$

Here we are thinking of all numbers as divided into two sets: $A = $ evens, and $B = $
odds. We need only an informal notion of set here, for we don't need to be given all
numbers at once to divide them up. All we need is that the following functions are
primitive recursive:

$$Odd(n) = \begin{cases} 1 & \text{if } n \text{ is odd} \\ 0 & \text{if } n \text{ is even} \end{cases}$$

and

$$Even(n) = \overline{sg}\,[Odd(n)] = \begin{cases} 0 & \text{if } n \text{ is odd} \\ 1 & \text{if } n \text{ is even} \end{cases}$$

(See Exercises 2 and 3).

The *characteristic function of a condition* (or, informally, of a set) A is:

$$C_A(x) = \begin{cases} 1 & \text{if } x \text{ satisfies the condition} \\ 0 & \text{if } x \text{ does not satisfy the condition} \end{cases}$$

We say a *condition* (*set*) is *primitive recursive* if its characteristic function is
primitive recursive.

Suppose we have n primitive recursive conditions $A_1, \ldots, A_n$ such that
every number x satisfies one and only one of these (e.g., odd/even). (Informally,
we have a disjoint (nonoverlap) partition (dividing up) of all natural numbers into
sets $A_1, \ldots, A_n$.) Suppose further that we have n primitive recursive functions
$h_1, \ldots, h_n$. We may define:

$$f(x) = \begin{cases} h_1(x) & \text{if } x \text{ satisfies } A_1 \\ \vdots & \vdots \\ h_n(x) & \text{if } x \text{ satisfies } A_n \end{cases}$$

which is then primitive recursive: f is $\sum_{i=1}^{n} h_i \cdot C_{A_i}$

Often we need nonconstructive proofs to *demonstrate* that every x satisfies exactly one of $A_1, \ldots, A_n$. But that's *outside* the system and does not affect whether a function is primitive recursive or not. Remember, we are viewing functions extensionally.

As an example, we can show that given a primitive recursive function g, the following function is primitive recursive:

$f(0) = x_0$
$f(1) = x_1$
$\vdots$
$f(n) = x_n$
and for $x > n$, $f(x) = g(x)$.

We may always specify the value of a function at some arbitrary number of places before we give a general procedure: That's like providing an accompanying table of values. From our extensional point of view:

$$FINITE \;=\; TRIVIAL$$
$$GENERAL\ METHOD \;=\; for\ all\ but\ finitely\ many$$

3. Predicates and logical operations

We can have conditions involving more than one number; for instance "$x < y$" or "*max*(x, y) is divisible by z". We call a condition which is either satisfied or is not satisfied by every k-tuple of numbers a *predicate* or *relation* of k variables. For instance $R(x, y)$ defined as $x < y$ is satisfied by $(2,5)$ and is not satisfied by $(5,2)$. We say that $R(2,5)$ *holds* (or is true), or we simply write "$R(2,5)$", and "not $R(5,2)$". Another example is the predicate $Q(x,y,z)$ defined as $x + y = z$. Then $Q(2,3,5)$ but not $Q(5,2,3)$. We usually let capital letters stand for predicates.

As for sets, we define *the characteristic function of predicate R* as:

$$C_R(\vec{x}) = \begin{cases} 1 & \text{if } R(\vec{x}) \\ 0 & \text{if not } R(\vec{x}) \end{cases}$$

W say that a predicate is *primitive recursive* if its characteristic function is. We can view sets as predicates of one variable.

Given two predicates we can form new ones; for example, from "x is odd" and "x is divisible by 7" we can form:

"x is odd *and* x is divisible by 7"

"x is odd *or* x is divisible by 7"

"x is *not* divisible by 7"

Given two predicates P, Q we write

$$P(\vec{x}) \wedge Q(\vec{x}) \equiv_{\text{Def}} \vec{x} \text{ satisfies } P \text{ and } \vec{x} \text{ satisfies } Q$$

$$P(\vec{x}) \vee Q(\vec{x}) \equiv_{\text{Def}} \vec{x} \text{ satisfies } P \text{ or } \vec{x} \text{ satisfies } Q,$$
$$\text{or } \vec{x} \text{ satisfies both } P \text{ and } Q$$

$$\neg P(\vec{x}) \equiv_{\text{Def}} \vec{x} \text{ does not satisfy } P$$

$$P(\vec{x}) \rightarrow Q(\vec{x}) \equiv_{\text{Def}} \vec{x} \text{ does not satisfy } P \text{ or } \vec{x} \text{ satisfies } Q$$

(In Chapter 19.C we will suggest that we can read $P \rightarrow Q$ as "if P, then Q".)
We needn't require that P and Q use the same number of variables. For example,
"x is odd and $x < y$" will be viewed as a predicate of 2 variables: (x, y) will
be said to satisfy "x is odd" if x does.

In Exercise 14 we ask you to show that if P, Q are primitive recursive then so
are all the above.

We may recast these ideas in terms of sets. Given A and B, define

$$A \cap B = \{ x : x \in A \wedge x \in B \}$$
$$A \cup B = \{ x : x \in A \vee x \in B \}$$
$$\overline{A} = \{ x : x \notin A \}$$

If A and B are primitive recursive, so are these sets.

4. Bounded minimization

If we have a computable function, we ought to be able to check what we know about
it up to some given bound. We say that *f is obtained from h by the operation of
bounded minimization* if

$$f(\vec{x}) = min \ y \leq n \ [\ h(\vec{x}, y) = 0 \ ]$$

where this means:

the least $y \leq n$ such that $h(\vec{x}, y) = 0$ if there is one; n otherwise

Note that n is fixed for all $\vec{x}$; that is, we have a different function for each n.

There are two ways we can show that this is a primitive recursive operation.
We could define a different function for each n, and then show by induction on n
that each is primitive recursive. But in general we want to show something more,
namely that there is one primitive recursive function which calculates them all.
We say that the functions $h_1, \dots, h_n, \dots$ are *uniformly* primitive recursive if there
is a primitive recursive function q such that for all n, $h_n(\vec{x}) = \lambda \vec{x} \ q(n, \vec{x})$
(cf. Exercise 12a).

In this case we define $min\ y \le n\ [\ h(\vec{x}, y) = 0\ ]$ as $q(n, \vec{x})$ where

$$q(0, \vec{x}) = 0$$
$$q(n + 1, \vec{x}) = q(n, \vec{x}) + sg(h(\vec{x}, q(n, \vec{x})))$$

Although the bound needs to be fixed for each $\vec{x}$, it needn't be the same for all $\vec{x}$: if h and g are primitive recursive, then so is f where $f(\vec{x}) = min\ y \le g(\vec{x})\ [\ h(\vec{x}, y) = 0\ ]$. Moreover, we can check more than just whether an output of h equals 0. If the function g and the predicate Q are primitive recursive, then so is f, where $f(\vec{x}) = min\ y \le g(\vec{x})\ [Q(\vec{x}, y)]$, usually written as $f(\vec{x}) = min\ y\ _{y \le g(\vec{x})}\ [Q(\vec{x}, y)]$ (Exercise 15). We also ask you to show the same when " $\le$ " is replaced by " $<$ " .

5. Existence and universality below a bound

We can view bounded minimization as a way to deal with questions of existence below a bound. For a predicate P we define:

$$\exists y \le n\ P(\vec{x}, y) \equiv_{\text{Def}} \text{there is a } y \le n \text{ such that } P(\vec{x}, y)$$

and

$$\forall y \le n\ P(\vec{x}, y) \equiv_{\text{Def}} \text{for all } y \le n,\ P(\vec{x}, y)$$

In Exercise 15 you're asked to show that these are primitive recursive predicates if P is.

6. Iteration

Iteration is the simplest form of definition by induction. Informally, the iteration of the function h is

$$f(n, x) = h^{(n)}(x) = \underbrace{h(h(...h(x)...))}_{n \text{ times}}$$

This way of describing f is only suggestive. Including $n = 0$, we define:

$$h^{(0)}(x) = x$$
$$h^{(n+1)}(x) = h(h^{(n)}(x))$$

Then we say that *f arises by iteration from h* if $f(0, x) = id(x)$ and

$$f(n + 1, x) = h^{(n+1)}(x) = h(P_3^{\ 1}(f(n, x), n, x))$$

which is in a correct form to demonstrate that it is primitive recursive.

7. Simultaneously defined functions

Sometimes we define two functions f and g together, so that at stage $n + 1$ the value of each depends on the previous values of both of them (cf. the prices on the stock exchanges in Chicago and New York). More precisely, let k, q, h, and t

be primitive recursive. We define:

$$f(0, \vec{x}) = k(\vec{x})$$
$$f(n + 1, \vec{x}) = h(f(n, \vec{x}), g(n, \vec{x}), n, \vec{x})$$
$$g(0, \vec{x}) = q(\vec{x})$$
$$g(n + 1, \vec{x}) = t(f(n, \vec{x}), g(n, \vec{x}), n, \vec{x})$$

We ask you to show that these are primitive recursive in Exercise 18.

8. Course-of-values induction

Up to this point we have only used the single previously calculated value $f(n)$ in calculating $f(n + 1)$. This corresponds to simple induction. Using any of the previously calculated values corresponds to a proof by *course-of-values induction*: Given a statement $A(n)$,

if $A(0)$, and for all n, if all $y \leq n\ A(y)$, then $A(n + 1)$;
then for all n, $A(n)$

Course-of-values induction can be reduced to simple induction by applying simple induction to $\forall y \leq n\ A(y)$.

Similarly, we wish to show that a definition of a function that can use all of its previously calculated values, which we call a *course-of-values recursion*, can be reduced to primitive recursion. To do that we code up the previous values of the function into one function, since we can't have a varying number of variables in the inductive step. Let p_n be the nth prime: $p_0 = 2$, $p_1 = 3$, $p_2 = 5$, ... Let f be the function that has a course-of-values definition. Define f^* by:

$$f^*(0, \vec{x}) = 1$$
$$f^*(n + 1, \vec{x}) = p_n^{f(n,\bar{x})+1} \cdot \cdots \cdot p_1^{f(1,\bar{x})+1} \cdot p_0^{f(0,\bar{x})+1}$$
$$= p_n^{f(n,\bar{x})+1} \cdot f^*(n, \vec{x})$$

Any definition of $f(n + 1, \vec{x})$ that uses the values $f(0, \vec{x}), f(1, \vec{x}), \ldots, f(n, \vec{x})$ in some primitive recursive auxiliary function can be defined by extracting those values from $f^*(n + 1, \vec{x})$. And we can simultaneously define f and f^*, so long as our coding and uncoding procedure on the primes is primitive recursive (see Exercise 18).

E. Prime Numbers for Codings

1. We first want to show that the function $p(n) = $ the nth prime p_n is primitive recursive. We begin by noting that

m divides n (written "$m|n$") iff $\exists i \leq n\ (m \cdot i = n)$

is a primitive recursive predicate by Section D.5; we denote its characteristic function as $d(m, n)$. Then

n is a prime iff $(1 < n) \wedge [\forall x < n\,(x = 1 \vee \neg\,(x \mid n))]$

is a primitive recursive predicate (via Section D.5). Denote its characteristic function:

$$prime\,(n) = \begin{cases} 1 & \text{if } n \text{ is prime} \\ 0 & \text{if } n \text{ is not prime} \end{cases}$$

By Euclid's theorem (Exercise 6 of Chapter 5) we know that if p is prime then there is another prime between p and $p! + 1$. We define the auxiliary function $h(z) = min\ y\ _{y \le z!+1}\,[\,z < y \wedge prime(y) = 1\,]$. Then we can define the function p by $p(0) = 2$, $p(n + 1) = h(p(n))$.

2. If we code along primes and are given a number, say 270 , we need to know the exponents of the primes in its decomposition: $270 = 2^1 \cdot 3^3 \cdot 5^1$. Let

$[\,x\,]_n = $ the exponent of the nth prime in the prime decomposition of x

That this is a well-defined function depends on the fact that every natural number has a unique decomposition into primes (Exercise 4 of Chapter 5). To show that $\lambda x\,[\,x\,]_n$ is primitive recursive we note that $p_n^{[x]_n}$ divides x, but $p_n^{[x]_n\,+1}$ does not. So $[\,x\,]_n = min\ y < x\ [\,d(\,p(n)^{y+1}, x) = 0\,]$.

3. We define the *length* of x to be

$lh(x) = min\ y < x\ (\,[\,x\,]_y = 0\,)$

This measures the number of different primes in a row beginning with 2 that have non-zero exponents in the prime decomposition of x. For example,

$lh(6) = lh(2 \cdot 3) = lh(p_0 \cdot p_1) = 2$
$lh(21) = lh(3 \cdot 7) = lh(p_1 \cdot p_3) = 0$
$lh(42) = lh(2 \cdot 3 \cdot 7) = lh(p_0 \cdot p_1 \cdot p_3) = 2$

4. We need to be able to code 0, but $p_n^0 = 1$ and we can't tell if p_n is "there" or not. So we code y into a number via $p_n^{y\,+1}$. To uncode we then need the function

$$(x)_n = [\,x\,]_n \dot{-} 1 = \begin{cases} 1 \text{ less than the exponent of the } n \text{ th prime} \\ \text{in the prime decomposition of } x \end{cases}$$

Note that for $x > 0$, $(x)_n < x$. We write $(x)_{n,m}$ for $((x)_n)_m$.

5. Now we have a way to code finite sequences of numbers into single numbers. We code $(a_0, a_1, \ldots, a_n)$ by:

$$\langle\,a_0, a_1, \ldots, a_n\,\rangle = p_0^{a_0+1} \cdot p_1^{a_1+1} \cdot\ \cdots\ \cdot p_n^{a_n+1}$$

For each n this is primitive recursive (Exercise 19).

With the convention that every number codes only up to its length, we also have a unique sequence assigned to each natural number:

x codes the sequence $((x)_0, (x)_1, \ldots, (x)_{lh(x) - 1})$,

where if $lh(x) = 0$, x codes the empty sequence

This is the coding we will use throughout this text.

We needn't have that $x = \langle (x)_0, (x)_1, \ldots, (x)_{lh(x) - 1} \rangle$, for example, $756 = 2^2 \cdot 3^3 \cdot 7$ which codes the sequence $(1, 2)$, but $\langle 1, 2 \rangle = 108$. Different numbers can code the same sequence.

6. We can, if we wish, give a 1-1 coding, though it's harder to construct and use. Recall the pairing function $J(x, y) = \frac{1}{2}[(x + y)(x + y + 1)] + x$ of Chapter 6.B.2, which you were asked to show is 1-1 and onto (Exercise 4 of Chapter 6). We can code $(a_0, a_1, \ldots, a_n)$ as $J(a_0, J(a_1, \ldots, J(a_{n-1}, a_n)) \ldots)$. That is, given the coding of n-tuples, J_n, the coding of $n + 1$-tuples is $J_{n+1}(a_0, a_1, \ldots, a_n) = J(a_0, J_n(a_1, \ldots, a_n))$. To uncode we define the unpairing functions

$$K(z) = min \ x \le z \ [\exists y \le z \ (J(x, y) = z)]$$
$$L(z) = min \ y \le z \ [\exists x \le z \ (J(x, y) = z)]$$

These are primitive recursive by Sections D.4 and D.5.

$$K(J(x, y)) = x, \quad L(J(x, y)) = y, \quad J(K(z), L(z)) = z$$

Uncoding longer sequences is left for you as Exercise 21.

Numbering the Primitive Recursive Functions

Here is a sketch of how we can computably number the primitive recursive functions.

First we give every initial function a number: $\#(Z) = 11$, $\#(S) = 13$, $\#(P_n^i) = (p_{n+5})^{i+1}$. Then to each operation under which the class is closed we will associate an arithmetical operation. If $\#(g) = a$ and $\#(h) = b$, then the composition $g \circ h$ will have number $2^a \cdot 3^b$. And if $\#(h_1) = a_1$, $\#(h_2) = a_2$, $\ldots$, $\#(h_m) = a_m$ and each is a function of k variables and g is a function of m variables, and $\#(g) = b$, then the function $g(h_1(\vec{x}), \ldots, h_m(\vec{x}))$ will have number $\#(f) = 2^b \cdot 3^{\langle a_1, \ldots, a_m \rangle}$. Lastly, if $\#(g) = a$ and $\#(h) = b$, and these are functions of the appropriate number of variables, then f defined by primitive recursion on h with basis g will have number $5^a \cdot 7^b$.

Given any primitive recursive definition, we can follow the steps above and obtain a number for the function, which we call an *index*. And given any number, we can decompose it into primes, further decompose the exponents into primes, and so on, until we have an expression consisting only of primes; then we can determine if it corresponds to a definition of a primitive recursive function. Thus, the conditions for a Gödel numbering are satisfied (Chapter 8.C). Moreover, we can check if the definition corresponds to a function of one variable. So we can make a computable list of the primitive recursive functions of one variable: $f_0, f_1, \ldots, f_n, \ldots$

where f_n is the function which has the nth index. Our list will have repetitions since every primitive recursive function has arbitrarily many different definitions (Exercise 9), and we are really numbering definitions.

G. Why Primitive Recursive $\neq$ Computable

Consider the function $g(x) = f_x(x) + 1$.

This is computable since our numbering is. Yet it can't be primitive recursive: if it were it would be f_n for some n, and then we would have $g(n) = f_n(n) + 1 = f_n(n)$. We have *diagonalized*.

$$
\begin{array}{ccccc}
f_0(0) + 1 & f_0(1) & f_0(2) & f_0(3) & \cdots \\[2mm]
f_1(0) & f_1(1) + 1 & f_1(2) & f_1(3) & \cdots \\[2mm]
f_2(0) & f_2(1) & f_2(2) + 1 & f_2(3) & \cdots \\
\vdots & & & & \\
f_n(0) & f_n(1) & f_n(2) & \cdots & f_n(n) + 1 \quad \cdots
\end{array}
$$

We've made g disagree with every primitive recursive function of one variable by making g disagree with each f_n on the diagonal. Hence, we have found a computable function which is not primitive recursive.

Here is another way to produce a computable function that isn't primitive recursive. Define

$$
\begin{aligned}
h(0) &= f_0(0) + 1 \\
h(1) &= f_0(1) + f_1(1) + 1 \\
&\;\vdots \\
h(n) &= f_0(n) + f_1(n) + \cdots + f_n(n) + 1 \\
&\;\vdots
\end{aligned}
$$

Again, h is computable since our numbering is. Yet h *dominates* all primitive recursive functions of one variable; that is, if f is a primitive recursive function of one variable, then $f = f_n$ for some n, so for all $x \geq n$, $h(x) > f(x)$. Thus h cannot be primitive recursive.

If we do not yet have all the computable functions, how can we obtain all of them? What further operations or initial functions do we need?

Exercises Part 2

11. For every $n \geq 2$ give proper primitive recursive definitions of

$$
\sum_{i=1}^{n} f_i(\vec{x}) \equiv_{\text{Def}} f_1(\vec{x}) + \cdots + f_n(\vec{x}) \qquad \text{and}
$$

$$\prod_{i=1}^{n} f_i(\vec{x}) \equiv_{\text{Def}} f_1(\vec{x}) \cdot \ \cdots \ \cdot f_n(\vec{x})$$

(*Hint:* It's easy for $n = 2$; then proceed by induction.)

12. a. Show that by holding one variable fixed in a primitive recursive function we obtain a primitive recursive function. That is, given that $\lambda n\lambda\vec{x}\, f(n, \vec{x})$ is primitive recursive, show that for every n, $\lambda\vec{x}\, f(n, \vec{x})$ is primitive recursive. (*Hint:* Use Section C.1.)

   b. Use part (a) and Exercise 11 to show that if f is primitive recursive, so are

   $$\lambda\vec{x} \sum_{i=1}^{n} f(i,\vec{x}) \quad \text{and} \quad \lambda\vec{x} \prod_{i=1}^{n} f(i,\vec{x})$$

13. Suppose we have countably many primitive recursive conditions $A_1, \ldots,$ $A_n, \ldots$ such that every x satisfies exactly one of these. And suppose we also have countably many primitive recursive functions $h_1, \ldots, h_n, \ldots$. Let f be defined by $f(x) = h_n(x)$ if A_n is satisfied by x.

   Is f necessarily primitive recursive? Give a proof or a counterexample with appropriate restrictions.

14. Show that if P and Q are primitive recursive conditions, then so are $P \wedge Q$, $P \vee Q$, $\neg P$, and $P \rightarrow Q$.

15. a. Show that if h and g are primitive recursive, then so is f, where $f(\vec{x}) = \min y\ _{y \le g(\vec{x})}\, [\, h(\vec{x}, y) = 0\,]$.

   b. Show that if the function g and predicate Q are primitive recursive, then so is f, where $f(\vec{x}) = \min y\ _{y \le g(\vec{x})}\, [Q(\vec{x}, y)]$.

   c. Show that if the predicate P and function g are primitive recursive, then so are the predicates $\exists y \le g(\vec{x})\, [P(\vec{x}, y)]$ and $\forall y \le g(\vec{x})\, [P(\vec{x}, y)]$.

   d. Repeat parts (a)–(c) with "$\le$" replaced by "$<$".

16. Give one application of definition by conditions and one by bounded minimization which show the utility of knowing that these operations are primitive recursive.

17. Show that the function $e(x) = x^x$ is primitive recursive. Describe the function $f(n, x)$ obtained by iteration of e. Calculate $f(3, 2)$, $f(3, 3)$, $f(10, 10)$. Try to describe in informal mathematical notation the function g that arises by iteration of $\lambda x\, f(x, x)$. Calculate $g(3)$.

†18. a. Show that if h, g, and t are primitive recursive then so is f defined by
   $$f(0, \vec{x}) = g(\vec{x})$$
   $$f(1, \vec{x}) = t(\vec{x})$$
   $$\text{and for } n \ge 1, \ f(n + 1, \vec{x}) = h(f(n - 1, \vec{x}), n, \vec{x})$$
   Use our codings, not simultaneous recursion.

   b. Show that simultaneous definition by recursion (Section D.7) is a primitive recursive operation.

Hint: Set up a new function $j(n, \vec{x}) = \begin{cases} f\left(\frac{n}{2}, \vec{x}\right) & \text{if } n \text{ is even} \\ g\left(\frac{n-1}{2}, \vec{x}\right) & \text{if } n \text{ is odd} \end{cases}$

19. Show that for each n, $\langle a_0, a_1, \ldots, a_n \rangle$ is primitive recursive.

20. Using our code (Section E.5), find $\langle 3,1,0 \rangle$, $\langle 0,0,2 \rangle$, and $\langle 2,1,0,2,2 \rangle$. What sequences are coded by 900 and by 19,600? By $2^{1047} - 1$?

21. In terms of K and L (Section E.6), give a function which outputs the ith element in the sequence represented by $J_{n+1}(a_0, a_1, \ldots, a_n)$.

†22. Let $f(n) =$ the nth digit in the decimal expansion of π; that is, $f(0) = 3$, $f(1) = 1$, $f(2) = 4$, ... Show that f is primitive recursive.

23. Compare the proof that there is a computable function which is not primitive recursive with:
    a. The proof that the reals are not countable.
    b. The proof that there is no set of all sets.
    c. The Liar paradox.

Further Reading
"Mathematical induction and recursive definitions" by R.C. Buck has a good discussion of definition by induction with lots of examples.

12 The Grzegorczyk Hierarchy

(Optional)

Though we've seen that the primitive recursive functions are not all the computable functions, they are nonetheless a very important class. Almost all the functions we normally study in number theory and all the usual approximations to real-valued functions (cf. Exercise 11.22) are primitive recursive. In this chapter we will study more deeply the nature of primitive recursion and induction and, in doing so, will develop an idea for how to define larger classes of computable functions. *Except for Section B* neither this material nor Chapter 13 is essential to our goal of investigating Hilbert's program.

A. Hierarchies and Bounded Recursion

We propose to analyze the primitive recursive functions by breaking them down into a hierarchy. What is a hierarchy? It's a classification system: of less important to more important; of more complex to less complex; or of any quality we believe can be stratified. The Elizabethans stratified all of existence into a hierarchy.

The Elizabethan Chain of Being

First there is mere existence, the inanimate class: the elements, liquids and metals. But in spite of the common lack of life there is a vast difference in virtue; water is nobler than earth, the ruby than the topaz, gold than brass: the links in the chain are there. Next there is existence and life, the vegetative class, where again oak is nobler than bramble. Next there is existence, life and feeling, the sensitive class. In it there are three grades. First the creatures having touch but not hearing, memory or movement. Such are shellfish and parasites on the base of trees. Then there are animals having touch, memory and movement but not hearing, for instance ants. And finally there are the higher animals, horses and dogs etc., that have all these faculties. The three classes lead up to man, who has not only existence, life and feeling, but understanding.

Tillyard, 1943, pp. 25–26

Amongst the primitive recursive functions we can easily recognize varying levels of complexity, say,

of definition: for instance, how many recursions are used

or

of calculation: surely adding is "simpler" than multiplying, and $x \cdot y$ is "simpler" to calculate than x^y, which is "simpler" still than

$$f(n,x) = x^{x^{\cdot^{\cdot^{x}}}} \left.\right\} n \text{ times}$$

or

of rate of growth: the values of the exponential function grow much more rapidly than do those of multiplication, which in turn grow more rapidly than those of addition.

How might we try to build classes of functions that reflect these intuitions?

We can base our first class, call it $\mathcal{E}^0$, on the simplest functions we have: Z (zero), S (successor), and P_n^i (the projection functions). What operations should we close it under? Composition doesn't seem to add to the complexity of the functions, at least compared to recursion. But if we allow recursion we can get addition, and indeed, all of the primitive recursive functions. So let's allow the use of recursion if we don't get a function that is more complicated, say, one that doesn't grow any faster than one we already have. This is what we call *bounded recursion*. Formally, the function f is defined by recursion on h with basis g bounded by k if

$$f(0, \vec{x}) = g(\vec{x})$$
$$f(n + 1, \vec{x}) = h(f(n, \vec{x}), n, \vec{x})$$
$$f(n, \vec{x}) \leq k(n, \vec{x})$$

Bounded recursion is ordinary recursion with an extra clause added that requires the function being defined to be smaller than one already obtained.

One full unbounded application of recursion (iteration) to the successor function yields addition. So we can let our second class, $\mathcal{E}^1$, be what we get when we add $\lambda x y (x + y)$ to the stock of initial functions of $\mathcal{E}^0$, closing under the same operations.

And $\mathcal{E}^2$ can be the class we get when we add $\lambda x y (x \cdot y)$ to the stock of initial functions, since multiplication is the result of one unbounded application of recursion (iteration) to addition.

And $\mathcal{E}^3$ can be the class we get when we add $\lambda x y (x^y)$ to the stock of initial functions, since exponentiation is the result of one unbounded application of recursion (iteration) to multiplication. This last class, defined originally by Kalmar, is important in its own right so we'll look at it first.

B. The Elementary Functions

We define the class of *elementary functions* as:

> $\mathcal{E}$ = the smallest class of functions containing Z, S, the projection functions, $\lambda x y \, (x^y)$, and which is closed under composition and bounded recursion.

We remind you that this is shorthand for an inductive definition of the label "elementary function" (see the end of Chapter 11.B).

   This isn't a very happy way to define the class because to apply bounded recursion we need to know ahead of time whether the function we shall get is going to be smaller than one we already have. The computation procedure would require an accompanying proof. Whenever we use recursion we need to know something about the functions, namely, that they have the right number of variables. But here we need to know in advance something about the calculations. We can show, however, that the operation of bounded recursion can be replaced by bounded minimization. Recall that a function f is defined by *bounded minimization* on functions g and h if $f(\vec{x}) = min \, y_{\, y \le g(\vec{x})} \, [\, h(\vec{x}, \, y) = 0 \,]$.

THEOREM 1   **a.**   Addition and the coding and uncoding functions are in $\mathcal{E}$.
   b.   In the definition of $\mathcal{E}$ the operation of bounded recursion can be replaced by bounded minimization.

Proof: a. The definition of addition in Chapter 11.C.2 is acceptable here: $\lambda x y \, (x + 2)^{y+2}$ is in $\mathcal{E}$ as it is a composition of functions in $\mathcal{E}$, and $(x + y) \le (x + 2)^{y+2}$. Now it is simply a matter of tracing through our definitions of the coding and uncoding functions (Chapter 11.E) and all the functions those depend on to see that they can be bounded by functions in $\mathcal{E}$. We leave that to you.

   b. To see that $\mathcal{E}$ is closed under bounded minimization refer to Chapter 11.D.4. All the recursions used there can be bounded by functions in $\mathcal{E}$, which you can check.

   Now define the class of functions C by the same definition as $\mathcal{E}$ except replace the operation of bounded recursion by the operation of bounded minimization. We already have that $C \subseteq \mathcal{E}$. We need to show that $\mathcal{E} \subseteq C$.

   So suppose we are given a function f defined by recursion on g and h bounded by k, where $g, h, k \in C$. Then note that

$$f(m, \vec{x}) = (min \, z \, [(z)_0 = g(\vec{x}) \land \forall i < m, (z)_{i+1} = h((z)_i, i, \vec{x}) \,] \,)_m$$

We leave to you to show that the characteristic function for the part in square brackets is in C. So to show that $f \in C$, it only remains to show that the minimization operator in this definition can be bounded by some function in C.

   We know that f is bounded by $\lambda m \, \vec{x} \, k(m, \vec{x}) \in C$. We may assume that k is increasing in m because each time we use bounded recursion we can choose such a function, beginning with $\lambda x y \, (x^y)$. So the least z in the definition above has the form:

$$z = \langle f(0, \vec{x}), \dots, f(m, \vec{x}) \rangle$$
$$\leq \langle k(0, \vec{x}), \dots, k(m, \vec{x}) \rangle$$
$$\leq \langle k(m, \vec{x}), \dots, k(m, \vec{x}) \rangle$$
$$< p_m^{m \cdot k(m, \vec{x})}$$

where p_m is the mth prime. Multiplication is in C because all the recursions needed to define it can be bounded by $\lambda xy(x^y)$. So we need only show that the function $p(m) = p_m$ is in C. For that we use the definition of p from Chapter 11.E.1 and the inequality $p_m \leq 2^{2^m}$ (Exercise 1); the function $\lambda x(2^{2^x})$ is in C because it is the composition of functions in C. ∎

In the proof of Theorem 1.b we see how important it is to have the coding and uncoding functions available to us. Of the classes described in the previous section, $\mathcal{E}$ is the smallest which contains the coding and uncoding functions and in which we can replace bounded recursion by bounded minimization. For this reason it's taken as the minimum base for investigating the primitive recursive functions. Yet at the same time it marks the limit of what many believe to be the "feasibly" computable functions since it contains exponentiation.

C. Iterating Iteration: The Ackermann-Péter Function

It should seem clear how to continue the classes $\mathcal{E}^0$, $\mathcal{E}^1$, $\mathcal{E}^2$, $\mathcal{E}^3$, ... But can we come up with a general way to describe all these classes? And by iterating often enough will we get all the primitive recursive functions? To answer these questions we look at what we get by successive iteration starting with the successor function.

1. The functions ψ_m and proof by double induction

Define the functions ψ_0, ψ_1, ..., ψ_m, ... by:

$$\psi_0(n) = n + 1$$

and

$$\psi_{m+1}(0) = \psi_m(1)$$
$$\psi_{m+1}(n + 1) = \psi_m(\psi_{m+1}(n))$$

We can prove by induction on m that each ψ_m is primitive recursive: ψ_0 is, and ψ_{m+1} arises by recursion on ψ_m. Indeed, ψ_{m+1} is a pure iteration of ψ_m:

$$\psi_{m+1}(n) = \psi_m^{(n+1)}(1)$$

as you can check (Exercise 2).

We are going to get functions that grow very fast (cf. Exercise 11.17). So fast, indeed, that given any primitive recursive function there will be some ψ_m which bounds it. In order to show that, we first need to establish various ways in which the ψ_m's grow.

THEOREM 2   **a.** $\psi_m(n) > n$

           b. $\psi_m(n + 1) > \psi_m(n)$

           c. $\psi_{m+1}(n) \geq \psi_m(n + 1)$

           d. $\psi_{m+1}(n) > \psi_m(n)$

We need a new proof technique for this theorem. A *proof by double induction* of a statement $P(m, n)$ has the following steps:

1. We prove $P(0, n)$ for all n by induction on n: first we prove $P(0,0)$, and then assuming $P(0, n)$ we show that $P(0, n + 1)$ follows.
2. We assume that $P(m, n)$ holds for all n (the induction step for m). Then we prove $P(m + 1, n)$ by induction on n: first we prove $P(m + 1, 0)$, and then assuming $P(m + 1, n)$ (the induction step for n), we prove $P(m + 1, n + 1)$.

This is repeated use of single induction, so it should be acceptable. And in a sense that can be made precise (see Chapter 13.B) it can be reduced to single induction. Note that at the induction step for m we can equally assume that $P(i, n)$ holds for all $i \leq m$, and at the induction step for n we can assume that $P(m + 1, i)$ holds for all $i \leq n$ (cf. course-of-values induction in Chapter 11.D.8) . We'll prove part (a) of the theorem by double induction and leave the rest of the proof as Exercise 3.

Proof: a. We can do the basis level all at one go, for $\psi_0(n) = n + 1 > n$.

For the induction step for m, suppose that for all n, $\psi_m(n) > n$. Then $\psi_{m+1}(0) = \psi_m(1) > 1 > 0$.

For the sub-induction step for n, suppose that $\psi_{m+1}(n) > n$. Then

$$\psi_{m+1}(n + 1) = \psi_m(\psi_{m+1}(n))$$
$$> \psi_{m+1}(n) \quad \text{by induction on } m$$
$$> n \quad \text{by induction on } n. \qquad \blacksquare$$

2. Dominating the primitive recursive functions

To establish our claim, we first have to make it precise. We say that g *strictly dominates* f if for all x, $g(x) > f(x)$. That definition only works for functions of one variable; so, more generally, we say that a function of one variable, $\lambda x\ g(x)$ *strictly dominates* $\lambda \vec{x}\ f(\vec{x})$ if for all $\vec{x}$, $g(\max \vec{x}) > f(\vec{x})$.

THEOREM 3

a. Each of the initial functions Z, S, P_n^i is strictly dominated by ψ_1.

b. If g is strictly dominated by ψ_a and $h_1, \ldots, h_r$ are strictly dominated by, respectively, $\psi_{a_1}, \ldots, \psi_{a_r}$, then $f(\vec{x}) = g(h_1(\vec{x}), \ldots, h_r(\vec{x}))$ is strictly dominated by ψ_{m+2}, where $m = max\ (a, a_1, \ldots, a_r)$.

c. If g is strictly dominated by ψ_a, h is strictly dominated by ψ_b, and f is obtained by primitive recursion on h with basis g, then f is strictly dominated by ψ_{m+2} where $m = max\ (a, b, 1)$.

d. If f is primitive recursive, then for some r, f is strictly dominated by ψ_r.

Proof: We leave parts (a), (b), and (c) as good (hard) exercises in proof by induction (Exercise 4).

For part (d) we're going to use induction, but in a new way. Recall our explanation of the inductive character of the definition of the label "primitive recursive" in Chapter 11.B. A function f gets the label primitive recursive if it has a definition starting with the initial functions using composition and primitive recursions. So we can *induct on the number of operations used in a definition* to prove a constructive version of part (d): if f has a definition that uses at most m applications of composition and primitive recursion starting with the initial functions, then f is strictly dominated by ψ_{2m+1}.

It is true for 0 operations, for in that case we have the initial functions which by part (a) are bounded by ψ_1. Therefore, suppose it is true for functions which use at most m operations. Suppose that f can be defined by recursion on g and h, where the latter use at most m applications of the operations and hence are strictly dominated by ψ_{2m+1}. Then by part (c), f is strictly dominated by $\psi_{(2m+1)+2} = \psi_{2(m+1)+1}$. The same argument works if f is defined by composition. ∎

3. The Ackermann-Péter function and nested double recursion

In working with the collection of ψ_m's you may have found that you could think of the whole sequence of functions as a description of how to compute $\psi_m(n)$ for any m, n. That is, you could think of $\psi_m(n)$ as a single function of two variables defined by

$$\psi(0, n) = n + 1$$

$$\psi(m + 1, 0) = \psi(m, 1)$$

$$\psi(m + 1, n + 1) = \psi(m, \psi(m + 1, n))$$

W. Ackermann first defined a function in 1928 to show that there is a computable procedure which is not primitive recursive. The function ψ is a variation on that, in its essentials due to Rózsa Péter in 1935.

But why are we justified in saying that ψ is a function, much less a computable one? It has the flavor of an inductive definition, but there seems to be too

much going on. That's because it's a *definition by induction on* 2 *variables*.
Consider all pairs of natural numbers arranged as follows:

	$(0,0)$	$(0,1)$	...	$(0,n)$	$(0,n+1)$	...
rows→	$(1,0)$	$(1,1)$	...	$(1,n)$	$(1,n+1)$	...
	⋮	⋮		⋮	⋮	
	$(m,0)$	$(m,1)$	...	(m,n)	$(m,n+1)$	...
	$(m+1,0)$	$(m+1,1)$	...	$(m+1,n)$	$(m+1,n+1)$	...

In ordinary (primitive) recursion, to determine the value of f at
$(m+1, n+1)$, that is, $f(m+1, n+1)$, we allow ourselves to look at values of
f at places in preceding rows only, $f(x, y)$ such that $x \leq m$. This seems an
arbitrary restriction: why shouldn't we allow ourselves to look at values of f at
places preceding $(m+1, n+1)$ on the *same row*, that is, $f(m+1, x)$ for
$x < n+1$?

Moreover, *nesting* causes no problems; that is, we can apply f to itself, for
example, $f(m+1, n+1) = f(m, f(m+1, n))$, for we are again only
thrown back to previous calculations of f. To calculate any such f at
$(m+1, n+1)$ we need only a finite number of values of f at places that
precede $(m+1, n+1)$:

> when we start at $(m+1, n+1)$ there are only finitely many
> places to go on the same row before $(m+1, n+1)$. Then we may go
> to an arbitrarily distant place on a preceding row, say $(m, n+400)$.
> But then again there are only finitely many places on *that* row to which
> we can be thrown back, ... continuing we must eventually reach $(0, 0)$
> for which a value is given.

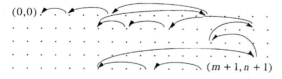

For example, for our function ψ how do we compute $\psi(2,1)$? Well,
$\psi(2,1) = \psi(1, \psi(2,0))$. So we need $\psi(2,0)$, which equals $\psi(1,1)$ and $\psi(1,1) =$
$\psi(0, \psi(1,0))$. And $\psi(1,0) = \psi(0,1) = 2$. Now trace the steps backwards.

So we conclude that nested double recursion (definition by induction on two
variables) is an appropriate operation to define a computable function. But it can't
be reduced to primitive recursion.

THEOREM 4 ψ is not primitive recursive.

Proof: Suppose ψ were primitive recursive. Then the *diagonal* $f(m) = \psi(m, m)$ would be primitive recursive, too. But by Theorem 3.d, for some r, $f(m) < \psi(r, m)$ for all m. In particular, we would have $f(r) < \psi(r, r) = f(r)$, which is a contradiction.   ∎

D. The Grzegorczyk Hierarchy

Now we can use the ψ_m's to build the hierarchy we began in Section A. Andrzej Grzegorczyk [pronounced G'-zhuh-gore´-chick], 1953, was the first to build a hierarchy based on successive recursions, and ours will be a variation on his, starting with the elementary functions.

Definition   For $m \geq 3$ define the inductive class :

$\mathcal{E}_m = \mathcal{E}(\psi_m)$

= the smallest class of functions containing Z, S, the projection functions, and ψ_m, and which is closed under composition and bounded recursion.

Thus the definition of $\mathcal{E}_m$ is the same as for $\mathcal{E}$ except that ψ_m replaces $\lambda xy\ (x^y)$.

THEOREM 5   **a.** $\mathcal{E} = \mathcal{E}_3$.

 b. In the definition of $\mathcal{E}_m$ the operation of bounded recursion can be replaced by bounded minimization.

Proof: a. It's enough to show that $\psi_3 \in \mathcal{E}$ and $\lambda xy\ (x^y) \in \mathcal{E}(\psi_3)$. Since $\psi_3(n) = 2^{n+3} - 3$ (Exercise 2), $\psi_3 \in \mathcal{E}$. From Exercise 5, $x \cdot y < 2^{x+y}$, so $x \cdot y < 2^{(x+y)+3} - 3$ and hence is in $\mathcal{E}(\psi_3)$. And similarly, $x^y \leq 2^{x \cdot y}$ so (with the details left to you) $\lambda xy\ (x^y) \in \mathcal{E}(\psi_3)$.

 b. Since $\lambda xy\ (x^y) \in \mathcal{E}_3$, $\lambda xy\ (x^y) \in \mathcal{E}_m$ for all m (via Theorem 2.d), so we can use the same proof as for Theorem 1.   ∎

Now we can establish that the $\mathcal{E}_m$'s give a hierarchy.

THEOREM 6   **a.** $\mathcal{E}_m \subseteq \mathcal{E}_{m+1}$.

 b. If f is primitive recursive then for some m, $f \in \mathcal{E}_m$.

 c. $\psi_{m+1} \notin \mathcal{E}_m$ and hence $\mathcal{E}_m \neq \mathcal{E}_{m+1}$.

Proof: a. This part follows from Theorem 2.d .

 b. This part follows from Theorem 3.d , since given a definition of a primitive recursive function, there is some m such that ψ_m dominates all the recursions used in that definition.

c. To establish this part we use the fact that if $f \in \mathcal{E}_m$ is a function of one variable, then for some a, for all x, $f(x) < \psi_m^{(a)}(x)$ (Exercise 6). So, for some a

$$f(x) \; < \; \psi_m^{(a)}(\psi_{m-1}^{(x+1)}(1)) \quad \text{by Exercise 2}$$

$$< \; \psi_m^{(a)}(\psi_m^{(x+1)}(1)) \quad \text{by Theorem 2}$$

$$= \; \psi_m^{(a+x+1)}(1) = \psi_{m+1}(a+x)$$

Now suppose that $\psi_{m+1} \in \mathcal{E}_m$. Then $\psi_{m+1}(x+x)$ would be too. But then for some a and all x, $\psi_{m+1}(x+x) < \psi_{m+1}(a+x)$. But this is impossible for $x = a$. So $\psi_{m+1} \notin \mathcal{E}_m$, and we have found a function in $\mathcal{E}_{m+1}$ that dominates all functions in $\mathcal{E}_m$. ∎

We have the following picture:

Note that we have built a *cumulative hierarchy*: Each class contains the preceding one, extending it with new elements. This is more common in mathematics (and usually easier) than making each class completely distinct from the preceding ones.

Every new class allows for one unbounded recursion (iteration) from the previous class. This yields a function that grows faster than any in the previous class and can be used to raise the bound for bounded minimization. The Ackermann-Péter function ties together these countably many new recursions into one function which grows faster than any primitive recursive function. The operation of nested double induction which we used to define the Ackermann-Péter function is computable, and that gives us our first clue for how to extend the primitive recursive functions in Chapter 13.

Exercises

Warning: Not every statement involving two variables requires a proof by double induction.

1. a. Prove by induction: $2^0 + 2^1 + \cdots + 2^n < 2^{n+1}$.
   b. Let p_n be the nth prime ($p_0 = 2$). Prove by induction on n that $p_n \leq 2^{2^n}$.
   (*Hint:* Replace $n!$ by a product of primes in Euclid's proof and use part (a).)

2. a. Prove that $\psi_{m+1}(n) = \psi_m^{(n+1)}(1)$.

   b. Calculate $\psi_1(3)$, $\psi_2(1)$, $\psi_4(2)$, $\psi_5(2)$, and $\psi_{421}(7)$.

   c. Express ψ_1, ψ_2, ψ_3, and ψ_4 as arithmetic functions using familiar operations, such as addition, multiplication, and exponentiation. In particular, show that $\psi_3(n) = 2^{n+3} - 3$. Try to devise a notation to describe ψ_m .

3. Complete the proof of Theorem 2 .

†4. Complete the proof of Theorem 3.

   (*Hint:* For part (c) show that $f(n, \vec{x}) < \psi_{m+1}(max(\vec{x}) + n)$. Then use the fact that $max(\vec{x}) + n < \psi_2(max(\vec{x}, n))$ and Theorem 2.c. We have not tried to get the best possible bounds and if you can't get ours, try to show that there are at least *some* bounds, for example, $m + 3$ instead of $m + 2$.)

5. a. Prove by induction that for all x, $x < 2^x$.

   b. Prove by induction on y that for all x, y, $x \cdot y < 2^{x+y}$.

   c. Prove by induction on y that for all x, y, $x^y \le 2^{x \cdot y}$.

†6. Prove that if $f \in \mathcal{E}_m$ and f has a definition which uses at most r compositions beginning with the initial functions of $\mathcal{E}_m$ then for all $\vec{x}$, $f(\vec{x}) < \psi_m^{(2^r+1)}(max(\vec{x}))$. (*Hint:* See the proof of Theorem 3. Remember that bounded recursion does not increase the value of the function beyond one we already have.)

†7. Iteration is just as strong as primitive recursion once we have the coding functions. Prove that the smallest class of functions which contains the initial functions S, Z, P_m^i for all m and $1 \le i \le m$, the coding functions $\lambda x (x)_m$ and $\langle x_0, x_1, \ldots, x_m \rangle$ for all m, and which is closed under composition and iteration is the class of primitive recursive functions. (*Hint:* If f is defined by recursion on h define $s(n, \vec{x}) = \langle f(n, \vec{x}), n, \vec{x} \rangle$ and show that s can be defined by iteration of a function t which gives $t(\langle a, n, \vec{x} \rangle) = \langle h(a, n, \vec{x}), n + 1, \vec{x} \rangle$.)

8. Show that we can't get all the primitive recursive functions by starting with some finite number of initial primitive recursive functions and closing under composition and bounded recursion.

Further Reading

Grzegorczyk's original paper, "Some Classes of Recursive Functions", is an excellent place to read more about his hierarchy. Rózsa Péter, in her book *Recursive Functions*, also develops the hierarchy and gives a good exposition of the Ackermann-Péter function. Odifreddi, in *Classical Recursion Theory* Volume 2, develops this material too, along with other hierarchies of the primitive recursive functions.

13 Multiple Recursion

(Optional)

In this chapter we will sketch how to extend the Grzegorczyk hierarchy in our quest for more computable functions and operations. This is only for motivation and is not needed later.

A. The Multiply Recursive Functions

1. Double recursion

In Chapter 12.C.3 we justified *nested recursion on two variables* as a computable operation. In its general form we define a function f of two variables from g, h, and k by the equations (highlighting the occurrences of f in bold)

$$\mathbf{f}(0, y) = g(y)$$
$$\mathbf{f}(x + 1, 0) = j(x, \mathbf{f}(x, a)) \quad \text{for a fixed number } a$$
$$\mathbf{f}(x + 1, y + 1) = h(x, y, \mathbf{f}(x + 1, y), \mathbf{f}(x, k(x, y, \mathbf{f}(x + 1, y))))$$

This is not really as complicated as it looks: for the lowest level we set f equal to another function. At a successor level we first define f for $y = 0$ using some earlier value of f. Then for successor stages of y it's almost as for primitive recursion, the main difference being that we now allow f to be applied to itself on an earlier value (nesting). To define a function $f(x, y, \vec{z})$ of more than 2 variables, simply insert the sequence $\vec{z}$ in all the "obvious" places.

By choosing h appropriately we can show that primitive recursion is a special form of double recursion. So we define inductively a new class of computable functions, the *doubly recursive functions*:

$\mathcal{M}_2$ = the smallest class containing the initial primitive recursive
functions Z, S, P_k^i and closed under composition and nested
double recursion

Calling the class of primitive recursive functions $\mathcal{P}$, we have $\mathcal{M}_2 \supseteq \mathcal{P}$. Since the
Ackermann-Péter function is not primitive recursive, $\mathcal{M}_2 \neq \mathcal{P}$.

Now do we have all the computable functions?

2. *n*-fold recursion

No, for we can computably number the doubly recursive functions just as we did $\mathcal{P}$
(Chapter 11.F), and then diagonalize or equally well define a function that dominates
them (Chapter 11.G).

With the primitive recursive functions, we turned that informal description into
a formal definition of a function ψ which uses nested recursion on one more variable
to dominate the primitive recursive functions (Chapter 12.C). We can do the same
here: We can define a function ρ that uses nested recursion on 3 variables and
dominates all the doubly recursive functions. Using ρ we can divide up the doubly
recursive functions into a hierarchy based on raising the bound in bounded minimization at each level. As you can well imagine after Chapter 12, the definition and
proofs that establish this are quite involved, although the ideas should be clear.
So we refer you to Péter's book, 1967, for the details.

Using a general form of nested recursion on 3 variables, we can define the class
of triply recursive functions $\mathcal{M}_3$. Then we can diagonalize with a function defined
by nested recursion on 4 variables, which establishes a hierarchy in $\mathcal{M}_3$ based on
raising the bound in bounded minimization. Generally, for every n we can define
the class of n-fold recursive functions, $\mathcal{M}_n$, with nested recursion on n variables
as a basic operation. And then we can diagonalize those functions with an $n + 1$ -
fold recursive function, establishing a hierarchy in $\mathcal{M}_n$ based on raising the bound
in bounded minimization.

By extending the bound under which we can do bounded
minimization we get larger and larger classes of computable functions.

3. Diagonalizing the multiply recursive functions

Once we've designated a general form of n-fold recursion for every n we've run out of variables to induct on. So now we've surely got all the computable functions.

No again, and for the same reason. We can give an inductive definition of the class $\mathcal{M}$ of all *multiply recursive functions* as the smallest class containing the initial functions Z, S, P_k^i and closed under the operations of composition and nested recursion on n variables for every n. Then it's only marginally more difficult to give a computable numbering of these functions and diagonalize, or equally well derive a computable function which dominates all of them.

How are we then to proceed?

B. Recursion on Order Types

Recall that we explained computing by double induction by referring to this picture:

	$(0,0)$	$(0,1)$	...	$(0,n)$	$(0,n+1)$	...
rows $\rightarrow$	$(1,0)$	$(1,1)$	...	$(1,n)$	$(1,n+1)$	...
	$\vdots$	$\vdots$		$\vdots$	$\vdots$	
	$(m,0)$	$(m,1)$	...	(m,n)	$(m,n+1)$	...
	$(m+1,0)$	$(m+1,1)$	...	$(m+1,n)$	$(m+1,n+1)$	...

To calculate $f(m+1, n+1)$ we could backtrack along the $m+1^{\text{st}}$ row, then be thrown back to some place on an earlier row, say to $f(m-16, 158107654)$, which, though a long way out on that row, would eventually lead us to a still earlier row since there are only 158107654 places preceding that one on that row; and that would lead to an earlier row still; and finally to the 0th row and to $f(0, 0)$. Since we knew that value, we could calculate $f(m+1, n+1)$.

In essence what we were doing was putting an ordering on pairs of numbers : $(m,n) < (x,y)$ iff $m < x$, or else $m = x$ and $n < y$. If we think of the natural numbers with their usual ordering, which we refer to as ω, then what we have here is a picture of ω^2. We can mock this ordering of pairs with a new ordering of the natural numbers.

For the 0th row put the odds : 1, 3, 5, 7, ...

For the first row put the numbers divisible by 2 but not by 4:
2, 6, 10, 14, ...

For the next row put the numbers divisible by 4 but not by 8:
4, 12, 20, 28, ...

For the nth row put the numbers divisible by 2^n but not by 2^{n+1}.

But we've left out 0. So subtract 1 from every entry above. More formally, first note that every natural number can be represented in the form $2^m(2n+1) - 1$ for some m and n. Then the ordering that mocks our square of pairs of natural numbers is:

$$a <_{\omega^2} b \quad \text{iff} \quad a = 2^m(2n+1) - 1 \quad \text{and} \quad b = 2^x(2y+1) - 1$$
$$\text{and } m < x, \text{ or } m = x \text{ and } n < y$$

Double recursion thus reduces to primitive recursion respecting this ordering instead of the usual ordering of the natural numbers. That is, the operation of double recursion can be replaced by the operation of recursion on $<_{\omega^2}$, where we define f from g, h, and q by

(‡)
$$f(0, \vec{x}) = g(\vec{x})$$
$$f(r+1, \vec{x}) = h(f(q(r+1, \vec{x}), \vec{x}), r, \vec{x})$$

where

$$q(0, \vec{x}) = 0$$

and

$$q(r+1, \vec{x}) <_{\omega^2} r + 1$$

In a similar way, proof by double induction can be reduced to proof by induction on this ordering.

For triple recursion what picture do we have? We need to use triples of natural numbers arranged in a cube, ω^3.

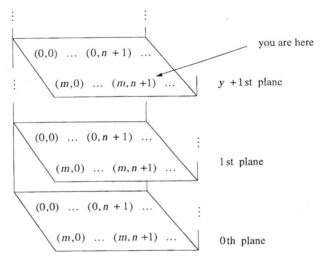

To calculate $f(y+1, m+1, n+1)$ we begin at that place in the mth row of the $y+1$st plane. Now you describe how we backtrack to eventually reach $(0, 0, 0)$. Triple recursion can then be reduced to single recursion on an ordering of the natural numbers that mocks this ordering of triples.

For induction on 4 variables, we use quadruples of natural numbers arranged in a line of cubes that look like ω^3, which altogether we call ω^4. Then for 5-fold recursion, we use quintuples of natural numbers arranged in a square of cubes like ω^3, which we call ω^5. Then for 6-fold recursion we use sextuples of natural numbers arranged in a cube of cubes like ω^3, which we call ω^6. And generally we can reduce recursion on n variables to recursion on an ordering $<_{\omega^n}$ as in (‡) above, which mocks the arrangement on n-tuples of natural numbers. Nested recursion on n variables can also be reduced to recursion on a single variable respecting the ordering ω^n, though its form is more complicated (see Odifreddi, volume 2).

These ω^n are examples of Cantor's "transfinite numbers" that Hilbert referred to in his paper (Chapter 7). But for our purposes they are simply ways to arrange the natural numbers in more complicated orderings. What is important about these orderings is that they are *well-orderings*: Every nonempty collection of natural numbers has a least element in the ordering. And they are computable, indeed primitive recursive. That's what we need to be certain that we have a computation procedure for a function defined by recursion on the ordering since we can always backtrack as we described above.

To continue beyond recursion on these order types, we concoct an ordering that corresponds to the collection of all n-tuples for all n (the $n + 1$ st-tuples following the n-tuples) followed by one more copy of the natural numbers, which together we denote $\omega^\omega + \omega$. Then we can define the class of functions which use the operation of nested recursion on $<_{\omega^\omega + \omega}$; using that operation we can diagonalize the multiply recursive functions. But even then we won't have all computable functions, for we can number these and diagonalize or create a function which dominates them and raises still higher the bound on bounded minimization.

No matter how far we go along this route, as long as we use well-orderings that we've previously justified as computable and are a "natural extension" of the previous ones, we can always computably diagonalize and raise the bound for bounded minimization.

Further Reading

For the multiply recursive functions consult Péter's book *Recursive Functions*. Hausdorff in *Set Theory* gives a basic exposition of the order types we have discussed, and Rogers in *Theory of Recursive Functions and Effective Computability*, pp. 219–222, presents the theory of computable order types as a series of good but hard homework exercises. Odifreddi, *Classical Recursion Theory* Volume 2, covers recursion on order types and relates it to various hierarchies of the computable functions.

14 The Least Search Operator

What happens if we eliminate the bound in bounded minimization? Would that be a computable operation?

A. The μ-Operator

Eliminating the bound leads to a problem of undefined points. Consider

$f(x)$ = the least y such that $y + x = 10$

For each $x > 10$, $f(x)$ is undefined. Yet f is still computable: for $f(12)$, for instance, we can check each y in turn to see that $y + 12 \neq 10$.

You might say that it's obvious that there is no such y that makes $f(12)$ defined. So why can't we use that fact to make a better function which is defined everywhere? That would be stepping outside the system. We'd need not only a program, an instruction for f that would tell us to calculate $y + x$ and verify whether it equals 10 or not, but also a *proof* that there is no such y if $x > 10$. In this case that would be easy. But consider the function

$h(0) = h(1) = 0$, and for $w > 1$,
$h(w)$ = the least $\langle x, y \rangle$ such that x and y are prime and $x + y = 2 \cdot w$

At present no one knows for which w this function is defined (the claim that it is defined for all w is called "Goldbach's Conjecture"). Yet for, say, $w = 4318$ we can constructively check in turn each pair $\langle x, y \rangle$ to see whether x and y are prime and $x + y = 2 \cdot 4318$.

We define the *least search operator*, also called the *μ-operator,* as:

$$\mu y\,[\,f(\vec{x},y) = 0\,] = z \quad \text{iff} \quad \begin{cases} f(\vec{x}, z) = 0 \text{ and} \\ \text{for every } y < z, \ f(\vec{x},y) \text{ is defined and} > 0 \end{cases}$$

B. The min-Operator

A comparison: Denote by " $\min_y [\,f(\vec{x}, y) = 0\,]$ " the smallest solution to the equation $f(\vec{x}, y) = 0$ if such exists, and undefined otherwise.

The following example shows that the min-operator is not the same as the µ-operator. Define the primitive recursive function:

$$h(x, y) = \begin{cases} x - y & \text{if } y \leq x \\ 1 & \text{otherwise} \end{cases}$$

Now define:

$$g(x) = \mu y [2 \div h(x, y) = 0]$$
$$g^*(x) = \min_y [2 \div h(x, y) = 0]$$

Then

$g(0), g(1)$ are undefined, $g(2) = 0$

$g^*(0), g^*(1)$ are undefined, $g^*(2) = 0$

But now define:

$$f(x) = \mu y [g(y) \cdot (x + 1) = 0]$$
$$f^*(x) = \min_y [g^*(y) \cdot (x + 1) = 0]$$

Then $f(x)$ is undefined for all x; but for all x, $f^*(x) = 2$.

C. The µ-Operator Is a Computable Operation

Why do we choose the µ-operator rather than the min-operator? We might not be able to predict for which x, $f(x, y) = 0$ has a solution. To say that $\min_y [f(x, y) = 0] = 1$ when $f(x, 0)$ is undefined due to an infinite search entails the completion of an infinite task.

But with the µ-operator if $f(x, 0)$ is undefined (that is, we're put into a search that never ends) then $\mu y [f(x, y) = 0]$ is undefined too, for we never get a shot at trying $f(x, 1) = 0$.

To calculate $g(x) = \mu y [f(x, y) = 0]$ we proceed in steps:

Step 0: calculate $f(x,0)$

      if defined $= 0$ output the number of this step;

      if defined > 0 we continue the search at the next step;

      if we never get an answer for $f(x,0)$ then $g(x)$ will be undefined, for it is in an unending search

Step 1: calculate $f(x,1)$ — proceed as in step 0

Step 2: calculate $f(x,2)$ — proceed as in step 0

    ⋮

This is a well-defined *computable procedure* for calculating $g(x)$, though it may not always give a result.

15 Partial Recursive Functions

A. The Partial Recursive Functions

1. Our investigations have led us to make the following definition:

> The *partial recursive functions* are the smallest class containing the zero, successor, and projection functions and closed under composition, primitive recursion, and the μ-operator.

Again, we remind you that even though phrased in terms of classes, this is really an inductive definition of the label "partial recursive" (see Chapter 11.B). We sometimes abbreviate "partial recursive" as *p. r.*

Since our functions may not be defined on all inputs, we will say that a function that is defined for all inputs is *total* and will continue to use lowercase Roman letters, f, g, h, etc. for total functions. Note that any total function we have so far investigated is partial recursive: We can define it just as before simply deleting any reference to a bound. The term *recursive* is reserved for total p. r. functions.

We call functions which may (for all we know) be undefined for some inputs *partial* functions and use lowercase Greek letters φ, ψ, ρ and so on, to denote them. We write, for example, $\varphi(x)$ to mean the function (thought of as a procedure) applied to x. We do not necessarily mean by this that there is an object called $\varphi(x)$, for φ applied to x may be undefined. We write

$\varphi(x)\downarrow$ for "φ applied to x is defined"

$\varphi(x)\updownarrow$ for "φ applied to x is not defined"

When are two partial functions (extensionally) the same? First, φ and ψ *agree on input* x if both $\varphi(x)\downarrow$ and $\psi(x)\downarrow$ and these are equal, or both $\varphi(x)$ and $\psi(x)$ are undefined. In that case, we write:

$\varphi(x) \simeq \psi(x)$

We say that φ and ψ are the same function if they agree on all inputs; that is, for all x, $\varphi(x) \simeq \psi(x)$. In that case, we write:

$\varphi \simeq \psi$

These conventions also apply to functions of several variables.

We say that a *set A* or *relation R is recursive* if its characteristic function, C_A or C_R, is recursive (see Chapter 11.D.2 and D.3). Note that every characteristic function is total (that's the law of excluded middle).

When we use the μ-operator we need to reverse the roles of 0 and 1 in the characteristic function, so we define the *representing function* for a relation R to be $\overline{sg} \circ C_R$.

2. It is not as restrictive as it may appear that the μ-operator requires us to search for a y such that $\varphi(\vec{x}, y) = 0$. Given a relation R, we write:

$$\mu y \leq g(\vec{x}) \, [R(\vec{x}, y)] \quad \text{to mean} \quad \mu y \, [y \leq g(\vec{x}) \wedge R(\vec{x}, y)] \,.$$

Lemma 1   If g and R are recursive, then the following functions are partial recursive.

$$\tau(\vec{x}) \simeq \mu y \, [\, \varphi(\vec{x}, y) = a \,]$$

$$\rho(\vec{x}) \simeq \mu y \, [R(\vec{x}, y)]$$

$$\psi(\vec{x}) \simeq \mu y \leq g(\vec{x}) \, [R(\vec{x}, y)]$$

$$\gamma(\vec{x}) \simeq \mu y < g(\vec{x}) \, [R(\vec{x}, y)]$$

We leave the proof as Exercise 1.

B. Diagonalization and the Halting Problem

Why can't we diagonalize out of this class?

Assume for the moment that we can effectively number all the partial recursive functions of one variable as φ_1, φ_2, ... , φ_n, ... (we'll indicate how in the next chapter, but from our previous experience with numbering this should be plausible). We can then define $\psi(x) = \varphi_x(x) + 1$. *But it won't diagonalize*, because some $\varphi_x(x)$ are *not defined*.

Can we avoid this and hence diagonalize by deciding if $\varphi_x(x)$ is defined?

THEOREM 2   There is no recursive function which can tell us whether $\varphi_x(x)$ is defined.

Proof: Suppose such a recursive function, f, exists:

$$f(x) \simeq \begin{cases} 1 & \text{if } \varphi_x(x) \downarrow \\ 0 & \text{if } \varphi_x(x) \not\downarrow \end{cases}$$

Then

$$\rho(x) \simeq \begin{cases} \not\downarrow & \text{if } \varphi_x(x) \downarrow \\ 0 & \text{if } \varphi_x(x) \not\downarrow \end{cases}$$

is partial recursive. For the first and last time we will formally define such a function: $\rho(x) \simeq \mu y \, [\, y + f(x) = 0 \,]$. So ρ must be φ_v for some y. But

$$\rho(y) = \begin{cases} \maltese & \text{if } \varphi_y(y)\!\downarrow \\ 0 & \text{if } \varphi_y(y)\!\maltese \end{cases}$$

This is a contradiction. So no such f exists.   ∎

Diagonalization bites the dust—but at the cost of introducing partial functions!

We've shown that $K \equiv_{\text{Def}} \{x : \varphi_x(x)\!\downarrow\}$ is not recursive. More generally, define $K_0 \equiv_{\text{Def}} \{\langle x, y \rangle : \varphi_x(y)\!\downarrow\}$. The characteristic function of K_0 is called the *halting problem* for the partial recursive functions since $\langle x, y \rangle$ is in K_0 iff the xth algorithm applied to y halts. We leave the proof of the following as Exercise 4.

COROLLARY 3 (The Halting Problem Is Unsolvable) K_0 is not recursive.

C. The General Recursive Functions

Remember when you were asked to say what you thought the criteria should be for a procedure to be computable (Chapter 8)? If you said it should terminate, you must be very dissatisfied with our introduction of partial functions.

Let's say a function $g(\vec{x}, y)$ is *regular* if it is total and for every $\vec{x}$ there is some y such that $g(\vec{x}, y) = 0$. The class of *general recursive functions* is defined exactly as the partial recursive ones except that the μ-operator may be applied only to regular functions. Clearly, every general recursive function is a total function, indeed, a total partial recursive function.

Church's Thesis (in one of its equivalent forms) asserts:
A function is computable iff it is general recursive.

But it is a false pleasure we get from creating a class all of whose functions are total. The operation of μ-operator applied to regular functions is "ill-defined" because for arbitrary x *we cannot decide* if $g(x, y) = 0$ has a solution. That might require calculating each of $g(x,0)$, $g(x,1)$, $g(x,2)$, ... none of which may equal 0 (and, note, it would entail solving the halting problem). So we *cannot effectively number* the general recursive functions (and hence can't computably diagonalize them either). The ambiguity of partial functions is essential in order to obtain a class of computable functions we can computably number and yet not computably diagonalize. In Chapter 16 we'll show:

the general recursive functions $\equiv$ the total partial recursive functions

D. Gödel on Partial Functions

Gödel points out that the precise notion of mechanical procedures is brought out clearly by Turing machines producing partial rather than general recursive functions. In other words, the intuitive notion does not require that a mechanical procedure should always terminate or succeed. A sometimes unsuccessful procedure, if sharply defined, still is a procedure, i.e. a well determined manner of proceeding. Hence we have an excellent example here of a concept which did not appear sharp to us but has become so as a result of a careful reflection. The resulting definition of the concept of mechanical by the sharp concept of 'performable by a Turing machine' is both correct and unique. Unlike the more complex concept of always-terminating mechanical procedures, the unqualified concept, seen clearly now, has the same meaning for the intuitionists [a brand of constructive mathematicians, see Chapter 26] as for the classicists. Moreover it is absolutely impossible that anybody who understands the question and knows Turing's definition should decide for a different concept.

Wang, p. 84

Exercises

1. Prove Lemma 1 (cf. Exercise 11.15).

2. Show that the function defined by
$$g(x) = \begin{cases} 0 & \text{if } x \text{ is even} \\ \updownarrow & \text{if } x \text{ is odd} \end{cases}$$
is partial recursive by giving a μ-operator definition of it using functions which we've already shown are partial recursive.

3. a. Applied to regular functions the min-operator and the μ-operator are the same. Suppose that φ is partial recursive and f is defined by
$f(\vec{x}) = \mu y[\varphi(\vec{x}, y) = 0]$. If f is total, show that for all $\vec{x}$,
$f(\vec{x}) = \min y[\varphi(\vec{x}, y) = 0]$.

   b. Show that the partial recursive functions are not closed under the min-operator.
   (*Hint:* Define $\psi(x, y) = 1$ if either $y = 1$, or both $y = 0$ and $\varphi_x(x) \downarrow$.)

4. Prove that K_0 is not recursive.

5. We say that a function ψ *extends* a function φ if whenever $\varphi(x) \downarrow$, $\psi(x) \downarrow = \varphi(x)$. Show that there is a partial recursive function φ which cannot be extended to a total recursive function.
   (*Hint:* Consider $\varphi \simeq \lambda x(\varphi_x(x) + 1)$.)

6. We said that we can't diagonalize out of the class of partial recursive functions. But doesn't the function f defined by
$$f(x) = \begin{cases} \varphi_x(x) + 1 & \text{if } \varphi_x(x) \downarrow \\ 0 & \text{otherwise} \end{cases}$$
diagonalize the class of partial recursive functions? Explain.

7. Explain why you do or do not agree that partial functions are computable. Are Gödel's remarks (via Wang) a convincing argument for accepting partial procedures as computable ?

16 Numbering the Partial Recursive Functions

A. Why and How: The Idea

We want to number the partial recursive functions for two reasons. First, we want to justify and make precise the comments we made in Chapter 15 about the halting problem and diagonalization. Second, as we number the functions we will code how each is built up, so that given the number of a function we can uncode it to compute the function on any input. In this way we will have a partial recursive procedure which simulates all partial recursive functions, what we call a "universal function" for the partial recursive functions.

The idea of the numbering isn't hard; it's no harder than the sketch we made of numbering the primitive recursive functions in Chapter 11.F. But writing it down gets a bit complicated, so we'll give the idea in rough form first. We're going to use the coding of sequences of numbers we presented in Chapter 11.E.5 with which you should be familiar.

The numbering is an inductive procedure. As basis, we number the initial functions, say Z gets number 0, S gets number $\langle 1 \rangle$, and P_k^i gets number $\langle 1, i \rangle$ (the number of variables will determine k). At the induction stage we assume that we have already numbered various functions, say φ_a, φ_b, φ_{b_1}, φ_{b_2}, $\dots$, φ_{b_k}. Then corresponding to each operation we can use to produce new functions, we will associate an arithmetical operation:

for composition, $\varphi_a(\varphi_{b_1}, \varphi_{b_2}, \dots, \varphi_{b_k})$ will get number
$\langle a, \langle b_1, b_2, \dots, b_k \rangle, 0 \rangle$

for primitive recursion, the function defined by recursion on φ_b with basis φ_a will get number $\langle a, b, 0, 0 \rangle$

for the least search operator, $\mu y \, (\varphi_a(\vec{x}, y) = 0)$ will get number
$\langle a, 0, 0, 0, 0 \rangle$

There are two complications, however, that make the numbering harder than this sketch. First, we will want every number to be the number of some function, so we will build in a lot of redundancy. For example, rather than assigning just $\langle a,0,0,0,0 \rangle$ to the function that arises by an application of the μ-operator to φ_a, we'll assign every n with $lh(n) \geq 5$ such that $(n)_0 = a$. Second, we need to number all functions of every possible number of variables at one go. Thus, when we come to the number n we will have to stipulate for every k what the nth function of k variables is. This is analogous to defining a Turing machine that works for inputs of any number of variables.

Now, given any number we can unpack it to see what function it corresponds to. For example, if we have $n = \langle a,0,0,0,0 \rangle$ then we know that it's the number of a function which arises by least search operator applied to φ_a, where $a < n$. We can continue to unpack, say $a = \langle 4796521, 814, 0, 0 \rangle$. In this case, we know that φ_a is obtained by primitive recursion on φ_{814} with basis $\varphi_{4796521}$. By unpacking n until we arrive at a complete description of it using only primes, 0, and 1, we can get a description of the function it indexes in terms of initial functions and operations on them. Hence, given any x we can describe how to calculate $\varphi_n(x)$. The formal description of that process amounts to a universal function for the partial recursive functions.

B. Indices for the Partial Recursive Functions

Recall that $\varphi \simeq \psi$ means that for all x, $\varphi(x)\downarrow$ iff $\psi(x)\downarrow$, and they are equal if both are defined.

We shall index all partial recursive functions of every number of variables at once. The nth function of k variables for $k \geq 1$ will be denoted φ_n^k, though we will drop the k if it's clear. The variables will always be labeled $x_1, x_2, \dots, x_k$. We will give a definition by inducting on n for all k at once.

By induction, we define $\varphi_n^k (x_1, x_2, \dots, x_k)$.

Case 1   (zero, successor) $lh(n) = 0$ or 1
      n codes the empty sequence or $(n)_0$
      If $lh(n) = 0$, then $\varphi_n^k \simeq$ the constant function Z
      If $lh(n) = 1$, then $\varphi_n^k \simeq$ the successor function, $x_1 + 1$

Case 2   (projections) $lh(n) = 2$
      n codes the sequence $((n)_0, (n)_1)$
      If $(n)_0 = 0$ or $(n)_1 = 0$, then $\varphi_n^k \simeq$ the constant function Z
      If $(n)_0 \geq 1$ and $1 \leq (n)_1 \leq k$, then $\varphi_n^k \simeq P_k^{(n)_1} (x_1, x_2, \dots, x_k)$
      If $(n)_0 \geq 1$ and $k < (n)_1$, then $\varphi_n^k \simeq P_k^{\ k} (x_1, x_2, \dots, x_k)$

Case 3   (composition of functions) $lh(n) = 3$
      n codes the sequence $((n)_0, (n)_1, (n)_2)$
      If $lh((n)_1) = 0$, then $\varphi_n^k \simeq \varphi_{(n)_0}^k$

      If $lh((n)_1) \geq 1$, then $\varphi_n^k \simeq \varphi_{(n)_0}^{lh((n)_1)} (\varphi_{((n)_1)_0}^k, \dots, \varphi_{((n)_1)_{lh((n)_1) - 1}}^k)$

Case 4 (primitive recursion) $lh(n) = 4$

n codes the sequence $((n)_0, (n)_1, (n)_2, (n)_3)$

If $k = 1$, then

$$\varphi_n(0) \simeq (n)_3$$
$$\varphi_n(x_1 + 1) \simeq \varphi^2_{(n)_1}(\varphi_n(x_1), x_1)$$

If $k > 1$, then

$$\varphi^k_n(0, x_2, \ldots, x_k) \simeq \varphi^{k-1}_{(n)_0}(x_2, \ldots, x_k)$$
$$\varphi^k_n(x_1 + 1, x_2, \ldots, x_k) \simeq$$
$$\varphi^{k+1}_{(n)_1}(\varphi^k_n(x_1, x_2, \ldots, x_k), x_1, x_2, \ldots, x_k)$$

Case 5 (μ-operator) $lh(n) \geq 5$

n codes the sequence $((n)_0, \ldots, (n)_{lh(n)_1 \dot{-} 1})$

$$\varphi^k_n(x_1, x_2, \ldots, x_k) \simeq \mu x_{k+1}[\varphi^{k+1}_{(n)_0}(x_1, x_2, \ldots, x_k, x_{k+1}) = 0]$$

This completes the numbering. If $\psi \simeq \varphi^k_n$, we call n an *index* of ψ .

Note that we have numbered programs, descriptions of functions. In Exercise 2 you're asked to prove that every partial recursive function has arbitrarily many different indices. This is not an accidental feature of the numbering: In Exercise 9 you're asked to show that there is no recursive way to determine whether two programs give the same function.

THEOREM 1 φ is partial recursive iff for some n, $\varphi \simeq \varphi_n$.

Proof: We leave to you to show by induction on n that for every n, φ_n is partial recursive.

To show that if φ is partial recursive there is some n such that $\varphi \simeq \varphi_n$, we use *induction on the number of applications of the basic operations in a definition of φ* .

If no operations are used in the definition then φ is an initial function: if φ is Z then $\varphi \simeq \varphi_0$; if φ is S then $\varphi \simeq \varphi_4$; and if φ is P^i_k then $\varphi \simeq \varphi_{4 \cdot 3^i + 1}$.

Suppose now that the theorem is true for every function which has a definition using at most m applications of the basic operations and φ has a definition which uses $m + 1$ applications. If φ is defined by an application of the μ-operator applied to ρ, and ρ has a definition with at most m applications of the basic operations, then for some r, $\rho \simeq \varphi_r$ and $\varphi \simeq \varphi_{2^r + 1 \cdot 3 \cdot 5 \cdot 7 \cdot 11}$. The other cases are similar and we leave them to you as Exercise 1.   ∎

C. Algorithmic Classes
(Optional)

Generally, we call any class of functions of natural numbers *algorithmic* if it contains the zero, successor, and projection functions, and is closed under the operations of composition, primitive recursion, and μ-operator.

The class of partial recursive functions is the smallest algorithmic class. In set theoretic terminology, it is the intersection of all algorithmic classes, which is to say that every partial recursive function is contained in every algorithmic class. Thus, the difference between a general algorithmic class and the partial recursive functions is what additional nonrecursive initial functions are chosen.

If $f_1, f_2, \ldots, f_n$ are total functions of the natural numbers, then we say that φ is *partial recursive in* $\{f_1, f_2, \ldots, f_n\}$ iff φ can be obtained from the initial functions zero and successor, the projections, and $f_1, f_2, \ldots, f_n$ by the operations of composition, primitive recursion, and μ-operator.

For Exercise 11 we ask you to give a numbering of the functions partial recursive in $\{f_1, f_2, \ldots, f_n\}$.

D. The Universal Computation Predicate

Using our numbering, we'd like to check whether $\varphi_n(b)\!\downarrow = r$. But we can't because the halting problem is unsolvable (Corollary 15.3) . However, *if* $\varphi_n(b)\!\downarrow = r$, we can tell that: since n codes the definition of the function, we can actually do the computation. But $\varphi_n(b)$ might be undefined due to an infinite search. We will get a recursive predicate by limiting the searches we can do to check whether $\varphi_n(b)\!\downarrow = r$. So "$C(n, b, r, q)$" will mean that $\varphi_n(b)\!\downarrow = r$ and q bounds the largest number used in that computation. Thus if $C(n, b, r, q)$ holds, so will $C(n, b, r, w)$ for any $w > q$ (intuitively, if you can compute in time q and $q < w$, then you can compute in time w). Moreover, because all the searches are bounded by q, the predicate that checks the computation will actually be primitive recursive. This is an important point: there is no effective procedure for determining whether $\varphi_n(b)\!\downarrow = r$, but given a purported computation, "$\varphi_n(b)\!\downarrow = r$ in time q", we can check it primitive recursively. Indeed, the checking process is elementary (Corollary 3).

When we write "$C(n, b, r, q)$" we mean to assert that $C(n,b,r,q)$ holds.

THEOREM 2 (The Universal Computation Predicate)

There is a primitive recursive predicate C such that
$$\varphi_n^k(b_1, \ldots, b_k) = r \text{ iff } \exists q\, C(n, \langle b_1, \ldots, b_k \rangle, r, q)$$
Moreover, if $C(n, b, r, q)$ and $q < w$, then $C(n, b, r, w)$.

Proof: We are going to define C by induction on n (the number of the function) by stipulating those cases in which it holds (and thus by implication in all others it fails). Thus, when we define $C(n,b,r,q)$ we may assume that we have already defined $C(m,x,y,z)$ for any $m < n$ and all x, y, z. We can do this because the only functions which are referred to in the definition of φ_n are φ_m for $m < n$. Also note that if $C(n,b,r,v)$ and $lh(b) = k \geq 1$, then b will code $(b_1, \ldots, b_k)$ so that k is the number of variables involved (the variables are numbered starting with 1, the primes starting with $p_0 = 2$).

Case 1 $lh(n) = 0$ and $r = 0$
    or $lh(n) = 1$ and $r = (b)_0 + 1$

Case 2 $lh(n) = 2$ and
    $(n)_0 = 0$ or $(n)_1 = 0$, and $r = 0$
    or $(n)_0 \geq 1$ and $1 \leq (n)_1 \leq lh(b)$ and $r = (b)_{(n)_1 \div 1}$
    or $(n)_0 \geq 1$ and $k < (n)_1$ and $r = (b)_{lh(b) \div 1}$

Note that q is irrelevant in cases 1 and 2 .

Case 3 $lh(n) = 3$ and
    $lh((n)_1) = 0$ and $C((n)_0, b, r, q)$
    or $lh((n)_1) \geq 1$ and
    $\exists d \leq q$ with $lh(d) = lh((n)_1)$ and
    $C(((n)_1)_i, b, (d)_i, q)$ for $0 \leq i \leq lh((n)_1) \div 1$
    and $C((n)_0, d, r, q)$

Case 4 $lh(n) = 4$ and
    $lh(b) = 1$ and $(b)_0 = 0$ and $r = (n)_3$
    or $lh(b) = 1$ and $(b)_0 \geq 1$ and $\exists e \leq q$,
    $C(n, \frac{b}{2}, e, q)$ and $C((n)_1, \langle e, (b)_0 \div 1 \rangle, r, q)$
    or $lh(b) > 1$ and $(b)_0 = 0$ and
    $C((n)_0, \langle (b)_1, \dots, (b)_{lh(b) \div 1} \rangle, r, q)$
    or $lh(b) > 1$ and $(b)_0 \geq 1$ and $\exists e \leq q$,
    such that $C(n, \frac{b}{2}, e, q)$ and
    $C((n)_1, \langle e, (b)_0 \div 1, (b)_1, \dots, (b)_{lh(b) \div 1} \rangle, r, q)$

Case 5 $lh(n) \geq 5$ and
    $C((n)_0, b \cdot p(lh(b))^{1+r}, 0, q)$ [recall that $p(x) =$ the xth prime]
    and $\forall i < r$, $\exists e$, $0 < e < q$, such that
    $C((n)_0, b \cdot p(lh(b))^{1+i}, e, q)$

This completes the description of C.

    Now we must prove that C does what we claim. First, note that C is primitive recursive since every condition is obtained by bounded existence on some primitive recursive condition.

    To show that $\varphi_n(b_1, \dots, b_k) = r$ iff $\exists q \, C(n, \langle b_1, \dots, b_k \rangle, r, q)$ we induct on n and subinduct on $b = \langle b_1, \dots, b_k \rangle$. For the basis, we note that for $lh(n) \leq 2$ it's clear.

    Suppose now it is true for all $a < n$ and for n for all $x < b$. We'll do only one direction of one case, and leave the rest to you. Suppose $\varphi_n(b_1, \dots, b_k) = r$

and $lh(n) = 5$. Then $\varphi_{(n)_0}(b_1, \ldots, b_k, r) = 0$ and hence, since $(n)_0 < n$, by induction there is some v such that $C((n)_0, \langle b_1, \ldots, b_k, r \rangle, 0, v)$. Moreover, for each $i < r$, $\varphi_{(n)_0}(b_1, \ldots, b_k, i) \downarrow > 0$. So by induction we know that there are $u_0, \ldots, u_{r-1}$ and $v_0, \ldots, v_{r-1}$ such that for all $i \le r - 1$, $u_i > 0$ and $C((n)_0, \langle b_1, \ldots, b_k, i \rangle, u_i, v_i)$. Take

$$q = max(u_0, \ldots, u_{r-1}, v_0, \ldots, v_{r-1}, v, r) + 1$$

Then $C(n, \langle b_1, \ldots, b_k \rangle, r, q)$.   ∎

We only claimed in Theorem 2 that C is primitive recursive, but actually we've proved more. Recall that a function is elementary if it is in $\mathcal{E}$ (Chapter 12.B).

COROLLARY 3   The universal computation predicate is elementary.

Proof: This is just a matter of tracing through the definition of C to see that every condition is obtained by bounded existence on some elementary condition.   ∎

Since the representing function (total) for the universal computation predicate is partial recursive it must have an index. We call the least one c, so that

$$\varphi_c^4(n, m, r, q) = \begin{cases} 0 & \text{if } C(n, m, r, q) \\ 1 & \text{if not } C(n, m, r, q) \end{cases}$$

E. The Normal Form Theorem

What we have done so far in this chapter may seem merely a tedious exercise in labeling and reading labels. But the names we have given code a lot of information. Using the numbering we can define a universal partial recursive function, one which calculates all others, analogous to one Turing machine which simulates all others.

THEOREM 4   For $\vec{x} = (x_1, x_2, \ldots, x_k)$ the function

$$\lambda n, \vec{x}\ (\mu q\,[\,C(n, \langle \vec{x} \rangle, (q)_0, q)\,])_0$$

is partial recursive and is universal for the partial recursive functions of k variables. That is, if φ is a partial recursive function of k variables, then for some n, all $\vec{x}$,

$$\varphi(\vec{x}) \simeq (\mu q\,[\,C(n, \langle \vec{x} \rangle, (q)_0, q)\,])_0$$
$$\simeq (\mu q\,[\,\varphi_c^4(n, \langle \vec{x} \rangle, (q)_0, q) = 0\,])_0$$

Proof: By Theorem 1, if φ is partial recursive then for some n, φ is φ_n. So by Theorem 2, if $\varphi(\vec{x}) \downarrow = r$, then there is some s such that $C(n, \langle \vec{x} \rangle, r, s)$. Hence $C(n, \langle \vec{x} \rangle, r, \langle r, s \rangle)$ and the theorem follows.   ∎

We know that every general recursive function is partial recursive, but in Chapter 15 we said that we could prove the converse. Using the Normal Form Theorem that's now easy.

COROLLARY 5   a. Every partial recursive function may be defined with at most one use of the μ-operator.

         b. A total function is partial recursive iff it is general recursive.

Proof: a. This part follows from Theorem 4 since φ_c^4 is primitive recursive.

    b. Given any total partial recursive function, by Theorem 4 there is a definition of it that uses only one application of the μ-operator applied to a primitive recursive function. Hence that primitive recursive function must be regular.   ∎

F. The *s-m-n* Theorem

Consider the partial recursive function

$$\varphi(x,y) \simeq x^y + [y \cdot \varphi_x^1 (y)]$$

(Exercise 4). Suppose we take y as a parameter and consider, for example,

$$\lambda x (x^3 + [3 \cdot \varphi_x^1(3)])$$

Then that's partial recursive too. Generally, by using the Normal Form Theorem we can show that if we start with a partial recursive function and hold one or several variables fixed, we get another partial recursive function. More importantly, though, we can find an index for the new function effectively in the index of the given one.

THEOREM 6 (The *s-m-n* Theorem)   For every $n, m \geq 1$ there is a recursive function S_n^m such that if we hold the first m variables fixed in

$$\varphi_x(a_1, \dots , a_m, y_1, \dots , y_n)$$

then an index for the resulting partial recursive function is $S_n^m (x, a_1, \dots , a_m)$.

$$\lambda y_1 \cdots y_n [\varphi_x(a_1, \dots , a_m, y_1, \dots , y_n)]$$
$$\simeq \varphi_{S_n^m(x, a_1, \dots , a_m)} (y_1, \dots , y_n)$$

Proof: The left-hand side of the equation is

$$(\mu q [\varphi_c (x, \langle \underbrace{a_1, \dots, a_m}_{}, \underbrace{y_1, \dots, y_n}_{} \rangle, (q)_0, q) = 0])_0$$

        view these as constants   view these as projections

To begin, the number 36 is an index for the identity function. In Exercise 5 we ask you to define a primitive recursive function h such that for all n, $\varphi_{h(n)}$ is the constant function $\lambda x (n)$. As before, $p(m) =$ the mth prime. We'll let you calculate an index d for the function $\lambda x ((x)_0)$. Using these we can define

$$S_n^m (x, a_1, \ldots , a_m) = 2^{d+1} \cdot 3^{a+1} \cdot 5$$

where $a = 2^{b+1} \cdot 3 \cdot 5 \cdot 7 \cdot 11$ and $b = 2^{c+1} \cdot 3^{e+1} \cdot 5$ where

$$e = p(0)^{h(x)+1} \cdot p(1)^{h(a_1)+1} \cdot \ldots \cdot p(m)^{h(a_m)+1} \cdot$$
$$p(m+1)^{1+(2^2 \cdot 3^{1+1})} \cdot \ldots \cdot p(m+n+1)^{1+(2^2 \cdot 3^{n+1})}$$

To see that this is correct requires going back to the numbering in Section B to check each part. That's a good way to get a grip on how the numbering works, and so we'll leave it to you.

Note that each S_n^m function is elementary since it involves only addition, multiplication, and composition on the functions p and h, all of which are elementary. ∎

G. The Fixed Point Theorem

Self-reference can be a problem, as we know from the liar paradox and our many uses of diagonalization. But it has also been a useful tool for us, since that is exactly what primitive recursion is based on: A function is defined in terms of itself. And we have seen other forms of recursion in which a function could be defined in terms of itself in Chapters 11–13. Here we will show that the immunity to diagonalization that the partial recursive functions enjoy can be put to good use to find fixed points and hence very general ways to define a function in terms of itself.

THEOREM 7 (The Fixed Point Theorem)
  If f is recursive then there is an e such that $\varphi_e \simeq \varphi_{f(e)}$.

Proof: Consider the function

$$\lambda x y \; \varphi_{\varphi_x (x)} (y) \simeq \begin{cases} \varphi_{\varphi_x(x)} (y) & \text{if } \varphi_x (x) \!\downarrow \\ \downarrow & \text{otherwise} \end{cases}$$

This is partial recursive since it can be defined as

$$\psi(x, y) \simeq (\mu q \; C(\varphi_x (x), \langle y \rangle, (q)_0, q))_0$$

So by the *s-m-n* theorem there is a function g such that $\varphi_{g(x)} \simeq \lambda y \; \psi(x,y)$. Now consider $f \circ g$, which is recursive. For some d we have $f \circ g \simeq \varphi_d$. Hence by the definition of g,

$$\varphi_{g(d)} \simeq \varphi_{\varphi_d(d)}$$

and by the definition of d,

$$\varphi_{f(g(d))} \simeq \varphi_{\varphi_d(d)}$$

Thus we may take $e = g(d)$. ∎

The Fixed Point Theorem is sometimes called the *Recursion Theorem*. Here is an example of its use, which we will need later. Although there are other ways to obtain the same result, this proof illustrates a typical application.

COROLLARY 8   If A is an infinite recursive set, then its elements can be enumerated in increasing order by a recursive function.

Proof: Let a be the least element of A. First note that the function we want is defined by the equations:

$$f(0) = a$$
$$f(x + 1) = \mu y \, [y > f(x) \wedge y \in A]$$

Define a function of two variables ρ by:

$$\rho(i, 0) = a$$
$$\rho(i, x + 1) \simeq \mu y \, [y > \varphi_i(x) \wedge y \in A]$$

We leave to you to show that ρ is partial recursive. The collection of functions $\lambda x \, \rho(i, x)$ gives us a matrix for which (instead of diagonalizing) we can find a fixed point. First, by the *s-m-n* theorem there is some recursive s such that $\varphi_{s(i)} \simeq \lambda x \, \rho(i, x)$. Then by the Fixed Point Theorem there is some e such that $\varphi_{s(e)} \simeq \varphi_e$. Hence

$$\varphi_e(0) = a$$
$$\varphi_e(x + 1) \simeq \mu y \, [y > \varphi_e(x) \wedge y \in A]$$ ∎

Exercises

1. Complete the proof of Theorem 1.

2. Prove:
   a. There are exactly countably many different partial recursive functions.
   b. Every partial recursive function has arbitrarily many different indices.

3. We now have a universal partial recursive function. So from now on you don't have to write any more programs, just hand in an index. Right?

4. Show that the example we gave in Section F, $\varphi(x, y) \simeq x^y + [y \cdot \varphi_x^1(y)]$, is partial recursive. (*Hint:* Use the Normal Form Theorem.)

5. Define a primitive recursive function h such that for all n, $\varphi_{h(n)}$ is the constant function $\lambda x \, (n)$.

6. a. Give an index for addition.
   b. Give an index for multiplication.
   Express your answers in decimal notation.

†7. a. Show that there is a primitive recursive predicate *Prim* such that f is primitive recursive iff for some n, $f = \varphi_n^k$ and *Prim*(n).
   (*Note:* There may be some indices of addition, for example, which do not satisfy *Prim*, but at least one will.)
   b. Using part (a) show that there is a total recursive function which is not primitive recursive (cf. Chapter 11.G).

†8. The *graph* of a function φ is $\{z : z = \langle x, \varphi(x) \rangle\}$. Show that there is a function with primitive recursive graph that is not primitive recursive.

9. Show that there is no recursive function which can identify whether two programs compute the same function. That is, show that no recursive function f satisfies

$$f(x, y) = \begin{cases} 1 & \text{if } \varphi_x \simeq \varphi_y \\ 0 & \text{if } \varphi_x \not\simeq \varphi_y \end{cases}$$

(*Hint:* Show that g such that

$$\varphi_{g(x)} \simeq \begin{cases} \lambda x\,(1) & \text{if } \varphi_x(x)\downarrow \\ \updownarrow \text{ on all inputs} & \text{if } \varphi_x(x)\Updownarrow \end{cases}$$

is recursive, and look for a solution to the halting problem.)

10. Let $A = \{x : \varphi_x^1 \text{ is total}\}$.
    a. Show that there is no recursive f such that $A = \text{range of } f$.
       (*Hint:* Diagonalize.)
    b. Show that A is not recursive. (*Hint:* Reduce this to part (a).)

11. Number the functions partial recursive in $\{f_1, \ldots, f_n\}$. Using that, produce a universal computation predicate for the functions partial recursive in $\{f_1, \ldots, f_n\}$ which is also partial recursive in $\{f_1, \ldots, f_n\}$. What is the significance of this if $f_1, \ldots, f_n$ are recursive?

†12. Give an index for the universal partial recursive function of k variables defined in Theorem 4 in terms of the indices c, d, and e, where φ_c^4 is the representing function for the universal computation predicate, d is an index for $\lambda n\,(n)_0$, and e is an index for $\lambda \vec{x}\,\langle \vec{x} \rangle$.

13. We may delete primitive recursion as a basic operation in defining the partial recursive functions if we expand the class of initial functions. Show that the partial recursive functions are the smallest class containing the zero, successor, projection, and exponentiation functions and which is closed under composition and μ-operator.
    (*Hint:* Use Corollary 3, Theorem 4, and Theorem 12.1.)

†14. *Rice's Theorem*
    Prove: If C is a collection of partial recursive functions, then $\{x : \varphi_x \text{ is in } C\}$ is a recursive set iff $C = \varnothing$ or $C = $ all p. r. functions.
    (*Hint:* Assume the contrary and $\varphi_a \in C$, $\varphi_b \notin C$. Define a recursive function f such that

$$\varphi_{f(x)} \simeq \begin{cases} \varphi_a & \text{if } \varphi_x \notin C \\ \varphi_b & \text{if } \varphi_x \in C \end{cases}$$

Then use the Fixed Point Theorem.)
    Why doesn't this contradict Exercise 7?

Further Reading

Virtually all the theorems in this chapter were originally proved by Kleene, the Normal Form Theorem in particular in 1936; see his *Introduction to Metamathematics* for details of the history. Kleene's original computation predicate is called the "Kleene T-predicate"; our version of the Normal Form Theorem is based on one from G. Sacks via Robert W. Robinson.

The theorems in this chapter don't depend on the particular numbering we gave but only that there is *some* effective numbering of the partial recursive functions. The criteria for an acceptable numbering are discussed by Rogers in his book *Theory of Recursive Functions and Effective Computability* (exercises 2–10, 2–11, and 11–10) and by Odifreddi in *Classical Recursion Theory*, Chapter II.5.

For a more general discussion of the Fixed Point Theorem and extensions of it, see Odifreddi, especially Chapter II.2.

For the study of algorithmic classes, see Odifreddi, Chapter II.3 and Chapter V.

7 Listability

A. Listability and Recursively Enumerable Sets

Complementary to the notion of computability is the idea of effectively listing the elements of a set. Starting with computability, we take a set to be effectively enumerable if it is the output of some computable function: $\{f(0), f(1), f(2), \dots\}$. But, equally, starting with the idea of effectively making a list, we can take a set A to be computable if both it and its complement are effectively enumerable: In that case to decide if x is in A list the elements of A: $a_0, a_1, a_2, \dots$; simultaneously list the elements of the complement of A: $b_0, b_1, b_2, \dots$, until x appears on one of the lists.

Since we have identified the notion of computable function with that of total recursive function (Church's Thesis), we will identify the notion of a collection being effectively enumerable with the collection being the output of some total recursive function. However, we have to take the list with no elements on it as a separate case, since that isn't the output of any recursive function.

Definition   A set is *recursively enumerable* (r.e.) iff it is empty or it is the range of some total recursive function of one variable.

We may relax the condition that f be total, which in applications is very convenient.

THEOREM 1 B is r.e. iff there is some partial recursive function ψ such that
$$B = \{y : \exists x \ \psi(x)\!\downarrow = y\}.$$

Proof: $\Rightarrow$ In this direction it's immediate, since every recursive function is p. r.
$\Leftarrow$ Suppose we are given such a ψ. If $B = \varnothing$ then we are done. So suppose B is nonempty.

In intuitive terms we think of our universal function as a machine. We crank it up on an index for ψ. Then we do one step of the calculation for $\psi(0)$. Of course, that might not halt. So we do another step of $\psi(0)$ and the first step of $\psi(1)$. Then we do another step of $\psi(0)$, another of $\psi(1)$, and the first step of $\psi(2)$, ... , then another step of $\psi(0)$ (if it hasn't yet halted), another of $\psi(1)$ (if it hasn't yet halted) , ... , another of $\psi(n-1)$, and the first step of $\psi(n)$. This process is called *dovetailing* the computations.

Dovetailing Computations

The first time we get some y such that $\psi(y)\downarrow$, we label that output $f(0)$; the second time we get some y such that $\psi(y)\downarrow$, we label that output $f(1)$; ... ; the nth time in this process we get some y such that $\psi(y)\downarrow$, we label that output $f(n-1)$. If B is infinite, this will result in f being total. And if $\psi(y)\downarrow$, then eventually that value will show up as $f(n)$ for some n. However, B may be finite, so instead of f "searching" for new outputs, we will have it output its previous value if nothing new has been defined at a stage.

Formally, there is some m such that $\psi \simeq \varphi_m$. So

$$y \in B \quad \text{iff} \quad \exists x \exists q [\, \varphi_c(m, \langle x \rangle, y, q) = 0 \,]$$

where φ_c is the representing function of the universal computation predicate (Chapter 16.D). Give the name h to the function defined by $h(x, y, q) = \varphi_c(m, \langle x \rangle, y, q)$. Then this is 0 iff $\varphi_m(x)\downarrow = y$ "in time q" ; otherwise it is 1 . Let $f(0) =$ the least element in B. Then

$$f(x+1) = [\, h((x)_0, (x)_1, (x)_2) \cdot f(x)\,] + \overline{sg}\,[\,h((x)_0, (x)_1, (x)_2)\,] \cdot (x)_1$$

You can check that this does what we claim. ∎

COROLLARY 2 B is r.e. iff B is the range of some elementary function.

Proof: Because φ_c is elementary (Corollary 16.3), the function h in the proof of Theorem 1 is elementary. ∎

The following theorem is the formal version of our description that a set is computable iff it and its complement are both effectively listable.

THEOREM 3 A is recursive iff both A and $\overline{A}$ are r.e.

Proof: $\Rightarrow$ This direction is Exercise 1.

$\Leftarrow$ Suppose both are r.e. Then there are recursive f and g such that $A =$ the range of f , and $\overline{A} =$ the range of g . Define a recursive function h by $h(x) = \mu y [\, f(y) = x \vee g(y) = x\,]$. Then the characteristic function of A can be expressed as $E[f(h(x)), x]$ where E is the equality function of Exercise 11.6 . ∎

Thus, we have an equivalent form of Church's Thesis:

A set is effectively listable iff it is recursively enumerable.

B. Domains of Partial Recursive Functions

We can also characterize r.e. sets as domains of partial recursive functions. More-over, we can go effectively from a description of an r.e. set as the range of a p.r. function φ to a description of it as a domain of another p.r. function ψ. But what is a description? It's nothing more than an index. So to go effectively from one description to another means we have a recursive function that transforms indices.

THEOREM 4   There are recursive functions *f* and *g* such that

$$\text{domain } \varphi_x = \text{range } \varphi_{f(x)}$$

$$\text{range } \varphi_x = \text{domain } \varphi_{g(x)}$$

Thus, a set *A* is r.e. iff there is some partial recursive ψ such that $A = \{ x : \psi(x)\!\downarrow \}$.

Proof: First we will exhibit an *f*. Given any *x* we define:

$$\psi(y) \simeq \begin{cases} y & \text{if } \varphi_x(y)\!\downarrow \\ \mathord{\not\downarrow} & \text{otherwise} \end{cases}$$

Then the range of ψ = the domain of φ_x. This procedure is *effective in x*, by which we mean that there is a recursive procedure to derive an index for ψ solely in terms of *x*. To demonstrate this we first define ψ formally via $\tau(x, y) \simeq (\mu q\, C(x, \langle y \rangle, (q)_0, q))_0 \cdot 0 + y$, and then $\psi \simeq \lambda y\, \tau(x, y)$. Since τ is p.r., there is some *d* (which you can evaluate if you wish) such that $\tau \simeq \varphi_d$; that is, $\psi \simeq \lambda y\, \varphi_d(x, y)$. An index for ψ is then given by the *s-m-n* theorem as $S_1^1(d, x)$. Thus we may take *f* to be $\lambda x\, S_1^1(d, x)$.

For *g*, suppose we are given *x*. In a dovetailing procedure, run φ_x for more and more inputs, and define

$$\rho(y) \simeq \begin{cases} 1 & \text{if } y \text{ shows up on the list of outputs} \\ \mathord{\not\downarrow} & \text{otherwise} \end{cases}$$

Then the domain of ρ = the range of φ_x. Again, this procedure is effective in *x*. That is, define γ by $\gamma(x, y) \simeq sg(\mu q\, C(x, \langle (q)_0 \rangle, y, q))$ and then $\rho \simeq \lambda y\, \gamma(x, y)$. Since for some *e*, $\gamma \simeq \varphi_e$, by the *s-m-n* theorem we may take *g* to be $\lambda x\, S_1^1(e, x)$.   ∎

The usual naming of r.e. sets in the literature is:

$$W_x = \text{the domain of } \varphi_x$$

We call *x* an r.e. *index* for W_x.

Now we can restate the halting problem (Theorem 15.2) in terms of r.e. sets. Define:

$$K = \{ x : \varphi_x(x) \downarrow \} = \{ x : x \in W_x \}$$

We know that K is not recursive. However, it is r.e.: It is the domain of ψ, where $\psi(x) \simeq \mu q\, C(x, \langle x \rangle, (q)_0, q)$. Thus, K is r.e. but not recursive. Moreover, $\overline{K}$ is not r.e. (and hence not recursive), for if it were, then by Theorem 3, K would be recursive. If, further, we are willing to countenance nonconstructive methods of proof, we can show that there are infinitely many r.e. sets which are not recursive: "Choose" any element a of $\overline{K}$ and $K \cup \{a\}$ will be r.e. but not recursive (Exercise 2); now choose any element of $\overline{K \cup \{a\}}$ and repeat the process, ... Exercise 5 has examples of r.e. and non-r.e. sets that arise more naturally.

C. The Projection Theorem

Our final characterization of recursive enumerability links it to expressibility in terms of arithmetic conditions.

THEOREM 5   A set A is r.e. iff there is some recursive predicate R such that
$$A = \{ x : \exists y\, R(x, y) \}.$$

Proof: $\Rightarrow$ This direction is easy via the universal computation predicate.
$\Leftarrow$ Given R, define a p.r. function ψ by $\psi(x) \simeq \mu y [R(x, y)]$. Then $\psi(x) \downarrow$ iff for some y, $R(x, y)$ holds. So $A =$ the domain of ψ, and by Theorem 4 we are done.   ∎

Exercises

For some of the exercises here you will need the new techniques of dovetailing or obtaining a function effectively in the indices of others (see the proof of Theorem 4).

1. Prove that if A is recursive then both A and $\overline{A}$ are r.e.

2. a. Show that if $a \in \overline{K}$, then $K \cup \{a\}$ is r.e. but not recursive.
   † b. Show that there is a set which is not r.e. and whose complement is also not r.e., but don't use simply a counting argument.

3. Show that if A is r.e. and can be enumerated in increasing order, then A is recursive.

4. A is *recursively enumerable without repetitions* if there is some recursive function f which is 1-1 and $A =$ range of f. Prove that if A is r.e. and infinite then it can be recursively enumerated without repetitions.

†5. Classify, where possible, the following sets as (i) recursive, (ii) recursively enumerable, and/or (iii) having recursively enumerable complement.
   a. $\{ x : x$ is even $\}$
   b. $\{ x :$ there is a run of exactly x 5's in a row in the decimal expansion of $\pi \}$
   c. $\{ x :$ there is a run of at least x 5's in a row in the decimal expansion of $\pi \}$
   d. $\{ x : W_x = \varnothing \}$
   e. $\{ x : \varphi_x$ is total $\}$ (*Hint:* See Exercise 16.10.)
   f. For a fixed n, $\{ x : W_x = W_n \}$
   g. $\{ x : W_x$ is infinite $\}$
   h. For arbitrary $z \in [0,1)$, $\{ x : x = \langle n, y \rangle$ where y is the nth digit in the decimal expansion of $z\}$

6. Show that the r.e. sets are effectively closed under union and intersection. That is, there are recursive functions f and g such that $W_x \cup W_y = W_{f(x,y)}$ and $W_x \cap W_y = W_{g(x,y)}$. Are they effectively closed under complementation?

Further Reading

An especially good development of computability based on the notion of an effective enumeration is "Diophantine decision problems" by Julia Robinson. You might also enjoy the seminal paper Post wrote on this subject, "Recursively enumerable sets of positive integers and their decision problems". For a further development of the subject see Odifreddi's *Classical Recursion Theory* and Soare's *Recursively Enumerable Sets and Degrees*.

18 Turing Machine Computable = Partial Recursive

(Optional)

In Chapter 10 we referred to the Most Amazing Fact that all systems which have been proposed as formalizations of the notion of computability have been shown to be equivalent. In this chapter we will prove the equivalence of two of them. Moreover, we will do it constructively: Given a definition of a partial recursive function φ, we will produce a Turing machine which calculates that function; and given a Turing machine which calculates a function φ, we will produce a partial recursive definition of φ. Nothing later in this text depends on this chapter.

A. Partial Recursive Implies Turing Machine Computable

In order to make the proof in this section go more smoothly we are going to show something a bit stronger than the title of this section. We say that a Turing machine that computes a function *uses a one-way tape* if at no point in any of its computations on any input does it go to the left of the first blank at the left of the input. That is, the initial configuration for input $(n_1, \ldots, n_k)$ is

never goes to the left of here   blank squares can be added here

$\downarrow$ $\downarrow$

$0 \; 1^{n_1+1} \; 0 \cdots 0 \; 1^{n_k+1}$

1

THEOREM 1   Every definition of a partial recursive function can be effectively converted into a Turing machine that computes that function and which uses a one-way tape and never halts in a nonstandard configuration on an input of the correct number of variables.

Proof: First note that all the machines of Chapter 9.C use one-way tapes, and the machines you defined as answers to the exercises in Chapter 9 can be modified to run on a one-way tape if they don't already.

Hence, we can conclude that the initial partial recursive functions (successor, the projections (Exercise 9.2), zero) as well as the equality function (Exercise 9.6), addition, and multiplication can all be computed on TM's that use a one-way tape. Moreover, for these total functions the machines always halt in standard configuration. We now need to show that the class of functions computed by TM's that use one-way tapes and never halt in a nonstandard configuration is closed under composition, μ-operator, and primitive recursion.

Composition   We showed this for functions of one variable in Chapter 9.C. Now suppose that $\varphi_1, \ldots, \varphi_m$ are functions of k variables, ψ is a function of m variables, and we have a machine for each which computes it using a one-way tape and which never halts in a nonstandard configuration. We will describe the operation of a machine that computes the composition of these, $\psi(\varphi_1, \ldots, \varphi_m)$, and leave to you the actual definition of the machine as a set of quadruples (no, we don't think it's easy to do that, but it's not very instructive either).

For input $\vec{x} = (n_1, \ldots, n_k)$, we begin with the tape configuration $0\,1^{n_1+1}\,0 \ldots 0\,1^{n_k+1}$, which we'll call $1^{\vec{x}+1}$. Here are the successive contents of the tape.

1. $0\,1^{\vec{x}+1}$

2. $0\,1^{\vec{x}+1}\,0\,0\,0\,1^{\vec{x}+1}$

3. $0\,1^{\vec{x}+1}\,0\,0 \cdots 0\,1^{\varphi_1(\vec{x})}$

We aren't operating the machine T_{φ_1} on a badly configured tape here. What we are doing is adding quadruples that allow for the simulation of the machine at the appropriate stage in the computation, and we know that the machine will be simulated correctly because it never goes to the left of the first blank to the left of its input. Similarly, we continue,

4. $0\,1^{\vec{x}+1}\,0\,0\,0\,1^{\varphi_1(\vec{x})+1}$

5. $0\,1^{\vec{x}+1}\,0\,0\,0\,1^{\varphi_1(\vec{x})+1}\,0\,0\,1^{\vec{x}+1}$

6. $0\,1^{\vec{x}+1}\,0\,0\,0\,1^{\varphi_1(\vec{x})+1}\,0\,0\,1^{\varphi_2(\vec{x})+1}$

7. $0\,1^{\vec{x}+1}\,0\,0\,0\,1^{\varphi_1(\vec{x})+1}\,0\,0\,1^{\varphi_2(\vec{x})+1}\,0\,0\,1^{\vec{x}+1}$

8. $0\,1^{\vec{x}+1}\,0\,0\,0\,1^{\varphi_1(\vec{x})+1}\,0\,0\,1^{\varphi_2(\vec{x})+1}\,0\,0 \cdots 0\,0\,1^{\varphi_m(\vec{x})+1}$

9. $0\,1^{\varphi_1(\vec{x})+1}\,0\,0\,1^{\varphi_2(\vec{x})+1}\,0\,0 \cdots 0\,0\,1^{\varphi_m(\vec{x})+1}$

10. $1^{\psi[\varphi_1(\vec{x})+1,\,\varphi_2(\vec{x})+1,\,\varphi_m(\vec{x})+1]}$

Note that if for some i the machine that calculates $\varphi_i(\vec{x})$ does not halt, then the composition machine does not halt on $\vec{x}$.

The μ-operator   Suppose we have a machine that computes a function φ that uses a one-way tape and never halts in a nonstandard configuration. We will describe the operation of a machine that computes $\mu y\,[\,\varphi(\vec{x}, y) = 0\,]$, the μ-operator applied to φ, and leave to you to define the machine as a set of quadruples. Here are the successive contents of the tape.

1. $0\,1^{\vec{x}+1}$

2. $0\,1^{\vec{x}+1}\,0\,1$

3. $0\,1^{\vec{x}+1}\,0\,1\,0\,1^{\varphi(\vec{x},0)+1}\,0\,1$

4. Use the equality machine, T_E, to determine if $\varphi(\vec{x},0) = 0$ by applying it to the string beginning to the right of $1^{\vec{x}+1}\,0\,1\,0$ (i.e., insert the appropriately relabeled quadruples of T_E).

5. If equal, erase the tape and stop.

6. If not equal, erase the tape back to $0\,1^{\vec{x}+1}\,0\,1$ and add a 1 to the right,
    $0\,1^{\vec{x}+1}\,0\,1\,1$.
    $\vdots$

7. $0\,1^{\vec{x}+1}\,0\,1^{n+1}$

8. $0\,1^{\vec{x}+1}\,0\,1^{n+1}\,0\,1^{\varphi(\vec{x},n)+1}$

9. $0\,1^{\vec{x}+1}\,0\,1^{n+1}\,0\,1^{\varphi(\vec{x},n)+1}\,0\,1$

10. Use the equality machine, T_E, to determine if $\varphi(\vec{x}, n) = 0$ by applying it to the string beginning to the right of $0\,1^{\vec{x}+1}\,0\,1^{n+1}\,0$.

11. If equal, erase all but 1^n and stop.

12. If not equal, erase the tape back to $0\,1^{\vec{x}+1}\,0\,1^{n+1}$ and add a 1 to the right, $0\,1^{\vec{x}+1}\,0\,1^{n+2}$, and repeat the process.

Primitive recursion   This is difficult, so difficult that we're tempted to leave it to you. But there's a way out. In Chapter 22.A we prove that the partial recursive functions comprise the smallest class containing the zero function, the successor function, the projections, addition, multiplication, and the characteristic function for equality and which is closed under composition and the μ-operator. That's just what we need to complete the proof here. You can read that section with the background you already have.   ∎

B. Turing Machine Computable Implies Partial Recursive

THEOREM 2   If a Turing machine calculates a function φ then the set of quadruples of the machine can be effectively converted into a partial recursive definition of φ .

Proof: Let M be a Turing machine that calculates the function φ. By coding what the machine does at each step of its computation, we will be able to derive a partial recursive definition of φ.

First we assign the following numbers to the operations:

Delete the current symbol, if any	0
Write the symbol 1	1
Move one square to the right	2
Move one square to the left	3

Let the highest numbered state of M be n. The quadruples of M can be seen as a function g from (states, symbols) to (operations, states). For example, $q_1 0 R q_2$ can be written as $g[(q_1,0)] = (R,q_2)$. If we assign the number j to state q_j, we can express this function numerically as

$$d(x) = \begin{cases} g[(x)_0,(x)_1] & \text{if } g \text{ is defined for this input} \\ 47 & \text{otherwise} \end{cases}$$

We'll let you make this more formal, but it should be clear that d is (primitive) recursive since it's a finite table.

To code the tape descriptions, suppose that after t steps in the computation on input $\vec{x}$ the machine M is in state $q(t)$ and the symbol being observed is

$$s(t) = \begin{cases} 0 & \text{if blank} \\ 1 & \text{if the symbol is 1} \end{cases}$$

And suppose that we have the following configuration:

blank from this point on   blank from this point on

$$\downarrow \qquad\qquad\qquad\qquad\qquad\qquad\qquad\qquad\qquad\qquad \downarrow$$

$$\cdots\, b_s \,|\, b_{s-1} \,|\, \cdots \,|\, b_1 \,|\, b_0 \,|\, s(t) \,|\, c_0 \,|\, c_1 \,|\, \cdots \,|\, c_{r-1} \,|\, c_r \, \cdots$$

$$\uparrow$$
$$q(t)$$

We should indicate that both s and q depend on $\vec{x}$, but for the sake of legibility we will suppress that until the end of the proof. And, as usual, we will notate that b_i or c_i is blank with a 0. (Note that neither c_r nor b_s can be 0 since they mark the point beyond which the tape becomes blank.)

We may describe this configuration by coding up the contents of the tape to the left and the right of the square being observed

$$b(t) = \langle b_0, b_1, \dots, b_s \rangle$$
$$c(t) = \langle c_0, c_1, \dots, c_r \rangle$$

where these are 0 if the tape is blank in that direction.

On input $\vec{x}$, the machine begins on a tape containing only $1^{\vec{x}+1}$ (see the proof of Theorem 1 for notation). So $b(0) = 0$, $s(0) = 1$, $q(0) = 1$; we'll let you write down what $c(0)$ is. Then at step $t+1$, call $a = d(\langle q(t), s(t) \rangle)$, and we have $q(t+1) = (a)_1$ and one of the following cases:

i. The machine moves right at this step, that is, $(a)_0 = 2$

$$s(t + 1) = (c(t))_0$$
$$b(t + 1) = \langle s(t), b_0, b_1, \ldots, b_s \rangle$$
$$c(t + 1) = \langle c_1, \ldots, c_r \rangle$$

ii. The machine moves left, that is, $(a)_0 = 3$

$$s(t + 1) = (b(t))_0$$
$$b(t + 1) = \langle b_1, \ldots, b_s \rangle$$
$$c(t + 1) = \langle s(t), c_0, c_1, \ldots, c_r \rangle$$

iii. The machine writes or deletes, that is, $(a)_0 = 0$ or 1

$$s(t + 1) = (a)_0$$
$$b(t + 1) = b(t)$$
$$c(t + 1) = c(t)$$

We leave to you to confirm that b, c, q, and s are primitive recursive functions (see Chapter 11.D.2 and D.7). Indeed, they are elementary.

To determine at stage t whether the machine has halted in a standard configuration we need to know at most how many squares of tape have been used up to that stage. Since at each stage the machine can add no more than one new square, an upper bound is $c(0) + t$. Then at stage t the machine has halted in standard configuration iff

$$d(\langle q(t), s(t) \rangle) = 47$$
$$b(t) = 0$$

and

$$\forall i < c(0) + t \; [\neg[(p_i)^2 \mid c(t)] \rightarrow \neg[(p_{i+1})^2 \mid c(t)]]$$

This is a primitive recursive condition, indeed elementary; call its characteristic function h. If the machine does halt in standard configuration then the output is $lh(c(t))$. Remembering now that each of the functions we have defined depends on $\vec{x}$, we have

$$\varphi(\vec{x}) = lh(c(\vec{x}, \mu t \; [h(\vec{x}, t) = 1])) \qquad \blacksquare$$

Combining Theorems 1 and 2 we have the following.

COROLLARY 3   A function is TM computable

    iff it is partial recursive

    iff it is computable on a TM that uses a one-way tape and never halts in a nonstandard configuration.

These computable correspondences allow us to translate facts about partial recursive functions into facts about Turing machines. For example, from Theorem 16.4 we can derive the following.

COROLLARY 4   There is a universal Turing machine.

In Exercise 9.8 we defined the halting problem for Turing machines and sketched a proof that it was not Turing machine computable. Now we can conclude that directly from Corollary 15.3.

COROLLARY 5 (The Halting Problem for Turing Machines)
The halting problem for Turing machines is not Turing machine computable.

III

LOGIC AND ARITHMETIC

19 Propositional Logic

A. Hilbert's Program Revisited

Now that we have a better idea of what we might mean by "computable" or "constructive", let's return to Hilbert's ideas (Chapter 7).

Recall that Hilbert believed that completed infinities were not real in the sense of a table or chair being real: they corresponded to nothing "in the world". Infinite collections, according to him, were ideal elements comparable to the square root of −1 in algebra, and like the square root of −1 they could be employed to obtain finitistic facts.

Of course, we can't add just any fictions to our mathematics. Hilbert's criteria for adding ideal elements to a mathematical system were, in short, (1) they must not lead to any contradictions, and (2) they must be fruitful. Clearly, infinite collections were fruitful in providing foundations for analysis. Fruitful, that is, so long as they didn't lead to contradictions.

Hilbert believed that infinite collections were justified in mathematics and led to no contradictions, but he didn't want to build his mathematics on faith. He sought to justify the use of infinite collections by proving they were consistent with finitistic mathematics. And he realized that for such a proof to remove doubt about the suitability of infinite collections, it should not use any infinitistic reasoning. His program to justify his views, as distinct from his views about the nature of the finite and the infinite, was to formalize mathematics as a logical system and, by constructive, finitistic methods, show that no contradiction lay therein. He proposed to begin with arithmetic, the natural numbers, for once he could axiomatize that subject he felt confident about the rest.

In the following chapters we will present one particular axiomatization of arithmetic. Perhaps contrary to your expectations, we won't try to formalize all of arithmetic but rather only a small part of it. Once we have studied the properties of that formal system (as opposed to the properties of the natural numbers themselves), we will find we have answers for Hilbert for any formalization of arithmetic that contains even that small fragment.

B. Formal Systems

Perhaps you recall proving theorems in geometry in high school. You started with some undefined notions, such as "line" and "point" and "lies on". Then you had some axioms such as, "Given any two points there is one and only one line on which they both lie." The axioms were all taken to be "intuitively obvious", acceptable and unchallengeable. They were the first "truths" from which you were to derive all others as theorems.

Here we will do the same, only our subject matter is arithmetic and we will be much more careful. For example, in high school geometry you were probably not told explicitly what methods of proof you were allowed to use.

We will choose a formal language for arithmetic, pick certain axioms, and make explicit what constitutes a proof. The methods of proof, the formal language, and certain of the axioms will be the *logical* part of the system, whereas the other axioms will be specific to the subject matter of arithmetic. The *logic* we will present will, we hope, be agreed to be suitable for studying arithmetic. We won't make any claims about its suitability for any other discipline.

In this chapter we begin by investigating the simplest part of the logic.

C. Propositional Logic

1. The formal language

Here we will formalize reasoning about mathematical statements as wholes: We will not be concerned with the internal structure of such statements. For instance, we might call **p** the statement "All numbers are even or odd" and **q** the statement "Every number is either 0, or is 1 greater than some other number." In this *first* analysis **p** and **q** are simple: Their internal form won't matter to us.

We may begin with whatever statements we choose, and then we may combine them to form new statements using words such as "and", "or", "not", "if...then...". For instance, "All numbers are even or odd *and* every number is either 0 or is 1 greater than some other number" ; or another example, "*Not* all numbers are even or odd." These words used to join statements are called *connectives*: They create new statements from given ones. The English connectives are too vague and informal for our purposes, so we will replace them by formal counterparts: ∧ will replace "and", ∨ replaces "or", ⌐ replaces "not", and → replaces "if... then..." . Thus corresponding to our two examples we have **p∧q** and ⌐**p** . In a sense which we'll make precise below, these four connectives will be the only ones we'll need.

Now we're ready to define a formal language. In this definition and in the rest of this chapter we will use unsubscripted, boldface capital Roman letters such as **A** and **B** as *metavariables* to range over words in the language.

The formal language of propositional logic

Variables: $\mathbf{p_0}$, $\mathbf{p_1}$, $\cdots$, $\mathbf{p_n}$, $\cdots$
Connectives: $\wedge$, $\vee$, $\neg$, $\rightarrow$
Parentheses: (,)
Inductive definition of a well-formed-formula (*wff*) :

  i. $(\mathbf{p_i})$ is a wff.
  ii. If **A** and **B** are wffs then so are $(\mathbf{A} \wedge \mathbf{B})$, $(\mathbf{A} \vee \mathbf{B})$, $(\neg \mathbf{A})$, and $(\mathbf{A} \rightarrow \mathbf{B})$.
  iii. A string of symbols is a wff iff it arises via applications of (i) and (ii).

It may seem obvious that this is a proper inductive definition and that there is only one way to read each wff, but that requires a proof, which we give in the appendix to this chapter.

Sometimes when we are discussing a wff we might *informally* use other kinds of brackets or delete the outermost parentheses or parentheses around variables to make it more readable. For example, we might informally write $(((\mathbf{p_1}) \rightarrow (\mathbf{p_2})) \rightarrow (\neg((\mathbf{p_1}) \rightarrow (\mathbf{p_2}))))$ as $(\mathbf{p_1} \rightarrow \mathbf{p_2}) \rightarrow [\neg(\mathbf{p_1} \rightarrow \mathbf{p_2})]$. Remember that the latter is only shorthand for the former.

A wff is a formal object and doesn't assert anything until we say what statements the variables in it stand for. For example, we might let $\mathbf{p_1}$ stand for "2+2 = 4" and $\mathbf{p_2}$ stand for "2 divides 9". Then $\mathbf{p_1} \rightarrow \mathbf{p_2}$ stands for "2+2 = 4 $\rightarrow$ 2 divides 9". We may read this colloquially as "If 2+2 = 4, then 2 divides 9", but our goal is to replace the words "if...then..." by a formal connective in order to avoid the imprecision of English. What we must do now is say exactly how we are going to understand the formal connectives.

2. Truth and falsity: truth-tables for the connectives

What properties of statements will matter to us as mathematicians? Certainly their most important property is that they are true or false, though not both. They have many other properties: their subject matter, how they can be verified, and so on. But as mathematicians let us agree that the only property we'll concern ourselves with is their truth or falsity—no more. So a *proposition* will be a mathematical statement which we take to be true or false, but not both.

From simple propositions about arithmetic we may create compound ones using our connectives. Then the only properties of these with which we'll be concerned are their truth-values and the connectives appearing in them. So to formalize our understanding of the connectives we only need to decide under what conditions a compound proposition will be true or false.

The formalization of "not" , called *negation*, is easy. If **A** is true, then $\neg \mathbf{A}$ should be false; if **A** is false, then $\neg \mathbf{A}$ should be true. In tabular form, letting "T" stand for "true", and "F" stand for "false", we have the *truth-table* for negation :

A	$\neg$ A
T	F
F	T

Similarly, the formalization of "and", which we call *conjunction*, is easy to agree on. When is **A** ∧ **B** true? Exactly when both **A** and **B** are true. So its truth-table is:

A	B	A∧B
T	T	T
T	F	F
F	T	F
F	F	F

For the formalization of "or", which we call *disjunction*, we have two choices. We can say that **A** ∨ **B** is true if either is true, including the possibility that both are true, called *inclusive or*. Or we can say that **A** ∨ **B** is true if either, but not both, is true, called *exclusive or*. It's the former that is generally agreed upon in mathematics, and it has the truth-table:

A	B	A∨B
T	T	T
T	F	T
F	T	T
F	F	F

The connective that arouses most debate is "if...then...", the formalization of which is called the *conditional*. Given **A** → **B**, if the *antecedent*, **A**, is true, then the conditional should be true if and only if the *consequent*, **B**, is true, for from truths we can only conclude truths, not falsehoods. (If that seems puzzling in an example such as "If dogs are mammals, then $2+2 = 4$," remember that we are concerned only with the truth or falsity of the antecedent and consequent, not their subject matter. In any case, we said we are designing this logic to reason about mathematics, not dogs.)

But what if the antecedent is false? For example, consider "If m and n are odd natural numbers, then $m + n$ is an even number." We're sure you'll agree this is true. In that case, any time we substitute numbers for the variables it must come out true. In particular, both "If 4 and 8 are odd natural numbers, then $4+8$ is an even number" and "If 4 and 7 are odd natural numbers, then $4+7$ is an even number" must be true. In both of these the antecedent is false: in the first the consequent is true; in the second the consequent is false. So our truth-table must be:

A	B	A→B
T	T	T
T	F	F
F	T	T
F	F	T

Our formalization of "if...then..." allows us to deal with cases in which the "antecedent does not apply" by treating them as vacuously true.

We said earlier that these four connectives suffice to do logic, at least on the assumption that the only properties of propositions that are of interest to us are their truth-values. The reason is that any other connective that depends only on the truth-values of the constituent propositions can be defined in terms of these connectives. For example, suppose we want to formalize exclusive "or". We can do that with $(A \lor B) \land \neg (A \land B)$ since that is true iff exactly one of **A, B** is true. A more important connective which we can define is the formalization of "if and only if": $(A \to B) \land (B \to A)$, which we abbreviate as $A \leftrightarrow B$. A fuller discussion of definability of connectives and the entire subject of propositional logic can be found in Epstein, 1990.

3. Validity

Consider the wff $\neg (p_1 \land \neg p_1)$. No matter what proposition we let p_1 stand for, this is going to be evaluated as true, as you can check. And there are many other such wffs that are always evaluated as true, for example, $\neg \neg p_1 \to p_1$, $p_2 \lor \neg p_2$. We call a formal wff *valid* or a *tautology* if it is always evaluated as true regardless of which propositions the variables stand for: Any proposition of that form will be evaluated as true *due only to its form*. Not all wffs are valid; for example, $p_1 \to p_2$ is not valid because p_1 may be a true proposition and p_2 a false one, yielding $p_1 \to p_2$ false. We are especially interested in valid propositions because they are the ones we are justified in using to derive theorems.

D. Decidability of Validity

1. Checking for validity

After you check a few wffs for validity (Exercise 1) it may seem obvious that we can effectively decide whether any given wff is valid or not. Let's investigate that more carefully.

What does the procedure involve? We have to check that no matter what propositions are assigned to the variables the evaluation is always T. But in propositional logic a proposition is reduced to just its truth-value. Thus, all we have to do is look at all the ways to assign T or F to the variables appearing in the wff and check that each assignment yields T. In other words, we make up a table. For example, consider the wff $\neg (p_1 \to p_2) \to (p_1 \to p_2)$. We have the following table:

p_1	p_2	$(p_1 \to p_2)$	$\neg (p_1 \to p_2)$	$\neg (p_1 \to p_2) \to (p_1 \to p_2)$
T	T	T	F	T
T	F	F	T	F
F	T	T	F	T
F	F	T	F	T

Just one line of the table, one assignment of truth-values, is evaluated as F. But thus if p_1 is "$4 + 4 = 8$" and p_2 is "$2 + 2 = 5$", then $\neg (p_1 \to p_2) \to (p_1 \to p_2)$ is false. Hence that wff is not valid: It's not always true due only to its form.

Here is another way to check validity that is often easier for conditionals. We attempt to falsify the wff; that is, we try to come up with an assignment of truth-values that makes the evaluation come out F. If we can, it's not valid; otherwise it is. For instance, $A \rightarrow B$ can be falsified iff there is an assignment that makes both A true and B false. For example,

$$\lnot(p_1 \land p_2) \ \rightarrow \ (p_2 \rightarrow p_1)$$

is false iff	T	F

which is iff	$p_1 \land p_2$	p_2	p_1
are	F	T	F

and that is the falsifying assignment. Hence the wff is not valid.

Similarly,

$$((p_1 \land p_2) \rightarrow p_3) \ \rightarrow \ (p_1 \rightarrow (p_2 \rightarrow p_3))$$

is false iff	T	F

iff	p_1	$p_2 \rightarrow p_3$
are	T	F

iff	p_2	p_3
are	T	F

But if p_1 is T, p_2 is T, and p_3 is F, then $(p_1 \land p_2) \rightarrow p_3$ is F, so there is no way to falsify the wff; hence it is valid.

2. Decidability

If the procedure for checking validity of wffs is entirely mechanical, as it appears to be, then we ought to be able to express it as a recursive function.

We say that a class C of questions each of which can be answered as "yes" or "no" is *decidable* if we can number C (see pp. 67–68) and the resulting set of Gödel numbers representing problems for which the answer is "yes" is *computable*; that is, there is a computable procedure to determine whether a number representing a question has answer "yes". (If the problems have three or more possible answers, we have to decide how to reduce them to yes–no questions.) *According to Church's Thesis*, then, a class of questions is decidable if the set of Gödel numbers of the problems with answer "yes" is *recursive*. The *decision problem* for the class C is the problem whether C is decidable or not. We say that the *decision problem is solvable* if C is decidable. In that case, a representation of C via Gödel numbering and the recursive presentation of the set of problems with answer "yes" are together a *decision procedure for C*, though often an informal description of how to do that is called the decision procedure. If we want to stress that we've identified *decidable* with *presentation as a recursive set*, we say the problem is *recursively decidable* and we have a *recursive decision procedure*. Informally, we sometimes say a decision problem is *recursive* when we mean recursively decidable.

We gave an informal decision procedure for the class of questions "Is the wff
A of our formal language for propositional logic valid?", so we should be able to
produce a formal decision procedure. We will do that now, though it would be better
for you to do it yourself, since you already know how to do Gödel numberings and
derive information recursively from them: That was what Chapter 16 and the
universal computation predicate was all about. If possible, you should read only the
outline of this, come up with your own numbering, and fill in the details yourself.

Gödel numbering of wffs

We present a function from wffs to natural numbers, notated $[\![\mathbf{A}]\!]$, read
"the Gödel number of **A**".

First, $[\![(\mathbf{p_i})]\!] = \langle i \rangle$

If $[\![\mathbf{A}]\!] = a$ and $[\![\mathbf{B}]\!] = b$, then:

$$[\![(\neg\mathbf{A})]\!] = \langle a, 0 \rangle$$
$$[\![(\mathbf{A} \wedge \mathbf{B})]\!] = \langle a, b, 0 \rangle$$
$$[\![(\mathbf{A} \vee \mathbf{B})]\!] = \langle a, b, 0, 0 \rangle$$
$$[\![(\mathbf{A} \rightarrow \mathbf{B})]\!] = \langle a, b, 0, 0, 0 \rangle$$

You can show that this numbering fulfills the conditions for a Gödel numbering, (1)–(3) of Chapter 8.C (p. 68). For our induction procedures, we need to note
that if $[\![\mathbf{A}]\!] = a$, then the Gödel number of any wff that is a part of **A** is less than a.

We could if we wish give a numbering such that every natural number is
the Gödel number of some wff, but that complicates matters and isn't necessary:
We can identify which numbers are Gödel numbers of wffs.

a. The set of Gödel numbers of wffs is decidable. That is, there is a recursive
function h such that

$$h(n) = \begin{cases} 1 & \text{if for some } \mathbf{A},\ [\![\mathbf{A}]\!] = n \\ 0 & \text{otherwise} \end{cases}$$

Proof: We define h by induction on n.

First, $h(0) = 0$.

Now given any $n > 0$, it falls into one of the following cases:

$lh(n) = 1$ and so n codes (i); then

if $n = \langle i \rangle$, set $h(n) = 1$;

if $n \neq \langle i \rangle$, set $h(n) = 0$.

$lh(n) = 2$ and so n codes (i,j); then

if $n = \langle i, j \rangle$ and $h(i) = 1$ and $j = 0$, set $h(n) = 1$ (since $i < n$,
$h(i)$ is defined);

otherwise, set $h(n) = 0$.

$lh(n) = 3$ and so n codes (i,j,k); then

if $n = \langle i, j, k \rangle$ and $h(i) = h(j) = 1$ and $k = 0$, set $h(n) = 1$;

otherwise, set $h(n) = 0$.

You should be able to complete the definition.   ■

Now we'll set out the decision procedure, which is just the truth-table method described above translated into arithmetic functions on the Gödel numbers of wffs.

b. First we show that we can list the indices of the propositional variables appearing in a wff.

There is a recursive function f such that if $n = [\![\mathbf{A}]\!]$ and $\mathbf{p}_{i_1}, \ldots, \mathbf{p}_{i_r}$ are the propositional variables appearing in $\mathbf{A}$, then $f(n) = \langle i_1, \ldots, i_r \rangle$; otherwise $f(n) = 0$.

Proof: Here is a recursive definition of f.

If $lh(n) \le 1$
>   and $n = \langle i \rangle$ then $f(n) = \langle i \rangle$;
>   otherwise $f(n) = 0$.

If $lh(n) = 2$
>   and $h(n) = h((n)_0) = 1$ [i.e., n and $(n)_0$ are Gödel numbers of wffs]
>>   then $f(n) = f((n)_0)$;
>   otherwise $f(n) = 0$.

If $lh(n) \ge 3$
>   and $n, (n)_0,$ and $(n)_1$ are Gödel numbers of wffs, then $f(n) = \langle i_1, \ldots, i_r \rangle$,
>   where
>> $i_1 = \mu z_{\le f((n)_0) \cdot f((n)_1)} \ [(\exists m \le lh(f((n)_0)) - 1$ and
>> $z = (f[(n)_0])_m)$ or $(\exists m \le lh(f((n)_1)) - 1$ and $z = (f[(n)_1])_m)]$
>>   (the bounds are there just to convince you that an infinite search isn't
>>   necessary), and for $j \ge 1$, $i_{j+1} =$ as for i_1 except $z > i_j$;
>   otherwise $f(n) = 0$.   ∎

c. Now we identify F with 0, and T with 1 and show that we can list all possible assignments of truth-values to the variables appearing in a wff.

There is a recursive function g such that if $n = [\![\mathbf{A}]\!]$ and $\mathbf{p}_{i_1}, \ldots, \mathbf{p}_{i_r}$ are the propositional variables appearing in $\mathbf{A}$, then $g(n) = \langle x_1, \ldots, x_{2^r} \rangle$, where $x_i = \langle$the sequence of 0's and 1's of length r representing i in binary notation$\rangle$. You can make this more explicit if you wish.

d. There is a recursive function $a(n, m)$ such that if $n = [\![\mathbf{A}]\!]$ and m represents an assignment of T's and F's to those variables appearing in $\mathbf{A}$ in terms of 0's and 1's, then $a(n, m) = 1$ iff $\mathbf{A}$ is evaluated T by the truth-tables for that assignment.

Proof: We give a recursive definition of $a(n, m)$.

If n is not the Gödel number of some wff, that is $h(n) = 0$, or if $m \ne (g(n))_k$ for some $k < lh(g(n))$, then $a(n, m) = 0$.

Otherwise, for some $\mathbf{A}$, $[\![\mathbf{A}]\!] = n$ and $f(n) = \langle b_1, \ldots, b_r \rangle =$ the list of indices of variables in $\mathbf{A}$, and $m = \langle j_1, \ldots, j_r \rangle$, where each j_k is 0 or 1.

If $lh(n) \le 1$, then $n = \langle p \rangle$, and $p = j_k$ some k, so $a(n, m) = j_k$.

If $lh(n) = 2$, then $a((n)_0, m) = i$ is already defined, so $a(n, m) = 1 \doteq i$.

If $lh(n) = 3, 4$, or 5, then $a((n)_0, m) = i$ and $a((n)_1, m) = j$ are already defined, so if $lh(n) = 3$, $a(n, m) = i \cdot j$; if $lh(n) = 4$, $a(n, m) = max(i, j)$; if $lh(n) = 5$, $a(n, m) = sg((1 - i) + j)$. ∎

e. Finally to check whether n is the number of a valid wff we need to know if there is an evaluation that comes out false. So we check whether $a(n, m) = 0$ for some $m = (g(n))_k$, where $k \le 2^{lh(g(n)) - 1}$. The recursive function that will do that is:

$$e(n) = \prod_{k=1}^{lh\,(g(n)\,-\,1)^{k=2}} a(n, (g(n))_k)$$

And **A** is valid iff $[\![A]\!] = n$ and $e(n) = 1$.

This decision procedure for validity though effective is practically unusable for wffs containing, say, 40 or more propositional variables, for we would have to check some 2^{40} assignments of 0's and 1's. There are many shortcuts we can take, but to date no one has come up with a method of checking validity that is substantially faster and could be run on a computer in less than (roughly) exponential time relative to the number of propositional variables appearing in the wff. It has been conjectured that it is not possible to find a faster decision method. This is of significance because many combinatorial problems can be shown to have the same complexity as the decision procedure for validity of the classical propositional calculus (see *Further Reading* at the end of the chapter).

E. Axiomatizing Propositional Logic

There is another way we can approach the valid wffs: We can axiomatize them. Axiomatize logic? If we can try to axiomatize the "truths" of geometry or of arithmetic, why can't we try to axiomatize the "truths" of logic?

We would like to find some small number of valid wffs that we can take as our basic logical principles and then prove all the other valid wffs. This we can do. But instead of taking specific wffs as axioms we will use forms of wffs. A *scheme* (plural *schema*) is a wff with the variables replaced by our metavariables, **A, B, C,** ... . For example, $(p_1 \wedge p_2) \to p_1$ is a valid wff; $(A \wedge B) \to A$ is a *valid scheme*: every instance of this is valid, no matter what wffs **A** and **B** stand for. Schema are the skeletal forms of wffs, as wffs are the skeletal forms of propositions. When we write down a scheme as an axiom we are to be understood as taking every instance of that scheme, that is, every wff which has that form.

The only rule that we will use is *modus ponens*: from **A** and $A \to B$ conclude **B**. This is a *valid method of proof*, for whenever both **A** and $A \to B$ are true, so then is **B**; and thus whenever both **A** and $A \to B$ are valid, so is **B**.

Our axiom system for propositional logic is the following, where we delete the outermost parentheses for legibility.

The Classical Propositional Logic (PC)

Every instance of each of the following schema is an axiom:

1. $\neg A \rightarrow (A \rightarrow B)$
2. $B \rightarrow (A \rightarrow B)$
3. $(A \rightarrow B) \rightarrow ((\neg A \rightarrow B) \rightarrow B)$
4. $(A \rightarrow (B \rightarrow C)) \rightarrow ((A \rightarrow B) \rightarrow (A \rightarrow C))$
5. $A \rightarrow (B \rightarrow (A \wedge B))$
6. $(A \wedge B) \rightarrow A$
7. $(A \wedge B) \rightarrow B$
8. $A \rightarrow (A \vee B)$
9. $B \rightarrow (A \vee B)$
10. $((A \vee B) \wedge \neg A) \rightarrow B$

Rule: From **A** and **A** → **B** conclude **B** (*modus ponens*) .

A *proof of* **B** in this axiom system is a sequence $B_1, \ldots, B_n = B$, where each B_i is either an axiom or is derived from two earlier wffs in the sequence, B_j and B_k with $j, k < i$, by the rule of modus ponens, that is B_k is $B_j \rightarrow B_i$ (here the B_i's are metavariables ranging over wffs). This corresponds to how we would prove something informally using just this one rule of proof. We say that **A** is a *theorem* of this axiom system if it has a proof. In that case we write "⊢**A**". Classical propositional logic is also called the classical propositional *calculus*, abbreviated *PC* .

Let's see how this system works by proving two theorems. The sequence of wffs constitute the proof; the comments to the right are there to help you see why we're justified in taking each wff in the sequence.

⊢ $p_1 \rightarrow p_1$

Proof:

1. ⊢ $p_1 \rightarrow ((p_1 \rightarrow p_1) \rightarrow p_1)$ an instance of axiom 2
2. ⊢ $p_1 \rightarrow (p_1 \rightarrow p_1)$ an instance of axiom 2
3. ⊢ $(p_1 \rightarrow ((p_1 \rightarrow p_1) \rightarrow p_1)) \rightarrow ((p_1 \rightarrow (p_1 \rightarrow p_1)) \rightarrow (p_1 \rightarrow p_1))$
       an instance of axiom 4
4. ⊢ $(p_1 \rightarrow (p_1 \rightarrow p_1)) \rightarrow (p_1 \rightarrow p_1)$ by modus ponens using wffs 1 and 3
5. ⊢ $p_1 \rightarrow p_1$ by modus ponens using wffs 2 and 4   ■

⊢ $(p_1 \wedge \neg p_1) \rightarrow p_2$

Proof:

1. $(p_1 \wedge \neg p_1) \rightarrow p_1$ axiom 6
2. $(p_1 \wedge \neg p_1) \rightarrow \neg p_1$ axiom 7
3. $(\neg p_1 \rightarrow (p_1 \rightarrow p_2)) \rightarrow [(p_1 \wedge \neg p_1) \rightarrow (\neg p_1 \rightarrow (p_1 \rightarrow p_2))]$ axiom 2
4. $\neg p_1 \rightarrow (p_1 \rightarrow p_2)$ axiom 1
5. $(p_1 \wedge \neg p_1) \rightarrow (\neg p_1 \rightarrow (p_1 \rightarrow p_2))$ modus ponens using wffs 3 and 4

6. $[(\mathbf{p_1} \wedge \neg \mathbf{p_1}) \rightarrow (\neg \mathbf{p_1} \rightarrow (\mathbf{p_1} \rightarrow \mathbf{p_2}))] \rightarrow$
   $[((\mathbf{p_1} \wedge \neg \mathbf{p_1}) \rightarrow \neg \mathbf{p_1}) \rightarrow ((\mathbf{p_1} \wedge \neg \mathbf{p_1}) \rightarrow (\mathbf{p_1} \rightarrow \mathbf{p_2}))]$ axiom 4
7. $[((\mathbf{p_1} \wedge \neg \mathbf{p_1}) \rightarrow \neg \mathbf{p_1}) \rightarrow ((\mathbf{p_1} \wedge \neg \mathbf{p_1}) \rightarrow (\mathbf{p_1} \rightarrow \mathbf{p_2}))]$
   modus ponens using wffs 5 and 6
8. $(\mathbf{p_1} \wedge \neg \mathbf{p_1}) \rightarrow (\mathbf{p_1} \rightarrow \mathbf{p_2})$ modus ponens using wffs 2 and 7
9. $[(\mathbf{p_1} \wedge \neg \mathbf{p_1}) \rightarrow (\mathbf{p_1} \rightarrow \mathbf{p_2})] \rightarrow [((\mathbf{p_1} \wedge \neg \mathbf{p_1}) \rightarrow \mathbf{p_1}) \rightarrow ((\mathbf{p_1} \wedge \neg \mathbf{p_1}) \rightarrow \mathbf{p_2})]$
   axiom 4
10. $((\mathbf{p_1} \wedge \neg \mathbf{p_1}) \rightarrow \mathbf{p_1}) \rightarrow ((\mathbf{p_1} \wedge \neg \mathbf{p_1}) \rightarrow \mathbf{p_2})$ modus ponens, wffs 8 and 9
11. $(\mathbf{p_1} \wedge \neg \mathbf{p_1}) \rightarrow \mathbf{p_2}$ modus ponens, wffs 1 and 10   ∎

The last example is important. If we could prove a proposition and its negation, then by using Axiom 5 we could prove their conjunction. Hence, by this example, we could prove any proposition, for $\mathbf{p_2}$ can stand for any one we like (in terms of schema, we can establish that every instance of $(A \wedge \neg A) \rightarrow B$ is a theorem). That is, in classical propositional logic, *if we have a contradiction in our system then we can prove all other propositions.*

Now try your hand at proving some formal theorems in Exercise 4.

Each of the axiom schema is valid, and since the rule is valid, every theorem must be valid (Exercise 3). But why choose these and not other axiom schema? One reason is that each is "intuitively obvious". But more, with these we can give a conceptually clear proof that every valid wff is a theorem and hence establish the following.

THEOREM 1 (Completeness of Classical Propositional Logic)
A is a valid wff iff $\vdash A$.

That conceptually clear proof can be found in Chapter II of Epstein, 1990. But it has a drawback: It is nonconstructive. In an appendix to this chapter we've included a more complicated constructive proof.

Now we can see why our one rule of proof is sufficient: We don't need any others to get all valid wffs. Moreover, we can simulate other rules. For example, the rule "from $A \rightarrow B$ and $A \rightarrow \neg B$ conclude $\neg A$" always leads from true wffs to true wffs. And we can derive that via the theorems, for if we have that $A \rightarrow B$ and $\neg A \rightarrow B$ are theorems, then we can produce the following proof of B:

> $(A \rightarrow B) \rightarrow ((\neg A \rightarrow B) \rightarrow B)$ is an axiom;
> insert here a proof of $A \rightarrow B$;
> conclude by modus ponens $(\neg A \rightarrow B) \rightarrow B$;
> insert here a proof of $\neg A \rightarrow B$;
> conclude by modus ponens B.

And to derive the rule of *modus tollens*, from $\neg B$ and $A \rightarrow B$ conclude $\neg A$, we use the fact that $\neg B \rightarrow ((A \rightarrow B) \rightarrow \neg A)$ is a theorem.

F. Proving As a Computable Procedure

After reading the two proofs we gave in this axiom system and trying to prove some theorems (Exercise 4), you may have decided that to prove that a particular valid wff is a theorem takes quite a bit of insight and creativity. But really it's completely mechanical. All we have to do is start with our axioms and list all possible proofs until we get that particular wff as the final one in a proof sequence. Mechanical? Can we come up with a recursive procedure? Yes, using our Gödel numbering of wffs, we can show that we can recursively list all the theorems. Here's how.

a. First we show that we can recognize whether a wff has the schematic form which makes it an axiom. For instance,

$$a_1(n) = \begin{cases} 1 & \text{if } n = [\![\mathbf{A}]\!] \text{ and } \mathbf{A} \text{ is an instance of axiom 1} \\ 0 & \text{otherwise} \end{cases}$$

is recursive. And more generally,

$$a_1(n) = \begin{cases} 1 & \text{if } n = [\![\mathbf{A}]\!] \text{ and } \mathbf{A} \text{ is an instance of an axiom} \\ 0 & \text{otherwise} \end{cases}$$

is recursive.

b. Then we note that we can recognize when a number codes a sequence of wffs the following function is recursive:

$$s(n) = \begin{cases} 1 & \text{if } n = \langle (n)_0, \dots, (n)_{lh(n)-1} \rangle \text{ and each } (n)_i \,, i < lh(n), \\ & \quad \text{is the Gödel number of a wff} \\ 0 & \text{otherwise} \end{cases}$$

c. Then we can determine whether a number codes a sequence of wffs which is a proof sequence. That is, the following function is recursive:

$$prf(n) = \begin{cases} 1 & \text{if some } n = \langle (n)_0, \dots, (n)_{lh(n)-1} \rangle \text{ and for } i < lh(n), \\ & \quad (n)_i \text{ is the Gödel number of a wff, say } \mathbf{B}_i \text{, and either } \mathbf{B}_i \text{ is} \\ & \quad \text{an axiom, or for some } j, k < i \; \mathbf{B}_j \text{ is } \mathbf{B}_k \to \mathbf{B}_i \\ 0 & \text{otherwise} \end{cases}$$

Thus the set of proofs in our axiom system is recursively decidable.

d. Finally, using part (c), we can define a recursive function t that lists the Gödel numbers of theorems. That is: if $t(n) = m$, then $m = [\![\mathbf{A}]\!]$ and $\vdash \mathbf{A}$; and if $\vdash \mathbf{A}$ and $[\![\mathbf{A}]\!] = m$, then some n, $t(n) = m$. All we have to do to calculate $t(n)$ is search for the nth largest number a such that $prf(a) = 1$ and take $t(n) = (a)_{lh(a)-1}$. You're asked to fill in the details in Exercise 6 .

You may well ask why we should go to this trouble since we know that $\mathbf{A}$ is a theorem iff $\mathbf{A}$ is valid, so we could list the theorems by listing the valid wffs for which we already have a recursive decision procedure. The reason is that we wanted to demonstrate that proving in a formal system is a computable (recursive) process. There will be times when there may be no decision procedure, or we may know of none, and the only way we can approach the "truths" of the subject we're axiomatizing is by proving theorems. In those cases it's important to know that we can list the theorems recursively. Note that our recursive procedure for listing theorems is not a decision procedure for theoremhood. In the language of Chapter 17, it only establishes that the set of (Gödel numbers of) theorems is recursively enumerable.

Note also that this recursive procedure for listing theorems in terms of their proofs doesn't distinguish between "interesting" theorems and uninteresting ones, nor does it distinguish between especially perspicuous proofs and tediously involved ones. The theory of mechanical theorem proving is a subject in itself, which you can read about in Chang and Lee.

Appendix (Optional)

1. The Unique Readability Theorem

THEOREM 2   There is one and only one way to read each wff.

Proof: If **A** is a wff, then there is at least one way to read it since it has a definition. To show that there is only one way to read it, we'll establish that no initial segment of a wff is a wff.

The idea is that if we begin at the left of a wff and subtract 1 for every left parenthesis and add 1 for every right parenthesis then we will sum up to 0 only at the end of the wff. More precisely, define a function f from any concatenation of primitive symbols $\sigma_1 \ldots \sigma_n$ of our formal language to the integers by

$$f(\neg) = 0 ; \quad f(\wedge) = 0 ; \quad f(\vee) = 0 ; \quad f(\rightarrow) = 0 ; \quad f(\mathbf{p}_i) = 0 ;$$
$$f(\,(\,) = -1 ; \; f(\,)\,) = +1 ;$$
$$f(\sigma_1 \ldots \sigma_n) = f(\sigma_1) + \cdots + f(\sigma_n)$$

To show that for every wff **A**, $f(\mathbf{A}) = 0$ we proceed by induction on the number of symbols in **A**. The wffs with fewest symbols are $(\mathbf{p}_i)$, $i = 0, 1, 2, \ldots$, and for them it is immediate. So now suppose it is true for all wffs with fewer symbols than **A**. We then have four cases, which we cannot yet assume are distinct cases:

Case i.   **A** arises as $(\neg \mathbf{B})$. Then **B** has fewer symbols than **A**, so by induction $f(\mathbf{B}) = 0$, and so $f(\mathbf{A}) = 0$.

Case ii.   **A** arises as $(\mathbf{B} \wedge \mathbf{C})$. Then **B** and **C** have fewer symbols than **A**, so $f(\mathbf{B}) = f(\mathbf{C}) = 0$ and hence $f(\mathbf{A}) = 0$.

Case iii.   **A** arises as $(\mathbf{B} \vee \mathbf{C})$ ⎫
Case iv.   **A** arises as $(\mathbf{B} \rightarrow \mathbf{C})$ ⎭ are done similarly.

Now we show that, if reading from the left α is an initial segment of a wff other than the entire wff itself, then $f(\alpha) < 0$; and if α is a final segment other than the entire wff itself, then $f(\alpha) > 0$. So no proper initial or final segment of a wff is a wff. To establish this we again use induction on the number of occurrences of connectives in the wff. We'll let you establish the base case for atomic wffs.

Now suppose the lemma is true for any wff that contains $\leq n$ occurrences of the connectives. If **A** contains $n + 1$ occurrences, then it must have (at least) one of the forms given in the definition of wffs. If **A** has the form $(\mathbf{B} \wedge \mathbf{C})$, then an initial segment of **A** must have one of the following forms:

    i.   (

    ii.   (β where β is an initial segment of **B**

    iii.   (**B** $\wedge$

    iv.   (**B** $\wedge$ γ where γ is an initial segment of **C**

For (ii), $f(`(') = -1$ and by induction $f(\beta) < 0$, so $f(`(\beta') < 0$. We leave (i), (iii), and (iv) to you. The other three cases follow similarly.

Now to establish the theorem, we proceed through a number of cases by way of contradiction. Suppose we have a wff that could be read as both $(\mathbf{A} \wedge \mathbf{B})$ and $(\mathbf{C} \rightarrow \mathbf{D})$. Then $\mathbf{A} \wedge \mathbf{B})$ must be the same as $\mathbf{C} \rightarrow \mathbf{D})$. Hence, either **A** is an initial part of **C**, or **C** is an initial part of **A**. But then $f(\mathbf{A}) < 0$ or $f(\mathbf{C}) < 0$, which is a contradiction, as we proved above that $f(\mathbf{A}) = f(\mathbf{C}) = 0$. Hence, **A** is **C**. But then we have that $\wedge$ **B**) is the same as $\rightarrow$ **D**), which is a contradiction.

The other cases are similar, and we leave them to you.   ■

For the purpose of using induction on wffs, we may now define inductively the *length of a wff*:

The length of $(\mathbf{p_i})$ is 1.

If the length of **A** is n, then the length of $(\neg\mathbf{A})$ is $n + 1$.

If the maximum of the lengths of **A** and **B** is n, then each of
   $(\mathbf{A} \wedge \mathbf{B})$, $(\mathbf{A} \vee \mathbf{B})$, and $(\mathbf{A} \rightarrow \mathbf{B})$ has length $n + 1$.

2. The Completeness Theorem for Classical Propositional Logic

To prove the Completeness Theorem we shall have to make some definitions and prove two lemmas. We will leave many details to you.

We first define **A** to be a *consequence* of a finite set of wffs Γ if there is a sequence of wffs $\mathbf{B_1}, \dots, \mathbf{B_n} = \mathbf{A}$ such that each $\mathbf{B_i}$ is either an axiom, is in Γ, or is derived from two earlier wffs in the sequence, $\mathbf{B_j}$ and $\mathbf{B_k}$ with $j, k < i$, by the rule of modus ponens. In that case we write "$\Gamma \vdash \mathbf{A}$".

We leave to you to prove (a) $\varnothing \vdash \mathbf{A}$ iff **A** is a theorem; (b) if $\mathbf{A} \in \Gamma$ then $\Gamma \vdash \mathbf{A}$; (c) if **A** is a theorem then $\Gamma \vdash \mathbf{A}$; and (d) if $\Gamma \vdash \mathbf{A}$ and $\Gamma \vdash \mathbf{A} \rightarrow \mathbf{B}$, then $\Gamma \vdash \mathbf{B}$.

Lemma 3 (The Deduction Theorem) $\Gamma \cup \{\mathbf{A}\} \vdash \mathbf{B}$ iff $\Gamma \vdash \mathbf{A} \rightarrow \mathbf{B}$.

Proof: The proof from right to left is immediate. So suppose there is a proof of **B** from $\Gamma \cup \{\mathbf{A}\}$, namely, $\mathbf{B_1}, \dots, \mathbf{B_n} = \mathbf{B}$. We will show by induction on i that for each i, $\Gamma \vdash \mathbf{A} \rightarrow \mathbf{B_i}$. Either $\mathbf{B_1}$ is an axiom, or is in Γ, or is **A** itself. In the first two cases, the result follows by using axiom scheme 2. For the latter, you can modify our proof in Section E that $\vdash \mathbf{p_1} \to \mathbf{p_1}$ to show that $\vdash \mathbf{A} \to \mathbf{A}$.

Now suppose that for all $k < i$, $\vdash \mathbf{A} \to \mathbf{B_k}$. If $\mathbf{B_i}$ is an axiom, or is in Γ, or is $\mathbf{A}$ we are done, as above. The only other case is when $\mathbf{B_i}$ is a consequence by modus ponens of $\mathbf{B_m}$ and $\mathbf{B_j}$ where $\mathbf{B_j}$, is $\mathbf{B_m} \to \mathbf{B_i}$ and $m, j < i$. But then by induction we have $\Gamma \vdash \mathbf{A} \to (\mathbf{B_m} \to \mathbf{B_i})$ and $\Gamma \vdash \mathbf{B_m}$, so by axiom scheme 4 we can conclude that $\Gamma \vdash \mathbf{A} \to \mathbf{B_i}$. ∎

Formally an *assignment* of truth-values to propositional variables is a function $v : \{ \mathbf{p_0}, \mathbf{p_1}, \dots \} \to \{\mathsf{T}, \mathsf{F}\}$. The Unique Readability Theorem justifies that every assignment can be extended in a unique way to all wffs via the truth-tables.

We will use the metavariables $\mathbf{q_1}, \mathbf{q_2}, \dots$ to stand for propositional variables.

Lemma 4 (Kalmár, 1935)   Let C be any wff and $\mathbf{q_1}, \dots, \mathbf{q_n}$ the propositional variables appearing in it. Let v be any valuation. Define, for $i \le n$,

$$\mathbf{Q_i} = \begin{cases} \mathbf{q_i} & \text{if } v(\mathbf{q_i}) = \mathsf{T} \\ \lnot\mathbf{q_i} & \text{if } v(\mathbf{q_i}) = \mathsf{F} \end{cases}$$

and define $\Gamma = \{ \mathbf{Q_1}, \dots, \mathbf{Q_n} \}$. Then   i. If $v(\mathbf{C}) = \mathsf{T}$, then $\Gamma \vdash \mathbf{C}$.
ii. If $v(\mathbf{C}) = \mathsf{F}$, then $\Gamma \vdash \lnot\mathbf{C}$.

Proof: In this proof there are several schema which must be shown to be schema of theorems. We will highlight each by an * and leave them for you to prove from the axioms with the aid of the Deduction Theorem.

We proceed by induction on the length of C. If C is $(\mathbf{p_i})$, then the proof devolves into showing that $\vdash \mathbf{p_i} \to \mathbf{p_i}$ and $\vdash \lnot\mathbf{p_i} \to \lnot\mathbf{p_i}$, which follow as in Section E above. Now suppose the lemma is true for all wffs of length $\le n$ and C has length $n + 1$. We have four cases:

Case i.   C is $\lnot\mathbf{A}$. If $v(\mathbf{C}) = \mathsf{T}$, then $v(\mathbf{A}) = \mathsf{F}$. Hence by induction $\Gamma \vdash \lnot\mathbf{A}$ as desired. If $v(\mathbf{C}) = \mathsf{F}$, then $v(\mathbf{A}) = \mathsf{T}$, so $\Gamma \vdash \mathbf{A}$. But $* \vdash \mathbf{A} \to \lnot\lnot\mathbf{A}$, hence $\Gamma \vdash \lnot\lnot\mathbf{A}$ as desired.

Case ii.   C is $\mathbf{A} \land \mathbf{B}$. If $v(\mathbf{C}) = \mathsf{T}$, then $v(\mathbf{A}) = v(\mathbf{B}) = \mathsf{T}$. So $\Gamma \vdash \mathbf{A}$ and $\Gamma \vdash \mathbf{B}$, and hence by axiom 5, $\Gamma \vdash \mathbf{A} \land \mathbf{B}$. If $v(\mathbf{C}) = \mathsf{F}$, then either $v(\mathbf{A}) = \mathsf{F}$ or $v(\mathbf{B}) = \mathsf{F}$. Suppose $v(\mathbf{A}) = \mathsf{F}$. Then $\Gamma \vdash \lnot\mathbf{A}$. Since $* \vdash (\mathbf{D} \to \mathbf{E}) \to (\lnot\mathbf{E} \to \lnot\mathbf{D})$, via axiom 6 we have $\vdash \lnot\mathbf{A} \to \lnot(\mathbf{A} \land \mathbf{B})$ and so $\Gamma \vdash \lnot(\mathbf{A} \land \mathbf{B})$. If $v(\mathbf{B}) = \mathsf{F}$ the argument is the same except we use axiom 7.

Case iii.   C is $\mathbf{A} \lor \mathbf{B}$. If $v(\mathbf{C}) = \mathsf{T}$, then $v(\mathbf{A}) = \mathsf{T}$ or $v(\mathbf{B}) = \mathsf{T}$. If $v(\mathbf{A}) = \mathsf{T}$ then $\Gamma \vdash \mathbf{A}$, so by axiom 8 we have $\Gamma \vdash \mathbf{A} \lor \mathbf{B}$, and similarly if $v(\mathbf{B}) = \mathsf{T}$. If $v(\mathbf{C}) = \mathsf{F}$ then $v(\mathbf{A}) = \mathsf{F}$ and $v(\mathbf{B}) = \mathsf{F}$, so $\Gamma \vdash \lnot\mathbf{A}$ and $\Gamma \vdash \lnot\mathbf{B}$. Since $* \vdash \lnot\mathbf{A} \to (\lnot\mathbf{B} \to \lnot(\mathbf{A} \lor \mathbf{B}))$ we have $\Gamma \vdash \lnot\mathbf{C}$.

Case iv.   C is $\mathbf{A} \to \mathbf{B}$. If $v(\mathbf{C}) = \mathsf{T}$, then $v(\mathbf{A}) = \mathsf{F}$ or $v(\mathbf{B}) = \mathsf{T}$. If $v(\mathbf{A}) = \mathsf{F}$ then $\Gamma \vdash \lnot\mathbf{A}$ and so by axiom 1 we have $\Gamma \vdash \mathbf{A} \to \mathbf{B}$. If $v(\mathbf{B}) = \mathsf{T}$ then use axiom 2. Finally, if $v(\mathbf{C}) = \mathsf{F}$ then $v(\mathbf{A}) = \mathsf{T}$ and $v(\mathbf{B}) = \mathsf{F}$. So $\Gamma \vdash \mathbf{A}$ and $\Gamma \vdash \lnot\mathbf{B}$ and hence since $* \vdash \mathbf{A} \to (\lnot\mathbf{B} \to \lnot(\mathbf{A} \to \mathbf{B}))$, we have $\Gamma \vdash \lnot\mathbf{C}$. ∎

THEOREM 5 (Completeness of Classical Propositional Logic)   A is a valid wff iff $\vdash$ A .

Proof: We remarked earlier that if **A** is a theorem then **A** is valid (see Exercise 3 below). To establish the converse, suppose **A** is valid and q_1 , ... , q_n are the propositional variables appearing in **A** . For every assignment v, $v(A) = T$. So by Lemma 4, for any collection $Q_1, \ldots, Q_n$ where for each i Q_i is either q_i or $\neg q_i$, $\{Q_1, \ldots, Q_n\} \vdash A$. Hence by the Deduction Theorem, for any such collection $\{Q_1, \ldots, Q_{n-1}\} \vdash Q_n \to A$. And so $\{Q_1, \ldots, Q_{n-1}\} \vdash q_n \to A$ and $\{Q_1, \ldots, Q_{n-1}\} \vdash \neg q_n \to A$. Hence via axiom 3, for any $Q_1, \ldots, Q_{n-1}$ we have $\{Q_1, \ldots, Q_{n-1}\} \vdash A$. Repeating this argument $n-1$ times we get $\vdash A$ (but notice that the i-th step of the argument really embodies 2^i formal derivations.)   ∎

Exercises

1. Write out the truth-tables for the following formulas, and check whether each is valid. You can write boldface letters by putting a squiggle under the letter.

   a. $p_1 \to (p_2 \to p_1)$

   b. $[p_1 \to (p_2 \to p_3)] \to [(p_1 \to p_2) \to (p_1 \to p_3)]$

   c. $[(p_1 \to p_2) \wedge \neg p_1] \to \neg p_2$

   d. $[p_1 \wedge \neg (p_1 \wedge p_2)] \to \neg p_2$

   e. $[((p_1 \wedge p_2) \to p_3) \vee p_1] \to (p_2 \vee p_3)$

   f. $[(p_1 \wedge p_2) \vee p_3] \leftrightarrow [(p_1 \wedge p_3) \vee (p_2 \wedge p_3)]$

2. Prove that if **A** is a theorem, then **A** has arbitrarily many proofs.

3. a. Prove that each axiom scheme is valid.

   b. Prove that if **A** is a theorem, then **A** is valid.
      (*Hint:* Induct on the number of steps in a proof.)

†4. Prove the following wffs of propositional logic using our axiom system:

   a. $(\neg p_1 \to p_1) \to p_1$ (*Hint:* Use a wff we proved earlier.)

   b. $\neg\neg p_1 \to p_1$ (*Hint:* Use part (a).)

5. Explain why the following rules can be replaced by derivations in our axiom system.

   Adjunction: from **A** and **B** conclude **A** ∧ **B**

   Distribution: from (**A** ∨ **B**) and (**A** ∨ **C**) conclude **A** ∨ (**B** ∧ **C**)

6. Fill in the details in the proof in Section F that we can recursively list all theorems of propositional logic. That is, give recursive definitions of the functions a_1, a, s, *prf*, and t .

Further Reading

For a comprehensive introduction to propositional logics, see Epstein's *Propositional Logics*. Garey and Johnson's *Computers and Intractability* is a thorough and quite readable textbook on the question of the complexity of the decision problem for the validity of propositional wffs. For a short introduction to the subject, see Book's review of that.

20 An Overview of First-Order Logic and Gödel's Theorems

Let's review what we've done and see where we're going. We were concerned about the use of infinitistic methods in mathematics. We looked at Hilbert's analysis and decided that we'd investigate whether the use of infinite sets could be justified by proving them consistent with ordinary finitistic mathematics. Any such proof, we agreed, would have to be finitistic itself if it weren't to beg the question.

So the first thing we did was see what we meant by "finitistic methods". We formalized the notion of computable function as being Turing machine computable and then saw that this notion was equivalent to being partial recursive. After a brief discussion, we adopted (at least as a working hypothesis) Church's Thesis that indeed the formalization was apt.

Then we turned to an analysis of the notion of proof. We presented the idea of a formal system and developed propositional logic. We explained how to connect our analysis of computable functions with finitistic methods in non-numerical situations by using Gödel numberings and the notion of decidability. As an example, we showed that the propositional logic we adopted was decidable and that we could view proving theorems in the formal system as a machine represented by a computable function.

Now it's time to see whether we can construct a formal system which is powerful enough to formalize much of arithmetic and yet which we can prove is consistent by finitistic means. Not to hide anything from you, the answer will be "No."

The fundamental idea is that for any formal system we may Gödel number the wffs and translate assertions about the system, such as "this wff is a theorem" or "the system is consistent" into assertions about the natural numbers. Then if our system can formalize even a small part of arithmetic we will be able to translate those statements back into the system itself. Thus, via our Gödel numberings we will be able to talk about the system within the system itself. Self-reference and a variation on the liar paradox will follow.

This is a long project which takes up the next four chapters. Since it's easy to lose your way, we'll outline the steps here.

1. The first step we need to take is to decide on a formal language for arithmetic. What will be the form of the propositions about the natural numbers for which the propositional variables will stand?

To begin, we will certainly want symbols for addition and multiplication: $+$ and $\cdot$. And we'll need symbols for equality, $=$, and for zero, $\mathbf{0}$. More, we need symbols for every natural number. The simplest way to do that is to have a symbol for the successor function: $\prime$. Using that we can write a numeral for any natural number in unary notation, for example, $\mathbf{0}''''$ would be a numeral for 4 . We'll abbreviate the numerals in shorthand using boldface italic: thus, $\mathbf{4}$ means $\mathbf{0}''''$.

What other functions will we want to have special symbols for? One of our goals is to come up with a system that is as simple as possible so that when we claim it is consistent we will have some confidence we are right. We don't want to build into our system more than we need.

We can already make up polynomials; for example, using $\mathbf{x}, \mathbf{y}, \mathbf{z}$ as variables for natural numbers, we can write the polynomial $x^2 + y^2 + z^3$ as $((\mathbf{x} \cdot \mathbf{x}) + (\mathbf{y} \cdot \mathbf{y})) + ((\mathbf{z} \cdot \mathbf{z}) \cdot \mathbf{z})$. From our experience with the primitive recursive functions, we know that starting with these functions and the projections, which we'll get for free from the logical notation, we can get much more complicated functions. All we need is that various operations on the functions are definable.

2. But what do we mean when we say we can "get" other functions from these? In just the same way as for the formal system of propositional logic, proving will be a computable procedure for the formal system of arithmetic. Indeed, that is an essential characteristic of proving. So we can view our formal system as a way to calculate functions if we know how to interpret the symbols, much as we interpreted Turing machine calculations as functions.

We will say (roughly) that a total function $f(x_1, \dots, x_k)$ is *representable* if there is some expression in the formal language, say $\mathbf{A}$, which uses variables $\mathbf{x}_1, \dots, \mathbf{x}_k, \mathbf{x}_{k+1}$ such that for any numbers $m_1, \dots, m_k$ and n, we have $f(m_1, \dots, m_k) = n$ iff we can prove in our formal system $\mathbf{A}(\boldsymbol{m_1}, \dots, \boldsymbol{m_k}, \boldsymbol{n})$ and we can't prove $\mathbf{A}(\boldsymbol{m_1}, \dots, \boldsymbol{m_k}, \boldsymbol{j})$ for any other number j. Since from a contradiction we can prove anything, this definition will be useful only if our system is consistent. In that case, any function that is representable will be computable (recursive): to find the value on input $m_1, \dots, m_k$ start searching through the list of theorems until you find one of the form $\mathbf{A}(\boldsymbol{m_1}, \dots, \boldsymbol{m_k}, \boldsymbol{n})$. Calculation in terms of proving in a formal system was one of the first analyses put forward to characterize computability.

If we take as axioms the inductive definitions of addition and multiplication and ones that say that successor is a 1-1 function whose range is everything except zero, which altogether we call axiom system Q, then we'll be able to establish that all the initial recursive functions are representable and the representable functions are closed under composition and the μ-operator. If we can show that they are also closed under primitive recursion, then we'll have that all the recursive functions are representable and thus have more than ample means to translate the numerical versions of assertions about our system back into the system itself.

In order to show that the representable functions are closed under primitive recursion, we take a slight detour through number theory to develop new coding and uncoding functions that don't depend on exponentiation. Thus we will prove in Chapter 22 that *a function is representable in our formal system iff it is recursive*, another characterization of the class of recursive functions.

3. Before we can even talk about representability of functions, however, we have to set up the formal language and the axioms and proof methods of our formal system of arithmetic. That's what we do in Chapter 21, but doing that will make a lot more sense if you keep in mind where we're going: we want to get a system just strong enough to be able to represent the recursive functions.

The only part of that system we haven't mentioned is that we need some way to say "for every" and "there exists". In Chapter 21 we discuss these *quantifiers* and explain that we'll only quantify over elements; for example, "for every x there is a y such that $x + y = x$." We won't need to quantify over sets of elements, as in "Every nonempty set of numbers has a least element." Quantification over elements is called *first-order* quantification, as opposed to quantification over sets, which is called *second-order*, and the logical tools we will use are called *first-order*.

4. Because the recursive functions are representable in the formal system Q, we will be able to show that the set of theorems of Q is undecidable. Roughly, here's how.

We have that the recursive sets are representable. So if we can diagonalize the representable sets, then we know the resulting set won't be recursive. To diagonalize the representable sets we need only to be able to recognize whether a wff has a particular form, namely, it has to have a variable in it for which we can substitute numerals. This we can do recursively in terms of the Gödel numbering; say these formulas are $A_1, A_2, \ldots$. Then on the assumption that Q is consistent, if a set is represented by some formula there must be some m for which it is $\{n : A_m(n)$ is a theorem of $Q\}$. The diagonalization of these sets is $S = \{m : A_m(m)$ is not a theorem of $Q\}$, which is not representable and hence not recursive. But if S isn't recursive, it can only be because the set of theorems of Q (via the Gödel numbering) is not recursive.

Thus, the set of theorems of our formal system is not recursively decidable. And this will be equally true for any formal axiomatic system that contains Q as long as it doesn't contain a contradiction, since all we're using is that we can get the recursive functions representable in the system. This is what we establish in Chapter 23. *On the assumption of Church's Thesis and that the formal system we've set up contains no contradiction, there's no computable procedure for deciding which propositions are theorems in any "reasonably strong" theory of arithmetic.*

5. But more, consider all the statements in our formal language that are true of the natural numbers. Suppose we could axiomatize these "truths". Then to decide whether a statement is a theorem, since every statement is either true or its negation is true but not both, we'd only have to set our proof machine going and wait until we got a proof of either the statement or its negation. That would be a computable procedure for deciding theoremhood, contrary to what we have just established. So *the statements true about the natural numbers in our formal language are not even axiomatizable.*

6. Finally, we turn to the question whether we can establish the consistency of infinitistic methods in arithmetic by finitary means.

Since "finitary means" can be converted by Gödel numbering into constructive procedures on the natural numbers, we look for a formal theory of arithmetic in which to try to capture all finitistic proof methods. It should extend *Q* so that we can talk about recursive sets via their representations. What we need beyond that is to be able to do proofs by induction. By adding the axiom scheme of induction to *Q* we have the theory called (*Elementary*) *Peano Arithmetic* (*PA*). It will seem plausible that any finitistic consistency proof could be formalized in this system. (We will not claim that *PA* formalizes *only* finitary proof procedures.)

So we ask, in particular, can we prove in *PA* that *PA* is itself consistent? To make that precise, we pick out a particular wff of the formal language which in terms of the Gödel numbering is true iff *PA* is consistent and call it **Consis** .

Then, finally, we use the power of the self-reference available to us in our system to construct a variation of the liar paradox: a wff **U** which in terms of the Gödel numbering *expresses that it itself is not provable*. If **U** were provable, then since it asserts its own unprovability, we'll find that we could also prove its negation. So if *PA* is consistent, then **U** is not provable and hence must be true. We have produced *a statement which is true but unprovable relative to our formal system.*

What we have shown is this: (*) if *PA* is consistent, then **U** is not a theorem of *PA* . But the proof of (*) is actually finitistic and we can formalize it within *PA* . That is, we can prove within *PA* the formalization of (*) which, since **U** asserts that it itself is unprovable, is **Consis** → **U** . If we could also prove **Consis** in *PA*, then we could prove **U**, which we know we can't. Thus, *if PA is consistent then we cannot prove it is consistent within PA* .

We could add **Consis** as an axiom, but small help that would be. Since the new system would be axiomatizable we could Gödel number everything again, in particular the new proof mechanism which uses that axiom, and repeat the whole process. That is, *there's no axiomatizable theory of arithmetic which can prove its own consistency*, at least if the theory is strong enough to allow us to give inductive definitions of addition and multiplication, characterize the successor function, and do proofs by induction. Thus, even if we haven't captured all possible finitistic proof methods within *PA* , adding them wouldn't help.

Here is how Gödel first announced his work in a discussion at a conference in Königsberg in 1930:

GÖDEL: According to the formalist view, one adjoins to the meaningful statements of mathematics transfinite (pseudo-)assertions which in themselves have no meaning, but serve only to round out the system, just as in geometry one rounds out the system by the introduction of infinitely distant points. This view assumes that if one adjoins to the system S of meaningful statements the system T of transfinite statements and axioms and then proves a theorem from S in a roundabout way via theorems from T, that that theorem is also contentually correct, and hence that no contentually false theorems become provable through the adjunction of the transfinite axioms. This demand is customarily replaced by that of consistency. Now I would like to point out that these two demands may in no way be regarded as equivalent without further justification. For if in a consistent system A (perhaps that of classical mathematics) a meaningful proposition p is provable with the help of transfinite axioms, it only follows from the consistency of A that *not-p* is not formally derivable *within* the system A. Nonetheless it remains conceivable that one could perceive [the truth of] *not-p* through some sort of contentual (intuitionistic) considerations which themselves can *not* be formally represented in A. In this case, despite the consistency of A, a proposition would be provable in A whose falsity one could perceive through finitary considerations. As soon as one interprets the concept of "meaningful statement" sufficiently narrowly (e.g., as restricted to finite numerical equations), something of that kind certainly cannot happen. On the other hand, it is quite possible that one could prove a statement of the form $(Ex)F(x)$ where F is a finitary property of natural numbers (the negation of Goldbach's conjecture has this form, for example) by transfinite methods of classical mathematics, and on the other hand could realize via contentual considerations that all numbers have the property *non-F*; indeed, I would like to point out that this would still remain possible even if one had demonstrated the consistency of the formal system of classical mathematics. For of no formal system can one affirm with certainty that all contentual considerations are representable within it.

[There is a comment by von Neumann, and Gödel continues.]

One can (assuming the consistency of classical mathematics) even give examples of propositions (and indeed, of such type of Goldbach or Fermat) which are really contentually true but are unprovable in the formal system of classical mathematics. Therefore if one adjoins the negation of such a proposition to the axioms of classical mathematics, one obtains a consistent system in which a contentually false proposition is provable.

from Dawson, 1984, pp. 125–126

21 First-Order Arithmetic

A. A Formal Language for Arithmetic

1. Variables

We will need variables that we can interpret as standing for numbers. Our formal variables will be x_0, x_1, ... , x_n, ... In our schema we will use x, y, z, w as metavariables that range over the formal variables.

2. Arithmetic functions and terms

As we discussed in Chapter 20, we choose to use symbols for just three functions: successor will be formalized with the symbol $'$; addition will be formalized with the symbol $+$; and multiplication will be formalized by $\cdot$.

We will also take the symbol **0** to be in our formal language, intending to interpret it as the number zero. Of course, all this talk about how we intend to interpret our symbols is just by way of guidelines. In the plane geometry you learned in school you couldn't assume anything about the undefined words "point" and "line" except what you officially postulated in the axioms. It's the same here. The symbols $+$, $\cdot$, and so on, are formal symbols, undefined, primitive. Our goal is to axiomatize the properties of the natural numbers using them. We always keep that goal in front of us, but we may not assume anything about our formal symbols except what we have explicitly postulated.

The first thing we must explicitly state is how we will form expressions for the result of applying these functions to either variables or to **0**. Informally, we might write $(x + y + 0) \cdot z$, but that won't do here because it's ambiguous as to which addition is to be done first. Here is the inductive definition of how to form the compound expressions, which we will call *terms*. We use the symbols **t**, **u**, and **v** as metavariables ranging over terms.

Terms
    i.   Every variable is a term, and **0** is a term.
    ii. If **t** and **u** are terms, then so are $(\mathbf{t})'$, $(\mathbf{t} + \mathbf{u})$, and $(\mathbf{t} \cdot \mathbf{u})$.
    iii. A string of symbols is a term iff it arises via applications of (i) and (ii).

3. Numerals in unary notation

One of the simplest arithmetic propositions we can make is "$2+2=4$". How can we say that formally? We will need a symbol for equality: $=$. But what about symbols for the numbers 2 and 4 ? We can use unary notation: a formal symbol for zero is **0**, for 1 is **(0)´**, for 2 is **((0)´)´**, for 3 is **(((0)´)´)´**, and for 4 is **((((0)´)´)´)´**. So "$2+2=4$" will be formalized as **((0)´)´ + ((0)´)´ = ((((0)´)´)´)´**. Did we say that we're using parentheses to lessen confusion? No, parentheses are there to ensure that our statements will be precise and unambiguous, though often it would be easier to read the statement without them. When we are talking informally, we will delete parentheses whenever it makes the statement more legible; in particular we'll write **t´** for **(t)´**. So *informally* we have $\mathbf{0}´´ + \mathbf{0}´´ = \mathbf{0}´´´´$.

But even that is hard to read, so we'll adopt the further convention of abbreviating the numeral in unary notation that has n occurrences of ´ by the decimal numeral for n in bold italic. For example, our new shorthand for the formalization of "$2+2=4$" is $\mathbf{\textit{2}} + \mathbf{\textit{2}} = \mathbf{\textit{4}}$.

4. Quantifiers: existence and universality

The point of using variables is to be able to express generality. Informally we may write $x + y = y + x$ as a statement of the commutative law of addition. But this statement is ambiguous between the assertion that for every x and y, $x + y = y + x$ and our having chosen some specific x and y for which $x + y = y + x$. It's much the same problem we resolved for functions by using the λ-notation. Here we can make our meaning explicit by using the *universal quantifier* $\forall$. Informally, we have $\forall x \, \forall y \, (x + y = y + x)$; a formal version could be $\forall \mathbf{x}_1 \, (\forall \mathbf{x}_2 \, (\mathbf{x}_1 + \mathbf{x}_2 = \mathbf{x}_2 + \mathbf{x}_1 \,) \,)$. We'll informally delete parentheses between quantifiers when the meaning is clear.

Now consider the formula $x + x = 4$. This isn't true for all natural numbers. However, there are some that make it true. Informally, we write $\exists x \, (x + x = 4)$ to mean that "there is an x such that $x + x = 4$". But for our formal language we want to be as parsimonious as we can, so we're going to show how to dispense with the existential quantifier. Consider the (informal) statement $\neg \forall x \neg \, (x + x = 4)$. What does this assert? "It's not the case that for every x it's not the case that $x + x = 4$." But if it's not the case that for every x, $x + x \neq 4$, then there must exist some x for which $x + x = 4$, *at least on the classical, nonconstructive understanding of existence*. Conversely, if there is some x for which $x + x = 4$, then it can't be that for every x we have $x + x \neq 4$. So from the classical, nonconstructive point of view (which after all is what we're investigating here), we don't need a new primitive symbol for existence. Instead, we will take $\exists \mathbf{x}$ as a defined symbol that abbreviates $\neg \forall \mathbf{x} \, \neg$. We'll nonetheless informally refer to such abbreviations as *existential quantifiers*.

It may seem that we've built into our system a highly nonconstructive assumption. But from a constructive viewpoint we can deny that this is a good abbreviation and hold that our formalism only has resources for quantifying universally.

There are other ways we quantify in mathematics, for instance when we quantify over sets as in "Every nonempty collection of natural numbers has a least element." But remembering that we want to keep things as simple as possible so we can have confidence in our system, we will take only universal quantifications over elements as primitive, and thus have no resources for formalizing quantification over sets or other kinds of quantification. This is what is meant by *first-order* logic.

A further caution about using quantifiers is necessary. Consider, informally, $\forall x \exists y \, (x + y = 0)$ and $\exists y \forall x \, (x + y = 0)$. The first is read "For all x there is a y such that $x + y = 0$ "; the second is read "There is a y such that for all x, $x + y = 0$." Taken as statements about the integers, the first is true, the second false. *The order of quantification matters and is read from left to right as in English.*

5. The formal language

A formal language for arithmetic

> Variables: $x_0, x_1, \ldots, x_n, \ldots$
> Constant: $\mathbf{0}$
> Function symbols: $', +, \cdot$
> Equality: $=$
> Connectives: $\wedge, \vee, \neg, \rightarrow$
> Quantifier: $\forall$
> Parentheses: $(\,,\,)$

The definition of *term* is given above. Here is the inductive definition of wffs in this language. As for propositional logic, we use $\mathbf{A}, \mathbf{B}, \mathbf{C}, \ldots$ to stand for wffs.

Well-formed-formula (wff)

> i. If $\mathbf{t}$ and $\mathbf{u}$ are terms, then $(\mathbf{t} = \mathbf{u})$ is a wff. It is an *atomic wff* .
> ii. If $\mathbf{A}$ and $\mathbf{B}$ are wffs, then so are $(\mathbf{A} \wedge \mathbf{B})$, $(\mathbf{A} \vee \mathbf{B})$, $(\neg \mathbf{A})$, and $(\mathbf{A} \rightarrow \mathbf{B})$.
> iii. If $\mathbf{A}$ is a wff and $\mathbf{x}$ is a variable, then $(\forall \mathbf{x} \, \mathbf{A})$ is a wff.
> iv. A string of symbols is a wff iff it arises via applications of (i), (ii), and (iii) .

These are inductive definitions, so you can Gödel number the terms, atomic wffs, and wffs of this language (Exercise 11).

Aside: The unique readability of wffs can be proved much as for propositional logic, pp. 165–166.

First we must show that each term has a unique reading. Define a function $g(+) = -1$, $g(\cdot) = -1$, $g(') = 0$, $g(\mathbf{0}) = 1$, $g(x_i) = 1$, $g(\,(\,) = 1$, $g(\,)\,) = -1$, and extend g to all concatenations of these symbols by setting $g(\sigma_1 \cdots \sigma_n) = g(\sigma_1) + \cdots + g(\sigma_n)$. Then you can show that for every term $\mathbf{t}$, $g(\mathbf{t}) = 1$, and if $\mathbf{t}^*$ is an initial segment of a term other than the entire term itself, then $g(\mathbf{t}^*) < 1$, and if a final segment then $g(t^*) > 1$. Proceeding as for wffs in propositional logic, you can show every term has a unique reading.

To show that there is only one way to read each wff, define a function f by: $f(\,(\,) = -1$, $f(\,)\,) = 1$, $f(\,\forall\,) = -1$, $f(=) = -2$, $f(t) = 1$ for any term t , and $f(\wedge) = f(\vee) = f(\rightarrow) = f(\neg) = 0$. Then proceed as for propositional logic.

Informal conventions: Here are some informal conventions we'll use to make the wffs more legible. Anything written using these conventions is to be understood as an abbreviation of the real wff in the formal language.

1. We abbreviate the numeral in unary notation that has n occurrences of ´ by the decimal numeral for n in bold italic.
2. We write $\exists x$ for $\neg \forall x \neg$.
3. We write $(\,t \neq u\,)$ for $\neg(\,t = u\,)$.
4. We delete the outermost parentheses.
5. We sometimes write] in place of) , and [in place of (.

Also, with considerable caution, and only when we're sure the meaning is clear, we will delete parentheses between quantifiers or even inner parentheses.

6. The standard interpretation and axiomatizing

As we said before, we can assume nothing about our formalism except what we explicitly postulate. So far we're free to interpret the variables as ranging over any things and the function symbols as functions on those things. For instance, we could use the wffs in this language to make assertions about the integers modulo 5, or the integers, or a ring where ´ could be the additive inverse. However, we've designed this symbolism to make assertions about the natural numbers where $+$ is interpreted as addition, $\cdot$ is interpreted as multiplication, ´ is interpreted as the successor function, and $\mathbf{0}$ is interpreted as zero. This is the *standard interpretation.* In the following discussions whenever we say a wff is *true of the natural numbers* or simply *true*, or we appeal to your intuition about the correctness of some principle, we have in mind the standard interpretation.

For propositional logic our task in axiomatizing was clear: we wanted axioms that would yield all valid wffs as theorems. Here we are not going to talk about validity, and in that sense our task will be easier. All we want are axioms that are true about the natural numbers and rules of proof that lead from true wffs to true wffs in order to axiomatize enough of arithmetic to be able to represent the recursive functions in the system. We first turn to the principles of reasoning that we'll codify.

B. Principles of Reasoning and Logical Axioms

1. Closed wffs and the rule of generalization

Informally, consider $x + y = z$. This is neither true nor false until we specify which numbers x, y, z are to stand for. For example, $2 + 3 = 5$ is true, whereas $2 + 3 = 7$ is false.

Now consider $\exists y\, (x + y = z)$. We're no longer free to let y be any number we want: this describes a two-place relation. Since $\exists y\, (2 + y = 5)$ is true, $(2, 5)$ is in the relation, whereas $(7, 2)$ is not. We say that x and z are *free* in this formula, whereas y is "bound". If we bind z by a universal quantifier we get $\forall z\, \exists y\, (x + y = z)$, which describes a property of the free variable x. Because $\forall z\, \exists y\, (5 + y = z)$ is false, 5 doesn't have the property; indeed, 0 is the only natural number which does.

If we further bind x by the existential quantifier, $\exists x\, \forall z\, \exists y\, (x + y = z)$, then none of the variables are free: we have a proposition which is either true or false. In this case it's true and expresses the idea that there exists a least natural number.

Only wffs in which every variable is bound can be true or false. To make this more precise, we first have to say which variables are affected by which quantifiers; for example, in $\forall x\, \exists y\, (x + y = 0) \rightarrow \forall y\, (x \cdot y = 0)$ the last x is not bound by a quantifier. We define the *scope of the quantifier* $\forall x$ in $(\forall x\, A)$ to be A. Thus, in $\forall x_1 \exists x_2\, (x_1 + x_2 = 0) \rightarrow \forall x_2 (x_1 \cdot x_2 = 0)$ the scope of $\forall x_1$ is $\exists x_2\, (x_1 + x_2 = 0)$.

An occurrence of a variable x *is bound in* A if it is either the variable immediately following the symbol $\forall$, that is, $\forall x$, or if it is within the scope of a quantifier $\forall x$ in the wff. Otherwise *the occurrence is free in* A. Thus in our example, the last occurrence of x_1 is free, whereas the first two are bound. Sometimes when we want to stress that there's a free occurrence of x in A we write $A(x)$.

Finally, we say that a wff is *closed* if it contains no free variables; otherwise it is *open*. Closed wffs are sometimes called *sentences*. It is closed wffs which we want as theses of our system.

In informal mathematics we often use open formulas to express laws, such as the commutative law of addition: $x + y = y + x$. When we do that we are implicitly assuming that x and y can be any numbers. That is, we are really saying $\forall x\, \forall y\, (x + y = y + x)$. This is a useful convention and by adopting it we'll simplify the formal proofs: we can assert any wff on the understanding that all the variables free in it are to be thought of as universally quantified.

The formal version of this convention is the *rule of generalization*: from A conclude $\forall x\, A$. For example, from $(x_1 + x_2 = x_2 + x_1)$ asserted as a thesis, we can conclude $\forall x_1 \forall x_2 (x_1 + x_2 = x_2 + x_1)$; we can also conclude $\forall x_{47} (x_1 + x_2 = x_2 + x_1)$, which isn't wrong though it may seem odd. We tolerate "superfluous" quantifiers because to avoid them would make the definition of wffs and many of our axioms and rules quite complicated.

Note that we can't replace this rule by the scheme $A(x) \rightarrow \forall x\, A(x)$ because that can be false; for example, $x + x = 2 \rightarrow \forall x\, (x + x = 2)$ is false if we let x stand for 1. On the other hand, $\forall x\, A(x) \rightarrow A(x)$ is acceptable: if the antecedent is true, then every instance of it is true, so in particular it will be true no matter what we let x stand for.

2. The propositional connectives

When we introduced the propositional connectives in Chapter 19 we said they connect propositions. What then are we to make of $\exists x \, \forall z \, [\, (\, x \cdot z = 0 \,) \, \wedge \, \exists y \, (\, x + y = z \,) \,]$ where $\wedge$ joins two formulas that are not closed? Think how we determine whether this is true about the natural numbers: we ask whether we can find a number to substitute for x, namely 0, such that no matter what number we let z stand for, their product is 0, and we can find a number to let y stand for such that the wff is true; for example, $(0 \cdot 5 = 0) \wedge (0 + 5 = 5)$. When the formulas are fully interpreted the propositional connectives connect propositions. In this sense we are justified in using the classical propositional logic of Chapter 19.

3. Substitution for a variable

A variable can stand for anything we are talking about. Suppose $A(x)$ is true of everything and t is a term. Then t stands for something, so if we put t for x in A, then that must be true too. The formal version of this is $\forall x \, A(x) \rightarrow A(t)$. But we must be careful how we use it. Consider several schematic examples:

$$\forall x \, \neg \, \forall y \, (\, x = y \,) \rightarrow \neg \, \forall y \, (\, 14 = y \,)$$
$$\forall x \, \neg \, \forall y \, (\, x = y \,) \rightarrow \neg \, \forall y \, (\, z + z = y \,)$$
$$\forall x \, \neg \, \forall y \, (\, x = y \,) \rightarrow \neg \, \forall y \, (\, x = y \,)$$

All these are true about the natural numbers. But we can get in trouble if we substitute a term that leads to a new variable being bound, as the following is false:

$$\forall x \, \neg \, \forall y \, (\, y = x \,) \rightarrow \neg \, \forall y \, (\, y = y \,)$$

Given a formula A and a variable x, we say that a term t *is free for an occurrence of* x *in* A (meaning free to be substituted) if that occurrence of x is free in A and does not lie within the scope of any quantifier $\forall y$ where y is a variable that appears in t . That is, by substituting t for that occurrence of x nothing new becomes bound. We write $A(t)$ for the result of substituting t for occurrences of x in A for which it is free, although when we use this notation we must specify which occurrences of which variable are being substituted for. If we restrict our substitution scheme to apply only when t replaces all occurrences of x for which it is free, then we have an axiom scheme every instance of which is true.

Note that to describe what axioms we want we used not a scheme in the sense of a wff with variables replaced by metavariables (see Chapter 19.E), but rather a scheme supplemented by a condition in English. From now on we'll also informally refer to a description of this sort as a *scheme*.

4. Distributing the universal quantifier

We need an axiom scheme to govern the relationship between the quantifiers and the propositional connectives.

If we have that $\forall x \, (A \rightarrow B)$ is true and A has no free occurrences of x, then

even though **A** is in the scope of ∀x, the quantifier really only affects the free
occurrences of **x** in **B**. So (**A**→∀x**B**) will be true (and conversely, although we
don't need to worry about that). Hence we're justified in taking as an axiom scheme:

$$\forall x(A \rightarrow B) \rightarrow (A \rightarrow \forall x B) \text{ where } A \text{ has no free occurrences of } x$$

In ∀x(y = z → y + x = z + x) the quantifier ∀x doesn't affect the antecedent at all,
so we can move it across the connective: y = z → ∀x (y + x = z + x).

5. Equality

We also need to assume some of the properties of equality for "=". The first is that
everything is equal to itself: ∀x (x = x). The other is that we can always substitute
equals for equals: (x = y) → [A(x) → A(y)]. Here, too, we have to restrict **y** to
being free for each occurrence of **x** which it replaces, although in this case we
needn't replace every such occurrence. From these alone we'll get all the usual
laws of equality (Theorem 2).

6. More principles?

There are many more principles of reasoning we could take as basic, but we're going
to stop here. With these we can feel confident that we've assumed only legitimate
principles of reasoning. And we'll see in the following chapters that these are
enough to derive the formal theorems we need. But they are also enough in a
stronger sense: any other principle of reasoning which can be formalized in our
language that would hold good for all possible interpretations of our symbols can be
derived from the ones we've chosen so far. This can actually be proved (non-
constructively) and is called *the completeness theorem for first-order logic* ("first-
order" refers to the fact that we only allow quantification over elements and not sets
of elements), although we won't need it here. See, for example, Epstein, 2006.

C. The Axiom System *Q*

1. The axioms

The axioms and rules given so far comprise the *logical* part of our axiom system.
To these we need to add axioms specific to arithmetic. But which ones? There are
many wffs true about the natural numbers, but which are really basic? We said that
one of our goals was to be able to represent the recursive functions in our system
starting with the functions we already have, namely, successor, addition, and
multiplication. So we need to assume enough about the symbols ´, + , and · to
ensure that goal. What we'll assume are the inductive definitions of addition in
terms of successor, and of multiplication in terms of addition and successor, and
that the successor symbol defines a 1-1 function whose range is everything but zero.
Here is the complete formal system.

Logical axioms and rules

Every wff in the formal language of arithmetic that is an instance of an axiom scheme of *classical propositional logic* (Chapter 19.E) is an axiom. As well, every instance of the following two schema is an axiom:

L1. *Substitution* $\forall x A(x) \rightarrow A(t)$
where $A(t)$ arises by substituting t for x in A, when t is free for every occurrence of x in A

L2. $\forall$- *distribution* $\forall x (A \rightarrow B) \rightarrow (A \rightarrow \forall x B)$
if A contains no free occurrences of x

Every instance of the equality schema is an axiom:

E1. $(x = x)$

E2. $(x = y) \rightarrow [A(x) \rightarrow A(y)]$
where $A(y)$ arises from $A(x)$ by replacing some, but not necessarily all, free occurrences of x by y, and y is free for the occurrences of x which it replaces.

The rules of proof are:

Modus ponens: From A and $A \rightarrow B$ conclude B.
Generalization: From A conclude $\forall x A$.

To the logical axioms and rules we add the following seven arithmetic axioms.

System Q

Q1. $(x_1' = x_2') \rightarrow x_1 = x_2$

Q2. $0 \neq x_1'$

Q3. $(x_1 \neq 0) \rightarrow \exists x_2 (x_1 = x_2')$

Q4. $x_1 + 0 = x_1$

Q5. $x_1 + (x_2)' = (x_1 + x_2)'$

Q6. $x_1 \cdot 0 = 0$

Q7. $x_1 \cdot (x_2)' = (x_1 \cdot x_2) + x_1$

Note that $Q1-Q7$ are (abbreviations of) wffs, not schema.

We define a *proof of* B to be a sequence $B_1, \ldots, B_n = B$, where each B_i is either an axiom, or is derived from two earlier wffs in the sequence, B_j and B_k with $j, k < i$ by the rule of modus ponens, that is, B_k is $B_j \rightarrow B_i$, or is derived by the rule of generalization, that is, B_i is $\forall x B_j$ for some $j < i$.

A wff A is a *theorem* of this formal system if there is a proof of A. In that case we write $\vdash_Q A$. The subscript is to remind us that the theorem depends on the arithmetic axioms we assumed. Whenever it is clear that we are referring to system Q, we will delete the subscript.

2. On consistency and truth

Our discussion in Section B was intended to convince you that the axioms of our formal system are true in the standard interpretation and that if the hypotheses of a rule are true, then so is its conclusion. Thus, all the theorems of Q are, we believe, true of the natural numbers.

That belief forms part of the motivation for our system, but it is not part of the system itself. Nor is it an assumption we need or want to make about the system. Except in one place (Theorem 3), we will use the notion of truth only informally until Chapter 23.C. As Hilbert enjoined us, we shall withhold judgment about questions of truth and meaning while we study the syntactic properties of the formal system.

What we shall assume instead, and state quite explicitly when we use it, is that Q has the syntactic property of being *consistent*: there is no wff A such that both $\vdash_Q A$ and $\vdash_Q \neg A$.

D. ∃-Introduction and Properties of = : Some Proofs in Q

To show you what proofs look like in this formal system, let's prove some theorems. The first is the formalization of a fundamental fact about existence.

THEOREM 1 (∃-Introduction)
   a. If **t** is free for **x** in A(x), then $\vdash A(t) \rightarrow \exists x A(x)$.
   b. If $\vdash A \rightarrow B$, then $\vdash (\exists x A) \rightarrow B$ whenever x is not free in **B** .

Proof: In order to prove that every instance of part (a) is a theorem we need to give a scheme of proofs. For example, in our proof take any wff with a free variable x_i in place of **x** as **A**, and any term **t** free for x_i in **A** and you will have an actual proof in Q.

a. 1. $(\forall x (\neg A(x)) \rightarrow \neg A(t))$ axiom L1
   2. $(\forall x (\neg A(x)) \rightarrow \neg A(t)) \rightarrow (A(t) \rightarrow \neg \forall x (\neg A(x)))$
          an instance of the valid propositional wff $(A \rightarrow \neg B) \rightarrow (B \rightarrow \neg A)$
          and hence a theorem, so insert its proof here
   3. $A(t) \rightarrow \neg \forall x (\neg A(x))$ modus ponens on (1) and (2).
And (3), in its abbreviated form, is what we set out to prove.

b. 1. $(A \rightarrow B) \rightarrow (\neg B \rightarrow \neg A)$ an instance of a valid propositional wff and hence a
          theorem, so insert its proof here
   2. $(A \rightarrow B)$ by hypothesis this is a theorem, so insert its proof
   3. $(\neg B \rightarrow \neg A)$ modus ponens on (1) and (2)
   4. $\forall x (\neg B \rightarrow \neg A)$ generalization
   5. $\forall x (\neg B \rightarrow \neg A) \rightarrow (\neg B \rightarrow \forall x \neg A)$ axiom L2
   6. $(\neg B \rightarrow \forall x \neg A)$ modus ponens on (4) and (5)
   7. $(\neg B \rightarrow \forall x \neg A) \rightarrow (\neg \forall x \neg A \rightarrow B)$ as for (1)
   8. $\exists x A \rightarrow B$ modus ponens on (6) and (7), and the definition of ∃ .   ∎

These proofs are relatively easy because of the completeness theorem for propositional logic: Any wff which is valid due to its propositional form is also a theorem, so we can insert its proof. We know it has a proof (Appendix 2 of Chapter 19), though we won't bother to produce it here, since all we want to do is convince ourselves that certain wffs have proofs, and for that it's enough to know that we can insert the proof if we wanted. From now on when we say *by propositional logic* (or *propositional calculus, PC*) we mean that the wff has a valid propositional form and that we can insert a proof of it.

Next, we'll show that the formalizations of the laws of reflexivity, symmetry, and transitivity of equality are provable.

THEOREM 2 (Properties of =)

a. For any term t , $\vdash t = t$.

b. $\vdash (t = u) \rightarrow [A(t) \rightarrow A(u)]$
for any wff $A(z)$ where both t and u are free for z in A, and $A(t)$ arises from $A(z)$ by replacing all occurrences of z by t, and $A(u)$ by replacing all occurrences of z by u .

c. $\vdash t = u \rightarrow u = t$.

d. $\vdash t = u \rightarrow (u = v \rightarrow t = v)$.

Proof: Again, we present schema of proofs.

a. 1. $\forall x_1 (x_1 = x_1)$ axiom E1
    2. $\forall x_1 (x_1 = x_1) \rightarrow t = t$ axiom L1
    3. $t = t$ modus ponens on (1) and (2).

b. Let x, y be distinct variables not appearing in $A(z)$ which are free for z in $A(z)$. Note then that y, t, u are all free for x in $A(x)$.
    1. $x = y \rightarrow (A(x) \rightarrow A(y))$ axiom E2
    2. $\forall x (x = y \rightarrow (A(x) \rightarrow A(y)))$ rule of generalization
    3. $t = y \rightarrow (A(t) \rightarrow A(y))$ by axiom L1 and modus ponens
    4. $\forall y (t = y \rightarrow (A(t) \rightarrow A(y)))$ rule of generalization
    5. $t = u \rightarrow (A(t) \rightarrow A(u))$ by axiom L1 and modus ponens, since u is free for y .

c. 1. $t = u \rightarrow (t = t \rightarrow u = t)$ by part (b)
    2. $t = t$ by part (a)
    3. $(t = t) \rightarrow ([t = u \rightarrow (t = t \rightarrow u = t)] \rightarrow (t = u \rightarrow u = t))$
       by propositional logic, since this is an instance of
       $B \rightarrow [(A \rightarrow (B \rightarrow C)) \rightarrow (A \rightarrow C)]$ which is valid
    4. $[t = u \rightarrow (t = t \rightarrow u = t)] \rightarrow (t = u \rightarrow u = t)$
       modus ponens on (2) and (3)
    5. $t = u \rightarrow u = t$ modus ponens on (1) and (4).

d. 1. $(u = t) \rightarrow (u = v \rightarrow t = v)$ by part (b)
    2. $t = u \rightarrow u = t$ by part (c)

3. $(t = u \rightarrow u = t) \rightarrow$
    $[((u = t) \rightarrow (u = v \rightarrow t = v)) \rightarrow ((t = u) \rightarrow (u = v \rightarrow t = v))]$
    by *PC* using $(A \rightarrow B) \rightarrow [(B \rightarrow C) \rightarrow (A \rightarrow C)]$
4. $((u = t) \rightarrow (u = v \rightarrow t = v)) \rightarrow [(t = u) \rightarrow (u = v \rightarrow t = v)]$
    modus ponens on (2) and (3)
5. $(t = u) \rightarrow (u = v \rightarrow t = v)$ modus ponens on (1) and (4).   ∎

The same method of substitution we use in the proof of Theorem 2 can be used to give formal proofs in Q of versions of $Q1$–$Q7$ with any variables x, y in place of x_1 and x_2; for example, $\vdash_Q (x' = y') \rightarrow x = y$ (Exercise 8).

E. Weakness of System Q

It may seem that if we've assumed enough about the natural numbers to be able to represent every recursive function in Q, then the system must be strong enough to prove almost all the basic facts of arithmetic. But as simple a wff as $x \neq x'$ cannot be proved in Q. How can we show that? We can show a wff is a theorem by exhibiting a proof. But how can we show that there is *no* proof?

Back to plane geometry. Do you remember hearing that there's no proof of Euclid's parallel postulate from his other axioms? Do you know how that was demonstrated? Beltrami, as well as Klein and Poincaré, exhibited a model of the other axioms in which the parallel postulate failed. Why was that enough?

Suppose we can exhibit something that satisfies all the axioms of the system Q, that is, a model of Q. The rules of proof never lead us from wffs that are true about something to ones that are false, so every theorem of Q must also be true in that model. Thus all we have to do is show something that satisfies all the axioms of Q and yet $x \neq x'$ is false in it. Then $x \neq x'$ cannot be a theorem of Q.

To present such a model we need two objects that are not natural numbers. Any two will do; for example, this book and your left ear, or a right angle and an obtuse angle. You choose, and label them α and β. The model then consists of the natural numbers supplemented by α and β with the following tables interpreting $'$, $+$, and $\cdot$:

$+$	n	α	β
m	$m + n$	β	α
α	α	β	α
β	β	β	α

x	successor of x
n	$n + 1$
α	α
β	β

$\cdot$	0	$n \neq 0$	α	β
0	0	0	α	β
$m \neq 0$	0	$m \cdot n$	α	β
α	0	β	β	β
β	0	α	α	α

To show that this is really a model of Q we have to assume that the axioms and hence theorems of Q are true of the natural numbers. Then it's easy to verify that they are also true when we have α and β. But the successor of α is α, and hence $x \neq x'$ cannot be a theorem of Q.

Here is a further list of simple wffs all of which are true of the natural numbers but cannot be proved in Q, as you can verify using the same model (Exercise 10).

THEOREM 3   If the theorems of Q are true of the natural numbers, then the following are not theorems of Q, where x, y, z are distinct variables (parentheses deleted for legibility):

a. $x \neq x'$

b. $x + (y + z) = (x + y) + z$

c. $x + y = y + x$

d. $0 + x = x$

e. $\neg [\exists x (x' + y = z) \land \exists x (x' + z = y)]$

f. $x \cdot (y \cdot z) = (x \cdot y) \cdot z$

g. $x \cdot y = y \cdot x$

h. $x \cdot (y + z) = (x \cdot y) + (x \cdot z)$

Remember, by our conventions these wffs are equivalent to their universally quantified forms.

F. Proving As a Computable Procedure

We are eventually going to show that there is no decision procedure for the set of theorems of Q. So it is even more important here than for propositional logic to establish that we can computably enumerate the theorems. The method is virtually the same as for propositional logic except that the structure of the wffs we are examining is more complicated. We are going to sketch the proof and leave the details to you, trusting that you understood how to do it for propositional logic.

To begin, we ask you to give a Gödel numbering of the terms and wffs (Exercise 11), since it will be easier for you to use your own numbering than one we cook up. We will write $[\![t]\!]$ for the Gödel number of a term t and $[\![A]\!]$ for the Gödel number of a wff A.

1. First we need to show that the set of Gödel numbers of terms and the set of Gödel numbers of wffs are recursive.

2. Next we show that we can decide whether a particular occurrence of a variable in a wff is free. That is, the following set is recursive:

$\{\langle n, m, p \rangle$: for some A and x $(n = [\![A]\!], m = [\![x]\!]$, and the pth
occurrence of x in A reading from the left is free in $A)\}$

Then we need to show that we can decide whether a term **t** is free for a particular occurrence of **x** in a wff. That is, the following set is recursive

$$\{ \langle n, m, q, p \rangle : \text{for some } \mathbf{A}, \mathbf{x} \text{ and } \mathbf{t} \ (n = [\![\mathbf{A}]\!], \ m = [\![\mathbf{x}]\!], \ q = [\![\mathbf{t}]\!] \text{ and}$$
$$\mathbf{t} \text{ is free for the } p^{\text{th}} \text{ occurrence of } \mathbf{x} \text{ in } \mathbf{A} \text{ reading from the left} \}$$

3. We can decide if a wff is an instance of one of the axiom schema. It shouldn't be too hard to modify your proof of this for propositional logic to take care of the propositional schema. And it's pretty straightforward to recognize the arithmetic axioms and the first equality axiom in terms of your Gödel numbering. Virtually all of the hard work in the whole decision procedure is to establish that we can recursively recognize whether a number is a Gödel number of an instance of one of the logical axioms or the second equality axiom, and it's for this that we have to go to all the trouble to do (2).

4. We establish that we can decide whether a wff is a consequence of two others by the rule of modus ponens or whether it is a consequence of another by the rule of generalization. That is, these sets are recursive:

$$\{ \langle n, m, p \rangle : \text{there are } \mathbf{A} \text{ and } \mathbf{B} \ (n = [\![\mathbf{A}]\!], \ m = [\![\mathbf{A} \rightarrow \mathbf{B}]\!], \text{ and } p = [\![\mathbf{B}]\!]) \}$$

and

$$\{ \langle n, m \rangle : \text{for some } \mathbf{A} \text{ and } i, \ (n = [\![\mathbf{A}]\!] \text{ and } m = [\![\forall x_i \mathbf{A}]\!]) \}$$

5. Then we can determine whether a number codes a sequence of wffs which is a proof sequence. That is, the following is recursive:

$$prf(n) = \begin{cases} 1 & \text{if some } n = \langle (n)_0, \ldots, (n)_{lh(n)-1} \rangle \text{ and for } i < lh(n) \\ & (n)_i \text{ is the Gödel number of a wff, say } \mathbf{B}_i, \text{ and} \\ & \text{either } \mathbf{B}_i \text{ is an axiom, or for some } j, k < i \ \mathbf{B}_j \text{ is } \mathbf{B}_k \rightarrow \mathbf{B}_i, \\ & \text{or some } j < i \ \mathbf{B}_i \text{ is } \forall x \, \mathbf{B}_j, \text{ for some variable } \mathbf{x} \\ 0 & \text{otherwise} \end{cases}$$

6. It is then easy to list the theorems: just search for the next n which codes a proof sequence, that is, $prf(n) = 1$, and output the number of the wff which it proves, namely $(n)_{lh(n)-1}$.

What we have shown is that we can recognize if a number codes a proof of a particular wff. That is, the predicate

$$Prf(x, y) \equiv_{\text{Def}} x \text{ codes a proof sequence which proves } \mathbf{A} \text{ where } y = [\![\mathbf{A}]\!]$$

is recursive, since $Prf(x, y)$ is $[prf(x) = 1 \text{ and } y = (x)_{lh(x)-1}]$.

Exercises

1. Which of the following are terms according to our formal definition? For those which are not, explain why.

a. $((x_1 \cdot x_2) + x_4)$

b. $(x_{47} + y)$

c. $(0'') \cdot (x_4 + x_1)$

d. $(t + u)$

e. $((((x_3)' \cdot x_1) + x_4) + ((0)')')$

f. $((x_1)^2 \cdot x_2)$

g. $((x_1 + x_2) = x_3)$

2. Give at least five different expressions in English that could be formalized by $\forall x$ and five that could be formalized by $\exists x$.

3. Write a wff or a scheme of wffs which you believe is an accurate formalization of each of the following informal statements about the natural numbers. You may use our conventions to make the wff more legible.

   a. $1 + 3 = 5$
   b. $8 = 2 \cdot 5$
   c. Every number is the sum of two numbers.
   d. Every even number is the sum of two odd numbers.
   e. 0 is the smallest natural number.
   f. For any n and m, either $n = m$, or $n > m$, or $m > n$.
   g. If an assertion is true for 0 and if whenever it's true for n it's also true for $n + 1$, then it's true for all n.
   h. If a set of natural numbers has no even numbers in it, then it must contain only odd numbers.

4. For each of the following wffs:
   —Identify the scope of each quantifier.
   —State which occurrences of which variables are free in it.
   —Try to reformulate it in English.
   —State whether it's true or false in the standard interpretation.

   a. $\forall x_1 (x_1 \neq 0 \rightarrow \exists x_3 (x_1 \cdot x_3 = 0'))$

   b. $\forall x_1 (\exists x_2 (x_1 \cdot x_2 = 0) \rightarrow \forall x_2 (x_1 \cdot x_2 = 0))$

   c. $\forall x_1 \exists x_2 (x_1 \cdot x_2 = 0) \rightarrow \forall x_2 (x_1 \cdot x_2 = 0)$

   d. $\exists x_1 (x_1' + x_2 = x_3) \rightarrow x_2 \neq x_3$

   e. $\forall x_1 \exists x_2 (x_1 + x_2 = x_1') \rightarrow \forall x_1 \exists x_2 [x_1 + (x_1 + 0') = x_1']$

   f. $\forall x_1 [(x_2 = x_3) \rightarrow (x_2 + x_1 = x_3 + x_1)]$

   g. $\forall x_1 (x_1 + 0 = x_1) \rightarrow 2 + 0 = 2$

   h. $\exists x_3 (2 + 2 = 4)$

   i. $[\forall x_1 \exists x_2 (x_1 = x_2') \rightarrow x_1 \neq 0] \rightarrow [\exists x_2 (x_1 = x_2') \rightarrow \forall x_1 (x_1 \neq 0)]$

5. a. In the axiom scheme of substitution why don't we allow t to replace some but not all occurrences of x for which it is free?

   b. Give an inductive definition of t *is free for x in* A.
   (*Hint:* Induct on the structure of A using the inductive definition of wff: if A is

atomic, then **t** is free for **x** in **A**; if **A** is of the form ¬**B**, then **t** is free for **x** in **A** iff ... .)

6. Give an informal justification of the schema involving the existential quantifier which were proved in Q in Theorem 1 .

7. Give an informal justification that the following are true of the natural numbers and then prove each using only the logical axioms and rules.
   a. ¬∃x**A**(x) ↔ ∀x¬**A**(x)
   † b. ∃x¬**A**(x) ↔ ¬∀x**A**(x)
   c. ¬∃x¬**A**(x) ↔ ∀x**A**(x)   (*Hint:* Use part (b).)

8. Show that every instance of the schematic forms of axioms Q1–Q7 is a theorem of Q . For example, show that for any **x**, **y** we have ⊢$_Q$(**x**´ = **y**´) → **x** = **y** . (*Hint*: cf. the proof of Theorem 2.)

† 9. Show that the following wffs are theorems of Q by exhibiting proofs of them:
   a. *3 = 4 → 2 = 3*
   b. *0 ≠ 4*
   c. *4 ≠ 0*
   d. *3 ≠ 4*
   e. *2 + 3 = 5*
   f. $∃x_1 (2 + (x_1)´ = 5)$

10. Complete the proof of Theorem 3 by verifying that all the axioms of Q are true for the system described there and that each of (a)–(h) fails to be true. Intuitively, what does (e) say?

†11. a. Give a Gödel numbering of the terms of the formal language.
   b. Give a Gödel numbering of the atomic wffs of the language.
   c. Give a Gödel numbering of the wffs of the language.
   d. Using your numberings, show that the set of terms, the set of atomic wffs, and the set of wffs of the language are recursively decidable.

12. Fill in the details of the proof in Section F that we can recursively enumerate the set of Gödel numbers of theorems of Q , and that $Prf(x,y)$ is recursive. (You will need to use Exercises 5 and 11.)

Further Reading

For a more thorough treatment of first-order logic as a formalization of mathematics see Epstein's *Classical Mathematical Logic*. Our logical axioms and rules are from Church, who derived them from Russell (see Church's *Introduction to Mathematical Logic*, p. 289). The arithmetic axioms are due to Raphael Robinson, 1950, though the best place to read about their history is in *Undecidable Theories* by Tarski, Mostowski, and Robinson, p.39. Mendelson in his *Introduction to Mathematical Logic* gives detailed solutions to Exercises 11 and 12 for essentially the same system as ours.

22 Functions Representable in Formal Arithmetic

In this chapter we're going to show that we can represent the recursive functions as those whose values we can compute using the proof machinery of Q. For later chapters all that is needed here are the definitions on p. 192 and the definitions and theorems starting with Corollary 21 (p. 200).

A. Dispensing with Primitive Recursion

It will be straightforward to show that the initial recursive functions (zero, successor, and the projections) can be represented in Q. Showing closure under the operations of composition and least search operator is fairly straightforward, though the latter is tedious since it requires several long proofs in Q. It is much more difficult to show that the representable functions are closed under primitive recursion.

One way we could show closure under recursion would be to add to Q infinitely many axioms corresponding to the recursion equations for all the primitive recursive functions. But that is inelegant and defeats the spirit of our program, which was to use a simple system of arithmetic whose consistency we could believe in.

More economical would be to use the full power of the least search operator, for by adding coding and uncoding functions we can dispense with primitive recursion as an initial operation (see Exercise 16.13). For example, we can define

$$x^y = (\mu z [(z)_0 = 1 \ \wedge \ (z)_{i+1} = x \cdot (z)_i \ \wedge \ lh(z) = y + 1])_y$$

But we used exponentiation to define our coding functions.

Actually all we needed to define the coding functions was exponentiation, the initial recursive functions, composition, and the least search operator (cf. Theorem 12.1). So we could get by with adding just one new function symbol, **exp**, to our language and axioms corresponding to the recursion equations for exponentiation in terms of multiplication. Still it would be a pity to do so, since exponentiation doesn't have the primacy and intuitive clarity of addition and multiplication. Fortunately, Gödel, 1931, showed how we may define a different coding function that doesn't depend on exponentiation. Then we can prove that the partial recursive functions can be characterized as

> C = the smallest class of functions containing zero, successor, the projections, addition, multiplication, and the characteristic function for equality, and closed under composition and the μ-operator

To prove this requires a diversion into number theory.

1. A digression on number theory

To define a function β in C that can do the coding we will need to use the Chinese Remainder Theorem.

Recall that $y \equiv z \, (mod \, x)$ (read "y is congruent to z modulo x") means that the integer difference between y and z is divisible by x. That is,

$$y \equiv z \, (mod \, x) \quad \text{iff} \quad x \mid (y - z)$$

Thus, if $y \equiv z \, (mod \, x)$ then y and z have the same remainder upon division by x. If we let

$$rem(x, y) = \text{the remainder upon division of } y \text{ by } x$$

then we have that for any y and x,

$$y \equiv rem(x, y) \, (mod \, x)$$

The *Chinese Remainder Theorem* (Exercise 1) gives sufficient conditions for finding a number z that simultaneously satisfies a set of congruence equations:

$$z \equiv y_1 \, (mod \, x_1), \quad z \equiv y_2 \, (mod \, x_2), \quad \ldots, \quad z \equiv y_n \, (mod \, x_n)$$

Namely, if the x_i's are relatively prime in pairs (that is, no two have a common factor except for 1), then there is such a z with $z \leq x_1 \cdot x_2 \cdot \cdots \cdot x_n$.

THEOREM 1 (Gödel's β-function)   There is a function $\beta \in C$ such that for any finite sequence of natural numbers $a_0, \ldots, a_n$ there exists a natural number d such that for every $i \leq n$, $\beta(d, i) = a_i$.

Proof: To construct β we need to show that a number of other functions are in C. First, we note that C is closed under the logical operations and the operations of bounded existential quantification and bounded universal quantification (which we leave as an exercise for you, cf. Chapter 11.D.3–5 and Exercises 11.14 and 11.15).

Thus the pairing function J and unpairing functions K and L of Chapter 11.E.6 are in C:

$$J(x, y) = \tfrac{1}{2} \, [(x + y)(x + y + 1)] + x$$
$$K(z) = min \; x \leq z \; [\exists y \leq z \, (J(x, y) = z)]$$
$$L(z) = min \; y \leq z \; [\exists x \leq z \, (J(x, y) = z)]$$

We'll let you define the following functions and (characteristic functions of) predicates in such a way that you can see they are in C (Exercise 2):

$m < n$ m divides n p is a prime

$m \div n$ n is a power of the prime p

Then the following is in C:

$$rem(x, y) = \mu z\,(\exists k \le y\,[(k \cdot x) + z = y])$$

Now we define the three-variable version of Gödel's β-function:

$$\beta^*(x, y, z) = rem(1 + (z + 1)\,y,\, x)$$

which we see is in C. All that's left to show is that for any sequence of natural numbers $a_0, \ldots, a_n$ there exist natural numbers b and c such that for every $i \le n$, $\beta^*(b, c, i) = a_i$. Then the function we want for our theorem is $\beta(d, i) = \beta^*(K(d), L(d), i)$.

Let $j = max(n, a_0, \ldots, a_n)$ and $c = j!$.

By Exercise 1.c below, the numbers $u_i = 1 + (i + 1)c$ for $0 \le i \le n$ have no factors in common except for 1. So by the Chinese Remainder Theorem the equations $z \equiv a_i \pmod{u_i}$ have a simultaneous solution $b < u_0 \cdot u_1 \cdot \;\cdots\; \cdot u_n$. But for $i \le n$, $a_i \le j \le j!$ and $j! = c < 1 + (i + 1)c = u_i$; that is, $a_i < u_i$ for all $i < n$. And hence the a_i's are the remainders upon division of b by the u_i's. That is, b is a number such that $a_i \equiv rem(u_i, b)$ for $i \le n$. So for that b, $\beta^*(b, c, i) = rem(1 + (i + 1)c, b)$.   ■

2. A characterization of the partial recursive functions

THEOREM 2   The partial recursive functions are the smallest class of functions containing zero, successor, the projections, addition, multiplication, and the characteristic function for equality and closed under composition and the μ-operator.

Proof: We have to show that this class is closed under primitive recursion. That's an easy variation on the example of exponentiation above. We'll show it for functions of one variable and leave the general case to you (Exercise 3).
Suppose

$f(0) = a$

$f(x + 1) = h(f(x), x)$ where h is in this class

Define the predicate

$$S(x, b) \equiv_{Def} \beta(b, 0) = a \;\wedge\; \forall i \le x\,[\beta(b, i + 1) = h(\beta(b, i), i)]$$

By Theorem 1 (and its proof) S is in this class of functions and for every x there is some b such that $S(x, b)$. Then for all x, $f(x) = \beta(\mu b\,[S(x, b)], x)$ and so f is in C.   ■

B. The Recursive Functions Are Representable in Q

Let f be a *total* function of k variables and $A(x_1, \dots, x_k, x)$ a wff with $k + 1$ free variables, where x is distinct from $x_1, \dots, x_k$. We say that f *is represented by* A *in* Q if:

$f(m_1, \dots, m_k) = n$ implies

1. $\vdash_Q A(m_1, \dots, m_k, n)$
2. $\vdash_Q A(m_1, \dots, m_k, x) \to x = n$

From this definition we will derive (Lemma 4) the further condition:

3. If $n \neq p$, then $\vdash_Q \neg A(m_1, \dots, m_k, p)$.

We say that f is *representable in* Q if there is some wff that represents it.

For example, we will show below that addition is represented in Q by the wff $x_1 + x_2 = x_3$. So we will have, for example, $\vdash_Q 0'' + 0''' = 0'''''$, $\vdash_Q 0'' + 0''' = x_3 \to x_3 = 0'''''$, and $\vdash_Q 0'' + 0''' \neq 0''$.

Given a function f that is represented by A in Q, we can calculate its values if Q *is consistent*: to calculate $f(m_1, \dots, m_k)$, start proving theorems until we get $\vdash_Q A(m_1, \dots, m_k, n)$ for some n. By condition (1) we will find one, and by condition (3) there will be only one such, for which $n = f(m_1, \dots, m_k)$.

This approach to calculation was originally taken as an explanation of what it means for a function to be computable (see Herbrand, 1931, Church, 1936, pp. 101–102; and the comments of Gödel below on p. 217). In this section we will show that every recursive function is representable.

To do this we are going to have to show that many specific wffs are theorems of Q. This is analogous to providing the Turing machines that compute the zero function, the successor function, and so on in the proof that every recursive function is Turing machine computable. If you wish, you can skip the details and go directly to the statement of Theorem 19.

Note: Other texts may define representability differently or use related notions such as "expressible", "strongly representable", or "weakly representable".

Our first goal is to establish condition (3), for which we prove the following lemma.

Lemma 3   For all natural numbers n and m, if $n \neq m$, then $\vdash_Q n \neq m$.

Proof: Here n and m are numbers, and n and m are numerals. So we are obliged to prove that infinitely many wffs are theorems of Q. To do this we induct on the number m, first assuming $n < m$.

Our basis is $n = 0$, $m = 1$: axiom Q2 is $\vdash_Q 0 \neq x_1'$, so by generalization, $\vdash_Q \forall x_1 (0 \neq x_1')$, and by L1 and modus ponens we have $\vdash_Q 0 \neq 0'$.

Suppose the lemma is true for m and all $n < m$. We will show it for all $n < m + 1$.

If $n = 0$, proceed as above to show that $\vdash_Q 0 \neq m'$. Suppose now it's true for n and $n + 1 < m + 1$. Hence $n < m$, and so by induction, $\vdash_Q n \neq m$. Axiom Q1 is $\vdash_Q x_1' = x_2' \rightarrow x_1 = x_2$, so by generalization and L1 we can substitute: $\vdash_Q n' = m' \rightarrow n = m$. Since $(A \rightarrow B) \rightarrow (\neg B \rightarrow \neg A)$ is a valid propositional scheme, $\vdash_Q (n' = m' \rightarrow n = m) \rightarrow (n \neq m \rightarrow n' \neq m')$, and hence by modus ponens twice, $\vdash_Q n' \neq m'$.

If $m < n$, then we have $\vdash_Q m \neq n$. By Theorem 21.2, $\vdash_Q x = y \rightarrow y = x$, so by generalization and L1 we have $\vdash_Q n = m \rightarrow m = n$, and hence as above, $\vdash_Q n \neq m$.   ∎

Several things are important to note in this proof:

1. The letters m, n, r, s, etc. stand for numbers; hence m, n, r, s stand for unary numerals, whereas x and y stand for variables in the formal language.

2. We used induction to prove something about Q, but not as a rule of proof within Q itself.

3. We used generalization to get the universally quantified version of Q1 which we then used to obtain the formula with n substituted for x_1. We do this often and from now on we'll simply say "by substitution" or "by L1". Also, we will use axioms Q1–Q7 with any variables x and y in place of x_1 and x_2, as justified by our remarks on p. 184 above and Exercise 21.8.

4. We specifically quoted the valid propositional scheme that justified one of our steps. In the future we'll just say "by *PC*".

5. We'll also leave out the subscript Q inside the proofs, though we'll keep it in the statements of the lemmas and theorems. We will refer to the equality axioms and Theorem 21.2 collectively as "properties of $=$", and Theorem 21.1 as "∃-introduction".

Lemma 4   Suppose f is represented by A in Q. If $f(m_1, \ldots, m_k) = n$ and $n \neq p$, then $\vdash_Q \neg A(m_1, \ldots, m_k, p)$.

Proof: If $p \neq n$, then by the previous lemma $\vdash p \neq n$. Since A represents f, $\vdash A(m_1, \ldots, m_k, x) \rightarrow x = n$. So $\vdash A(m_1, \ldots, m_k, p) \rightarrow p = n$, and so by *PC*, $\vdash \neg A(m_1, \ldots, m_k, p)$.   ∎

The rest of this section will be devoted to showing that the recursive functions are representable in Q. To do this we will first show that each of the functions zero, successor, the projections, addition, multiplication, and the characteristic function for equality are representable, and then show that the representable functions are closed under the operations of composition and the μ-operator.

Lemma 5   The zero function is represented in Q by $(x_1 = x_1) \wedge (x_2 = 0)$.

Proof: By E1, $\vdash 0 = 0$ and for every n, $\vdash n = n$. Therefore by *PC*,

$\vdash (n = n) \wedge (0 = 0)$. Calling the formula in the lemma $\mathbf{A}$, we have just proved $\vdash \mathbf{A}(n, 0)$ for every n, which shows that $\mathbf{A}$ satisfies condition (1).

To show that it satisfies condition (2) is easy, for by PC:
$$\vdash [(n = n) \wedge (x_2 = 0)] \rightarrow x_2 = 0.$$ ∎

Lemma 6   The successor function is represented in Q by $x_1{'} = x_2$.

Proof: For condition (1), suppose the successor of m is n. Then $m{'}$ and n are identical terms, so by the properties of $=$, $\vdash m{'} = n$. Condition (2) is exactly Theorem 21.2.c. ∎

Lemma 7   The projection function P_k^i where $1 \leq i \leq k$ is represented in Q by the formula $(x_1 = x_1) \wedge \cdots \wedge (x_k = x_k) \wedge (x_{k+1} = x_i)$.

We leave the proof of this as Exercise 4.

Lemma 8   **a.**   For all natural numbers n and m, if $n + m = k$,
then $\vdash_Q n + m = k$.

   b. $\vdash_Q n + 1 = n{'}$.

Proof: a. We prove this by induction on m. If $m = 0$, we have $\vdash n + 0 = n$ by axioms $Q4$ and $L1$. So suppose it's true for r and $m = r + 1$. Then for some s, $k = s + 1$, and $n + r = s$, so by induction, $\vdash n + r = s$. By $Q5$, we have $\vdash (n + r){'} = n + r{'}$, so by the properties of $=$, $\vdash s{'} = n + r{'}$, and by the same theorem, $\vdash n + r{'} = s{'}$; that is, $\vdash n + m = k$.

   b. This part is easy, and we leave it to you. ∎

Lemma 9   Addition is represented in Q by $x_1 + x_2 = x_3$.

Proof: By Lemma 8 condition (1) is satisfied. To show condition (2), suppose $n + m = k$. Then by the properties of $=$, since $\vdash n + m = k$ we have $\vdash (n + m = x_3) \rightarrow k = x_3$, and using the symmetry of $=$ and PC, we have $\vdash (n + m = x_3) \rightarrow x_3 = k$. ∎

Lemma 10   Multiplication is represented in Q by $x_1 \cdot x_2 = x_3$.

The proof is similar to the one for addition, and we leave it as Exercise 6.

Lemma 11   The characteristic function for equality is represented in Q by
$[(x_1 = x_2) \wedge (x_3 = 1)] \vee [(x_1 \neq x_2) \wedge (x_3 = 0)]$.

Proof: Call the characteristic function E. If $n \neq m$, then $E(n, m) = 0$ and $\vdash n \neq m$ by Lemma 4. By the properties of $=$ we have $\vdash 0 = 0$, so by PC, $\vdash [(n = m) \wedge (0 = 1)] \vee [(n \neq m) \wedge (0 = 0)]$. We'll leave to you to show that condition (1) is satisfied for the case $E(n,n)$.

For condition (2), call the wff $\mathbf{A}$. Then since $\vdash n = n$, by PC we get $\vdash \mathbf{A}(n, n, x_3) \rightarrow x_3 = 1$, and similarly if $n \neq m$ we get $\vdash \mathbf{A}(n, m, x_3) \rightarrow x_3 = 0$. ∎

Lemma 12 If f and $g_1, \ldots, g_k$ are representable in Q and f is a function of k variables, and all the g_i are functions of the same number of variables, then the composition $f(g_1, \ldots, g_k)$ is representable in Q.

Proof: We will prove this for $f \circ g$, where f and g are functions of one variable, leaving the generalization to you.

Suppose f is represented by $\mathbf{B}(\mathbf{x_1}, \mathbf{w})$ and g by $\mathbf{A}(\mathbf{x_1}, \mathbf{y})$. Let $\mathbf{z}$ be a variable that does not appear in either $\mathbf{A}$ or $\mathbf{B}$. By repeated use of generalization and L1, you can show that conditions (1) and (2) for the representability of f hold also for $\mathbf{B}(\mathbf{z}, \mathbf{w})$, and conditions (1) and (2) for g hold for $\mathbf{A}(\mathbf{x_1}, \mathbf{z})$. We claim that $f \circ g$ is represented by $\exists \mathbf{z}\, (\mathbf{A}(\mathbf{x_1}, \mathbf{z}) \wedge \mathbf{B}(\mathbf{z}, \mathbf{w}))$, which we will call $\mathbf{C}(\mathbf{x_1}, \mathbf{w})$.

If $(f \circ g)(n) = m$, then for some a, $g(n) = a$ and $f(a) = m$. So $\vdash \mathbf{A}(n,a)$ and $\vdash \mathbf{B}(a,m)$. So by PC, $\vdash \mathbf{A}(n,a) \wedge \mathbf{B}(a,m)$, and hence by $\exists$-introduction, $\vdash \mathbf{C}(n,m)$.

For condition (2), we have $\vdash \mathbf{A}(n, \mathbf{z}) \rightarrow \mathbf{z} = a$, so by PC, $\vdash [\,\mathbf{A}(n, \mathbf{z}) \wedge \mathbf{B}(\mathbf{z}, \mathbf{w})\,] \rightarrow \mathbf{z} = a$. We also have by the properties of $=$, $\vdash \mathbf{z} = a \rightarrow (\mathbf{B}(\mathbf{z}, \mathbf{w}) \rightarrow \mathbf{B}(a, \mathbf{w}))$ and by the representability of g, $\vdash \mathbf{B}(a, \mathbf{w}) \rightarrow \mathbf{w} = m$. So by PC, we get $\vdash [\,\mathbf{A}(n, \mathbf{z}) \wedge \mathbf{B}(\mathbf{z}, \mathbf{w})\,] \rightarrow \mathbf{w} = m$. By $\exists$-introduction, we have $\vdash \mathbf{C}(n, \mathbf{w}) \rightarrow \mathbf{w} = m$. ∎

To show that the representable functions are closed under the μ-operator we need the following lemmas.

Lemma 13 For every variable $\mathbf{x}$ and numeral n, $\vdash_Q \mathbf{x}' + n = \mathbf{x} + n'$.

Proof: The proof is by induction on n.

If $n = 0$, then by $Q4$, $\vdash \mathbf{x} + 0 = \mathbf{x}$, and $\vdash \mathbf{x}' + 0 = \mathbf{x}'$. Therefore, by the properties of $=$, $\vdash \mathbf{x}' + 0 = (\mathbf{x} + 0)'$. By $Q5$, $\vdash \mathbf{x} + 0' = (\mathbf{x} + 0)'$, so by the properties of $=$ again, we have $\vdash \mathbf{x}' + 0 = \mathbf{x} + 0'$.

For our induction hypothesis we suppose that $n = m + 1$, and $\vdash \mathbf{x}' + m = \mathbf{x} + m'$. By $Q5$, $\vdash \mathbf{x}' + m' = (\mathbf{x}' + m)'$, and the properties of $=$, $\vdash \mathbf{x}' + m' = (\mathbf{x} + m')'$. Again by $Q5$, we get $\vdash (\mathbf{x} + m')' = \mathbf{x} + m''$, and by the properties of $=$, we have $\vdash \mathbf{x}' + m' = \mathbf{x} + m''$; that is, $\vdash \mathbf{x}' + n = \mathbf{x} + n'$. ∎

For any terms $\mathbf{t}$ and $\mathbf{u}$ which do not contain the variable $\mathbf{x_3}$, take $\mathbf{t} < \mathbf{u}$ to be an abbreviation of the formula $\exists \mathbf{x_3}\, (\mathbf{x_3}' + \mathbf{t} = \mathbf{u})$.

Lemma 14 If $n < m$, then $\vdash_Q n < m$.

Proof: If $n < m$, then for some k, $(k + 1) + n = m$. Hence by Lemma 9, $\vdash k' + n = m$. So by $\exists$-introduction, $\vdash \exists \mathbf{x_3}\, (\mathbf{x_3}' + n = m)$; that is, $\vdash n < m$. ∎

Lemma 15 For any variable $\mathbf{x}$ other than $\mathbf{x_3}$,

    a. $\vdash_Q \neg (\mathbf{x} < 0)$.

    b. If $n = p + 1$, then $\vdash_Q \mathbf{x} < n \rightarrow (\mathbf{x} = 0 \vee \cdots \vee \mathbf{x} = p)$.

Proof: a. We will sketch the proof. Let **y** be a variable other than **x** or x_3 .

1. $x = 0 \rightarrow (x_3{'} + x = 0 \rightarrow x_3{'} = 0)$ properties of =, *Q4*, and *PC*
2. $x_3{'} \neq 0$ *Q2* and properties of =
3. $x = 0 \rightarrow \neg (x_3{'} + x = 0)$ from (1) and (2) by *PC*
4. $x = y{'} \rightarrow (x_3{'} + x = 0 \rightarrow x_3{'} + y{'} = 0)$ properties of =
5. $x_3{'} + y{'} = (x_3{'} + y)^{'}$ *Q5*
6. $x = y{'} \rightarrow (x_3{'} + x = 0 \rightarrow (x_3{'} + y)^{'} = 0)$ properties of = using (4) and (5)
7. $(x_3{'} + y)^{'} \neq 0$ *Q2* and properties of =
8. $x = y{'} \rightarrow \neg (x_3{'} + x = 0)$ from (6) and (7) by *PC*
9. $(x = 0 \vee x = y{'}) \rightarrow \neg (x_3{'} + x = 0)$ from (3) and (8) by *PC*
10. $(x = 0 \vee x = y{'}) \rightarrow \forall x_3 \neg (x_3{'} + x = 0)$ L1 and L2
11. $\exists x_3 (x_3{'} + x = 0) \rightarrow \neg (x = 0 \vee x = y{'})$ *PC* and definition
12. $x < 0 \rightarrow (x \neq 0 \rightarrow x \neq y{'})$ definition and *PC*
13. $x < 0 \rightarrow [x \neq 0 \rightarrow \forall y (x \neq y{'})]$ L1 and L2
14. $x \neq 0 \rightarrow \neg \forall y (x \neq y{'})$ *Q3* and definition
15. $\neg (x < 0)$ from (13) and (14) by *PC*

b. We will prove this by induction on n. In what follows it will be easier to read **z** for x_3 ; for the basis $n = 1$ we wish to prove $\vdash \exists z (z{'} + x = 1) \rightarrow x = 0$.

1. $\exists y (x = y{'}) \vee x = 0$ *Q3* and *PC*, where **y** is any variable other than **x** or **z**
2. $x = y{'} \rightarrow (z{'} + x = 1 \rightarrow z{'} + y{'} = 1)$ properties of =
3. $z{'} + y{'} = (z{'} + y)^{'}$ by *Q5*
4. $x = y{'} \rightarrow (z{'} + x = 0{'} \rightarrow (z{'} + y)^{'} = 0{'})$ by (2) and (3) using the properties of =, and the definition of *1*
5. $(z{'} + y)^{'} = 0{'} \rightarrow (z{'} + y = 0)$ axiom *Q1*
6. $(x = y{'} \wedge z{'} + x = 0{'}) \rightarrow (z{'} + y = 0)$ from (4) and (5) by *PC*
7. $(x = y{'} \wedge z{'} + x = 0{'}) \rightarrow \exists z (z{'} + y = 0)$ ∃-introduction
8. $\neg \exists z (z{'} + y = 0)$ by part (a)
9. $\neg (x = y{'} \wedge z{'} + x = 0{'})$ from (7) and (8) by *PC*
10. $(z{'} + x = 0{'}) \rightarrow \neg (x = y{'})$ by *PC*
11. $\exists z (z{'} + x = 0{'}) \rightarrow \neg (x = y{'})$ ∃-introduction
12. $(x = y{'}) \rightarrow \neg \exists z (z{'} + x = 0{'})$ by *PC*
13. $\exists y (x = y{'}) \rightarrow \neg \exists z (z{'} + x = 0{'})$ ∃-introduction
14. $\exists z (z{'} + x = 0{'}) \rightarrow \neg \exists y (x = y{'})$ by *PC*
15. $\exists z (z{'} + x = 0{'}) \rightarrow x = 0$ from (1) and (14) by *PC*

Now we have the induction step. Assume $n = p + 1$ and also
$\vdash x < n \rightarrow (x = 0 \vee \cdots \vee x = p)$. We wish to show
$\vdash x < n{'} \rightarrow (x = 0 \vee \cdots \vee x = p \vee x = n)$.

1. $(x = y{'} \wedge z{'} + x = n{'}) \rightarrow y < n$ as for the basis step
2. $(x = y{'} \wedge z{'} + x = n{'}) \rightarrow (y = 0 \vee \cdots \vee y = p)$

    by the induction hypothesis

3. $t = u \rightarrow t' = u'$ properties of $=$

4. $(x = y') \rightarrow [z' + x = n' \rightarrow (y' = 1 \vee \cdots \vee y' = n)]$

                                     by (2), (3), PC, and $\exists$-introduction

                                     (cf. the proof for the basis stage)

5. $(x = y') \rightarrow [z' + x = n' \rightarrow (x = 1 \vee \cdots \vee x = n)]$

                                     properties of $=$ and PC

6. $\exists y (x = y') \rightarrow [z' + x = n' \rightarrow (x = 1 \vee \cdots \vee x = n)]$ $\exists$-introduction

7. $[\exists y (x = y') \vee x = 0] \rightarrow [z' + x = n' \rightarrow (x = 0 \vee \cdots \vee x = n)]$

                                     PC on (6)

8. $\exists y (x = y') \vee x = 0$ $Q3$ and PC

9. $z' + x = n' \rightarrow (x = 0 \vee \cdots \vee x = n)$ modus ponens on (7), (8)

10. $[\exists z (z' + x = n')] \rightarrow (x = 0 \vee \cdots \vee x = n)$ $\exists$-introduction.   ■

Lemma 16   For every numeral n and every variable x except x_3,

$\vdash_Q n < x \rightarrow [(n' = x) \vee (n' < x)]$

Proof: In what follows we will write z for x_3. We also will need two variables y and w other than x and z.

1. $(z' + n = x \wedge z = 0) \rightarrow (0' + n = x)$ properties of $=$

2. $0' + n = x \rightarrow n' = x$ Lemma 13, $Q4$, and properties of $=$

3. $(z' + n = x \wedge z = 0) \rightarrow n' = x$ PC

4. $z = 0 \rightarrow (z' + n = x \rightarrow n' = x)$ PC

5. $z = y' \rightarrow (z' + n = x \rightarrow y'' + n = x)$ properties of $=$

6. $y'' + n = x \rightarrow y' + n' = x$ Lemma 13

7. $(z = y' \wedge z' + n = x) \rightarrow y' + n' = x$ PC

8. $(z = y' \wedge z' + n = x) \rightarrow \exists z (z' + n' = x)$ $\exists$-introduction

9. $z = y' \rightarrow [z' + n = x \rightarrow \exists z (z' + n' = x)]$ PC on (8)

10. $\exists y (z = y') \rightarrow [z' + n = x \rightarrow \exists z (z' + n' = x)]$

                                     $\exists$-introduction

11. $[z = 0 \vee \exists y (z = y')] \rightarrow [z' + n = x \rightarrow (n' = x \vee \exists z (z' + n' = x))]$

                                     PC on (4) and (10)

12. $z' + n = x \rightarrow [n' = x \vee \exists z (z' + n' = x)]$ $Q3$, (11), and PC

13. $\exists z (z' + n = x) \rightarrow [n' = x \vee \exists z (z' + n' = x)]$

                                     $\exists$-introduction   ■

Lemma 17   For every numeral n and every variable x except x_3,

$\vdash_Q (x < n) \vee (x = n) \vee (n < x)$.

Proof: The proof is by induction on n. We first consider $n = 0$.

1. $w' = x \rightarrow w' + 0 = x$ $Q4$ and properties of $=$

2. $w' + 0 = x \rightarrow \exists z (z' + 0 = x)$ $\exists$-introduction

3. $w' = x \rightarrow 0 < x$ PC and definition

4. $x = w' \rightarrow 0 < x$ properties of $=$ and PC

5. $x = z' \rightarrow 0 < x$ generalization and L1

6. $\exists z\,(x = z') \rightarrow 0 < x$	$\exists$-introduction
7. $x = 0 \lor \exists z\,(x = z')$	$Q3$ and PC
8. $(x = 0) \lor (0 < x)$	PC using (6) and (7)
9. $(x < 0) \lor (x = 0) \lor (0 < x)$	PC

Now for the induction stage, assume $n = p + 1$ and the lemma holds for p.

1. $(x < p) \lor (x = p) \lor (p < x)$	hypothesis
2. $p < n$	Lemma 14
3. $x = p \rightarrow (p < n \rightarrow x < n)$	properties of $=$ (twice)
4. $x = p \rightarrow x < n$	PC on (2) and (3)
5. $x < p \rightarrow x < n$	via Lemma 15.b and Lemma 14
6. $p < x \rightarrow (x = n \lor n < x)$	Lemma 16
7. $n = x \rightarrow x = n$	properties of $=$
8. $(x < n) \lor (x = n) \lor (n < x)$	PC using (4), (5), (6), and (7) ■

Now we are ready to show that the representable functions are closed under the μ-operator.

Lemma 18   If $g(\vec{x}, y)$ is representable in Q, and $f = \lambda \vec{x}\ \mu y\,[\,g(\vec{x}, y) = 0\,]$ is total, then f is representable in Q.

Proof: We will do this for the case where f is a function of one variable and let you generalize it.

Suppose that g is represented in Q by $A(x_1, x_2, x)$ and z is a variable not appearing in A. Then we claim that f is represented in Q by the following ·formula, which we call $C(x_1, x_2)$:

$$A(x_1, x_2, 0) \land \forall z\,[\,z < x_2 \rightarrow \neg A(x_1, z, 0)\,]$$

Suppose $f(n) = m$. We will first show that condition (1) is satisfied. We have two cases. First suppose $m = 0$.

1. $\vdash A(n, m, 0)$	by the representability of g, since $g(n, m) = 0$
2. $\neg(z < 0)$	by Lemma 15.a
3. $z < 0 \rightarrow \neg A(n, z, 0)$	PC
4. $\forall z\,(z < 0 \rightarrow \neg A(n, z, 0))$	generalization
5. $C(n, 0)$	PC on (1) and (4)

Now suppose $m > 0$.

6. $\vdash A(n, m, 0)$	as for (1)
7. for $k < m$, $\neg A(n, k, 0)$	as $g(n, k)\downarrow \neq 0$ [condition (3)]
8. for $k < m$, $z = k \rightarrow (\neg A(n, k, 0) \rightarrow \neg A(n, z, 0))$	Theorem 21.2
9. for $k < m$, $z = k \rightarrow \neg A(n, z, 0)$	PC
10. $z < m \rightarrow (z = 0 \lor \cdots \lor z = p)$	where $p + 1 = m$ by Lemma 15.b
11. $z < m \rightarrow \neg A(n, z, 0)$	PC on (9) and (10)

12. $\forall z\,[\,z< m \;\rightarrow\; \neg A(n,z,0)\,]$ generalization
13. $C(n,m)$ *PC* on (6) and (12)

        Now we turn to condition (2) for both $m=0$ and $m>0$. We need to prove
$C(n,x_2) \rightarrow x_2 = m$.

14. $C(n,x_2) \;\rightarrow\; \forall z\,[\,z< x_2 \;\rightarrow\; \neg A(n,z,0)\,]$ *PC*
15. $C(n,x_2) \;\rightarrow\; A(n,x_2,0)$ *PC*
16. $A(n,x_2,0) \;\rightarrow\; \neg(x_2 < m)$ *PC* on (11) and substitution of x_2
17. $C(n,x_2) \rightarrow \neg(x_2 < m)$ *PC* on (15) and (16)
18. $[\,m< x_2 \wedge C(n,x_2)\,] \;\rightarrow\; [\,m< x_2 \wedge A(n,m,0)\,]$
                                         (1) and *PC*
19. $[\,m< x_2 \wedge A(n,m,0)\,] \;\rightarrow\; \exists z\,[\,z< x_2 \;\wedge\; A(n,z,0)\,]$
                                         $\exists$-introduction
20. $[\,m< x_2 \wedge C(n,x_2)\,] \;\rightarrow\; \neg\forall z\neg[\,z < x_2 \;\wedge\; A(n,z,0)\,]$
                                         PC on (18) and (19)
                                         and definition
21. $[\,m< x_2 \wedge C(n,x_2)\,] \;\rightarrow\; \neg\forall z\,[\,z< x_2 \;\rightarrow\; \neg A(n,z,0)\,]$
                                         PC
22. $\neg[\,m< x_2 \wedge C(n,x_2)\,]$ *PC* on (14) and (21)
23. $C(n,x_2) \rightarrow \neg(m < x_2)$ *PC* on (22)
24. $(x_2 < m) \vee (x_2 = m) \vee (m < x_2)$ Lemma 17
25. $C(n,x_2) \rightarrow (x_2 = m)$ *PC* using (24), (23), and (17)   ■

        The results of this section, Theorem 2, and Corollary 16.5 bring us to our goal.

THEOREM 19   **a.**   Every general recursive function is representable in Q.
                b.   Every total partial recursive function is representable in Q.
                c.   If Q is consistent and f is represented in Q by A, then
                    $f(m_1, \ldots, m_k) = n$ iff $\vdash_Q A(m_1, \ldots, m_k, n)$.

C. The Functions Representable in Q Are Recursive

THEOREM 20   If Q is consistent, then every total function which is representable in Q is partial recursive.

Proof: We already described the calculation procedure informally when we defined representability. To proceed formally, suppose f is a total function of k variables that is represented in Q by A. First, we need to show that we can recursively list the wffs of the form $A(m_1, \ldots, m_k, n)$. That is, there is a recursive function h such that for all w, if $m_1 = (w)_0$, $\ldots$, $m_k = (w)_{k-1}$, and $r = (w)_k$, then $h(w) = [\![\,A(m_1, \ldots, m_k, r)\,]\!]$. We leave the proof of this to you (Exercise 7) based on your Gödel numbering.

Now recall from Chapter 21.F that there is a recursive predicate *Prf* such that $Prf(n, m)$ iff n codes a proof of the wff which has Gödel number m. So to calculate $f(m_1, \ldots, m_k)$ we look for the least w and y such that $lh(w) = k$ and for $i < k$, $(w)_i = m_{i+1}$ and $Prf(y, h(w))$. Then $f(m_1, \ldots, m_k) = (w)_k$. That is,

$$f(m_1, \ldots, m_k) = \left(\mu z \left[lh((z)_0) = k \wedge (i < k \rightarrow (z)_{0, i} = m_{i+1}) \right. \right.$$
$$\left. \left. \wedge Prf[(z)_1, h((z)_0)] \right] \right)_{0, k}$$

∎

COROLLARY 21   If Q is consistent, then for any function f each of the following is equivalent to f being representable in Q:
a. f is general recursive
b. f is total partial recursive
c. f is total and Turing machine computable

D. Representability of Recursive Predicates

Recursive predicates are representable in terms of their characteristic functions. For later reference we need to make some observations about them.

Suppose that C is a recursive set. Then we know that its characteristic function is representable in Q. That is, there is some **A** such that

if $n \in C$, then $\vdash_Q A(n, 1)$ and $\vdash_Q A(n, y) \rightarrow y = 1$
if $n \notin C$, then $\vdash_Q A(n, 0)$ and $\vdash_Q A(n, y) \rightarrow y = 0$

By Lemma 3, $\vdash_Q 0 \neq 1$; so we can conclude, with details of the proof left to you,

if $n \in C$, then $\vdash_Q A(n, 1)$
if $n \notin C$, then $\vdash_Q \neg A(n, 1)$

We can summarize this by saying that *the set C is represented by* $A(x)$ *in* Q, where we understand $A(x)$ to be $A(x, 1)$. By repeated use of generalization and substitution we can assume that x is x_1.

COROLLARY 22   If Q is consistent, then for any set C
a. C is representable in Q iff C is recursive.
b. If C is represented by A in Q, then $C = \{ n : \vdash_Q A(n) \}$.

Similarly we say that a *predicate* of natural numbers, R, of k variables *is representable in* Q if there is a formula $A(x_1, \ldots, x_k)$ such that

if $R(n_1, \ldots, n_k)$, then $\vdash_Q A(n_1, \ldots, n_k)$
if not $R(n_1, \ldots, n_k)$, then $\vdash_Q \neg A(n_1, \ldots, n_k)$

Beware: Other authors discuss a closely related notion of *definability* for predicates that is (usually) different from representability.

COROLLARY 23   If Q is consistent, then for any predicate R
  a. R is representable in Q iff R is recursive.
  b. If R is represented by $\mathbf{A}$ in Q, then
    $R(n_1, \ldots, n_k)$ iff $\vdash_Q \mathbf{A}(\boldsymbol{n}_1, \ldots, \boldsymbol{n}_k)$.

Exercises

†1. a. Show that if a and b are relatively prime natural numbers, then there is a natural number x such that $a\,x \equiv 1 \ (mod\ b)$. (This amounts to showing that there are integers u and v such that $1 = au + bv$.)
   b. Prove the *Chinese Remainder Theorem*:
      If $x_1, \ldots, x_n$ are relatively prime in pairs, and $y_1, \ldots, y_n$ are any natural numbers, then there is a natural number z such that $z \equiv y_i\,(mod\ x_i)$, for $1 \le i \le n$. Moreover, any two such z's differ by a multiple of $x_1 \cdot\ \cdots\ \cdot x_n$. (*Hint:* Let $x = x_1 \cdot\ \cdots\ \cdot x_n$ and call $w_i = x/x_i$. Then for $1 \le i \le n$, w_i is relatively prime to x_i, and so, by part (a), there is some z_i such that $w_i \cdot z_i \equiv 1 \ (mod\ x_i)$ for $1 \le i \le n$. Now let
      $$z = (w_1 \cdot z_1 \cdot y_1) + (w_2 \cdot z_2 \cdot y_2) + \cdots + (w_n \cdot z_n \cdot y_n)$$
      Then $z \equiv w_i \cdot z_i \cdot y_i \equiv y_i\,(mod\ x_i)$. In addition, the difference between any two such solutions is divisible by each x_i, and hence by $x_1 \cdot\ \cdots\ \cdot x_n$. And on the other hand, if z is a solution, so is $z - (x_1 \cdot\ \cdots\ \cdot x_n)$. Hence there must be a solution $z < x_1 \cdot\ \cdots\ \cdot x_n$.)
   c. Given any sequence of natural numbers $a_0, \ldots, a_n$, let $j = max(n, a_0, \ldots, a_n)$ and $c = j!$. Show that the numbers $u_i = 1 + (i + 1)c$ for $0 \le i \le n$ have no factors in common except for 1. (*Hint:* If a prime p divides $1 + (i + 1)c$ and $1 + (j + 1)c$ for $i \le j \le n$, then p divides $(j - i)c$. But p does not divide c, for otherwise it would divide 1, and p does not divide $j - i$, because then it would divide $n!$ which divides c.)

2. Complete the proof of Theorem 1 by showing that the following predicates and function are in C. (Compare Exercises 11.6 and 11.19.)
   a. $m < n$ d. p is a prime
   b. m divides n e. n is a power of the prime p
   c. $m \doteq n$

3. Complete the proof of Theorem 2 for functions of more than one variable defined by primitive recursion.

4. Prove that the projection functions are representable in Q (Lemma 7).

5. Prove $\vdash_Q n + 1 = n'$.

6. Prove that multiplication is representable in Q (Lemma 10).

7. Prove that we can recursively list all wffs of the form $\mathbf{A}(m_1, \ldots, m_k, n)$ as needed in the proof of Theorem 20.

23 The Undecidability of Arithmetic

A. *Q* Is Undecidable

We can diagonalize the sets representable in *Q* by diagonalizing the wffs with one free variable. This will give us a set which is not representable, and hence not recursive (Corollary 22.23). Since we can recursively distinguish these wffs, the only part of the diagonalization process that could fail to be recursive is the decision procedure for theoremhood in *Q*, so we can conclude that *Q*, viewed as the collection of its theorems, is recursively undecidable (Chapter 19.D.2).

THEOREM 1   If *Q* is consistent, then *Q* is recursively undecidable.

Proof: Recall that the provability predicate *Prf* was defined as $Prf(x, y)$ iff *x* codes a proof sequence which proves **A**, where $y = [\![A]\!]$, and that this is recursive (Chapter 21.F.6). Define a predicate *W* on the natural numbers by

$$W(a, b, x) \equiv_{\text{Def}} \text{ some } \mathbf{A} \text{ with exactly one free variable } \mathbf{x_1},$$
$$a = [\![\mathbf{A}]\!] \text{ and } Prf(x, [\![\mathbf{A}(b)]\!])$$

This is also recursive (Exercise 1). Now define a predicate *R* by

$$R(a, b) \equiv_{\text{Def}} \exists x\, W(a, b, x)$$

That is,

$$R(a, b) \text{ iff for some } \mathbf{A} \text{ with exactly one free variable } \mathbf{x_1},$$
$$a = [\![\mathbf{A}]\!] \text{ and } \vdash_Q \mathbf{A}(b)$$

Not every **A** with just one free variable $\mathbf{x_1}$ represents a set: only those for which, for every *n*, either $\vdash_Q \mathbf{A}(n)$ or $\vdash_Q \neg\mathbf{A}(n)$. But if **A** does represent a set in *Q*, then on the assumption that *Q* is consistent we have:

$$R(a, b) \text{ iff } b \text{ is in the set represented by } \mathbf{A} \text{ where } [\![\mathbf{A}]\!] = a$$

Now consider the diagonalization of *R*, namely,

$$S = \{ n:\ \text{not } R(n, n) \}$$

Suppose that S is representable in Q, say by $\mathbf{A}$ and $[\![\mathbf{A}]\!] = a$. Then

$$a \in S \quad \text{iff} \quad \vdash_Q \mathbf{A}(a)$$
$$\text{iff} \quad R(a, a)$$
$$\text{iff} \quad a \notin S$$

This is a contradiction, so S is not representable in Q. So by Corollary 22.23, S is not recursive, and hence R can't be recursive. But that can only be because the set of Gödel numbers of theorems of Q is not recursive. That is, Q is recursively undecidable.   ∎

COROLLARY 2   If Q is consistent then assuming Church's thesis, Q is undecidable.

We will postpone our historical remarks to the end of Chapter 24.

B. Theories of Arithmetic

We have established that there is no decision procedure for this one particular fragment of arithmetic, Q. But what about other fragments? And what about all of arithmetic?

1. Fragments simpler than Q

Let's first look at fragments that are simpler than Q.

Suppose we delete the symbol for multiplication from the formal language. Then the set of all wffs of that language that are true of the natural numbers (Chapter 21.A.6 and 21.C.2) is decidable (see Chapter 21 of Boolos and Jeffrey). The same is true if instead we delete the symbols for both addition and successor: $+$ and $'$. Thus, we also have finitistic proofs of the consistency of arithmetic without multiplication and of arithmetic without addition and successor by using the decision procedure in each case to show that some one wff is not a theorem.

Alternatively, if we keep the same language and delete even one of the seven axioms $Q1$–$Q7$ then, assuming that what we have is consistent, we can no longer represent all the recursive functions (see Tarski, Mostowski, and Robinson, particularly their Theorem 11, p. 62). So from now on Q will be the simplest fragment of arithmetic in which we'll be interested.

2. Theories

What do we mean by "fragment of arithmetic" ?

We've already said that we won't get anything more by adding any logical axioms or rules (Chapter 21.B.6) . So, keeping the same language, the only option is to add further arithmetic axioms. If we do that we'll want to look at the theorems we can generate, so we make the following definitions.

Given a collection T of wffs in our formal language, we write $\vdash_T A$ to mean that there is a proof of A using the set T as axioms (instead of Q) and the logical axioms and rules we adopted previously (Chapter 21.C.1). We write $\nvdash_T A$ to mean that there is no proof of A from T. If $\vdash_T A$, we say that A is a *consequence of T*.

The *theory of T* is the collection $\{A : \vdash_T A\}$. A collection T of wffs in our formal language is a *theory* if it contains all of its consequences: if $\vdash_T A$ then $A \in T$. We say that *theory T extends theory S* just in case $T \supseteq S$.

As for Q we say a theory T is *consistent* if there is no wff A such that both $\vdash_T A$ and $\vdash_T \neg A$. Otherwise it is *inconsistent*. There is only one inconsistent theory: the collection of all wffs (see Chapter 19.E), and that is certainly decidable.

From this point on when we refer to a collection of axioms as a theory we mean the collection of its theorems.

3. Axiomatizable theories

If we add more wffs to Q as axioms, we'd like to be able to prove further theorems from them. This will be pretty hard if we don't know which wffs are axioms. That is, we ought to require that the collection of new axioms be decidable.

We say that a theory T is *axiomatizable* if there is a decidable collection of wffs S such that $T =$ the theory of S. If we wish to stress that we have identified decidability with recursiveness, we will say a theory is *recursively axiomatizable*.

4. Functions representable in a theory

The definition of *representability* for functions and predicates (Chapter 22.B and D) carries over to any theory T by replacing Q by T.

THEOREM 3   If T is a theory that extends Q, then all the recursive functions are representable in T.

Proof: Suppose f is recursive. Then it is representable in Q; that is, there is some A such that whenever $f(m_1, \ldots, m_k) = n$, then $\vdash_Q A(m_1, \ldots, m_k, x)$ and $\vdash_Q A(m_1, \ldots, m_k, x) \rightarrow x = n$. Since T extends Q, these are also theorems of T, which is just to say that f is representable in T.   ∎

For any axiomatizable theory T that extends Q and any particular recursive axiomatization of T, we can establish just as for Q that proving is a computable procedure. Hence if T is consistent then every function representable in it must be recursive. So we have the following theorem.

THEOREM 4   a. If T is an axiomatizable extension of Q, then the set of (Gödel numbers of) theorems of T is recursively enumerable.
  b. If T is a consistent axiomatizable extension of Q, a total function is representable in T iff it is recursive.

5. Undecidable theories

What was the essential ingredient in proving that Q is undecidable? We needed that the recursive functions are representable in Q and that Q is consistent. We did not need that *only* the recursive functions are representable in Q.

THEOREM 5   a. Every consistent theory in which the recursive functions are
representable is recursively undecidable.

                                 b. Every consistent theory that extends Q is recursively undecidable.

Proof: a. Let T be such a theory. If T is axiomatizable the proof is as for Q. If T is not axiomatizable, then since the entire collection of theorems of T could serve as axioms, it must not be decidable. Note that we do not need to assume that T extends Q.

    b. This part follows from (a) and Theorem 3.   ■

C. Peano Arithmetic (*PA*) and *Arithmetic*

Induction is the most powerful tool to prove theorems about the natural numbers that we have used in this book. That proof method is not available in Q: within Q we have only the inductive definitions of addition and multiplication given by the axioms.

The *principle of induction* in its strongest form says:

For every set of natural numbers X: if $(0 \in X$ and if for every n,
if $n \in X$ then $n + 1 \in X$), then every natural number is in X.

We can't formalize this statement in our formal language since we've only allowed ourselves to quantify over natural numbers, not over sets of natural numbers. But the informal proofs we have given required something weaker: for the proposition P in question that depends on n, if $[P(0)$, and for every n, if $P(n)$ then $P(n + 1)$], then for all n, $P(n)$. This proof procedure we can formalize by taking the counterpart to such a proposition to be a wff with one free variable.

Peano Arithmetic (*PA*) is the theory obtained by adding every instance of the *first-order scheme of induction* to Q:

$$[A(0) \wedge \forall x (A(x) \rightarrow A(x'))] \rightarrow \forall x \, A(x)$$

We leave to you (Exercise 2) to show that the set of all instances of the induction scheme is decidable, and hence that *PA* is axiomatizable.

The first-order scheme of induction adds tremendous power to Q. For instance, we can now prove that addition and multiplication are commutative and associative, as well as all the other wffs of Theorem 21.3. Here is an example.

$$\vdash_{PA} x + (y + z) = (x + y) + z$$

Proof: Call this wff we want to prove $A(z)$. Here is a sketch of how to give a formal proof.

1. $y + 0 = y$ $Q4$
2. $x + (y + 0) = x + y$ properties of $=$ (Theorem 21.2)
3. $(x + y) + 0 = x + y$ $Q4$ and L1
4. $A(0)$ properties of $=$
5. $y + z' = (y + z)'$ $Q5$
6. $x + (y + z') = x + (y + z)'$ properties of $=$
7. $x + (y + z)' = (x + (y + z))'$ $Q5$ and L1
8. $(x + y) + z' = ((x + y) + z)'$ $Q5$ and L1
9. $A(z) \rightarrow A(z')$ (6), (7), (8), and properties of $=$
10. $\forall z\, (A(z) \rightarrow A(z'))$ generalization
11. $[A(0) \wedge \forall z\, (A(z) \rightarrow A(z'))] \rightarrow \forall z\, A(z)$ the induction scheme
12. $\forall z\, A(z)$ PC using (4), (10), and (11)
13. $x + (y + z) = (x + y) + z$ L1   ∎

Is *PA* so powerful that we can now prove everything true of the natural numbers that can be expressed in our formal language? On the assumption that that question even makes sense (Chapter 21.C.2), we make the following nonconstructive definition.

> *Arithmetic* is the collection of all wffs in our formal language
> that are true about the natural numbers

We needn't accept that such a collection exists, but to investigate whether the truths of arithmetic can be reduced to a formal system of proof we shall adopt as a working hypothesis that there is such a collection.

Given that there is a collection called *Arithmetic*, then it is a theory, since the consequences of true wffs are true. And it is consistent since no wff is both true and false. Moreover, as long as we've gone this far we may as well assume that it extends *Q*, for those axioms are as good candidates as we'll ever have for general truths of the natural numbers. But please note, we are not making these assumptions for *PA* nor for any of the other work that follows, except for discussions involving *Arithmetic* .

Now we can pose our question in the following terms: Is *PA* = *Arithmetic* ? And while we're at it: Is *PA* decidable? Is *Arithmetic* decidable?

For decidability we have the following as a Corollary to Theorem 5.

COROLLARY 6   **a.** If *PA* is consistent, then it is recursively undecidable.
               b. *Arithmetic* is recursively undecidable.

Thus, assuming Church's Thesis, *Arithmetic* is undecidable and so there is no constructive procedure for determining whether an arbitrary wff is true.

Let's now turn to the question of whether *PA* = *Arithmetic* .

THEOREM 7   *Arithmetic* is not axiomatizable.

Proof: Suppose ***Arithmetic*** were axiomatizable. Then it would be recursively decidable: To decide whether a wff **A** is true we begin the proof machinery that recursively enumerates the theorems, in this case all wffs true about the natural numbers. Since either **A** or ⅂**A** is true, we must eventually find one of them on the list. If we find **A** , then **A** is true. If we find ⅂**A** , then **A** is not true. Hence ***Arithmetic*** would be recursively decidable. But that contradicts Corollary 6, so ***Arithmetic*** is not axiomatizable.   ∎

COROLLARY 8   *Arithmetic* ≠ *PA*

Proof: *PA* is axiomatizable, but ***Arithmetic*** is not.   ∎

COROLLARY 9 (Gödel, 1931)   If T is a consistent axiomatizable theory in the language of first-order arithmetic, then there is some wff true of the natural numbers which cannot be proved in T .

Proof: By Theorem 7, ***Arithmetic*** ≠ T . We cannot have $T \supset$ ***Arithmetic*** by the same argument as for Theorem 7. Hence there must be some **A** which is true and is not a theorem of T .   ∎

Consider the theory T consisting of Q plus the scheme ⅂$\forall x (0 + x = x)$. In Theorem 21.3 we showed that T has a model and hence is consistent (assuming that the theorems of Q are true of the natural numbers). So for that theory Corollary 9 is trivial. What we are concerned with now are theories where $T \subseteq$ *Arithmetic* .

COROLLARY 10   If T is a consistent axiomatizable extension of Q all of whose theorems are true of the natural numbers, then there is some sentence **A** true of the natural numbers such that $\nvdash_T$ **A** and $\nvdash_T$ ⅂**A** .

A closed wff **A** such that neither $\vdash_T$ **A** nor $\vdash_T$ ⅂**A** is called *formally undecidable relative to* T .

What can we now conclude? The truths of arithmetic in the formal language we have chosen cannot be axiomatized. Even the very powerful theory *PA* cannot capture all the truths of arithmetic expressible in our formal language. Moreover, any interesting fragment of arithmetic that we can axiomatize (i.e., one extending Q) can be used to characterize the recursive functions. The power of this makes it undecidable.

But Corollaries 9 and 10 are unsatisfying. We have had to make some very strong assumptions about the meaning and truth of wffs. Moreover, for a given theory T we have not actually produced a sentence that is formally undecidable relative to T . For the theory Q we have some examples from Theorem 21.3 (Exercise 3 below), but how about other theories? In Chapter 24 we will show that assuming only the consistency of a theory extending Q we can construct a sentence formally undecidable relative to that theory along the lines of the liar paradox.

Exercises

1. (From the proof of Theorem 1)
   a. Show that the predicate *W* is recursive.
   b. Show that the predicate *R* is recursively enumerable.

†2. Show that the collection of all instances of the induction scheme of *PA* is recursively decidable.

3. a. Why are the wffs of Theorem 21.3 formally undecidable relative to *Q* ?
   (*Hint:* We've shown that they aren't provable. Why aren't their negations provable?)
   b. Show that we can deduce *Q*3 from the induction scheme of *PA* .
   †c. Prove that all the wffs of Theorem 21.3 are theorems of *PA* .
   (*Hint:* For wff (b) we've shown this above. Do (a), (d), (h), (f), (c), (g) in that order. Prove each informally first, and then convert your informal proof into a formal one. For part (e) use *PC* to reduce it to a wff with no quantifiers.)

4. We say that a theory *T* is *complete* if for every closed **A**, either $\mathbf{A} \in T$ or $\neg \mathbf{A} \in T$.
   a. Why do we require **A** to be closed? (*Hint:* Consider $\exists y (2 \cdot y = x)$.)
   b. What theories discussed in this chapter are complete?
   c. Is the inconsistent theory complete?
   d. Generalize the argument of Theorem 7 to show that any axiomatizable complete theory is decidable. Note that you will have to use two different methods, one if the theory is consistent, and another if it is not. Is there a decision procedure that will tell you which method to use?
   e. Suppose that *T* extends *Q*, and *T* is complete. Is *T* decidable? Is *T* axiomatizable?
   f. Is *PA* complete?

5. Show that the set of Gödel numbers of wffs in **Arithmetic** is not recursively enumerable (*Hint:* cf. Theorem 17.3 or Theorem 7.)

†6. a. Show that there is a function representable in **Arithmetic** that is not recursive.
   (*Hint:* Use any r.e. nonrecursive set and the Projection Theorem 17.5.)
   b. Prove that there is a function on the natural numbers that is not representable in **Arithmetic** . Do not simply use a counting argument.

†7. Prove *Church's Theorem* (1936a):
   The collection of theorems that can be proved using only the logical axioms and rules is (recursively) undecidable, if it is consistent.
   [*Hint:* Modify the definition of "consequence" from Chapter 19.Appendix.2 to apply to first-order logic without the axioms of *Q*. Then modify the proof of the Deduction Theorem there to show that for any closed wff **A** :

$\Gamma \cup \{A\} \vdash B$ iff $\Gamma \vdash A \rightarrow B$

Letting $A_1, \ldots, A_7$ be the universally quantified forms of $Q1$–$Q7$, show that:

$\vdash_Q B$ iff $\vdash_Q A_1 \rightarrow (A_2 \rightarrow (A_3 \rightarrow (A_4 \rightarrow (A_5 \rightarrow (A_6 \rightarrow (A_7 \rightarrow B))))))$.]

8. Is Corollary 10 a blow to Hilbert's program? (See the last paragraph on p. 57.)

Further Reading

For a full exposition of undecidability that augments both this chapter and the next, with many examples of both decidable and undecidable theories, see Epstein's *Classical Mathematical Logic*.

24 The Unprovability of Consistency

A. Self-Reference in Arithmetic: The Liar Paradox

There is no computable procedure for deciding whether an arbitrary wff of arithmetic is true (Corollary 23.6). Nor can we axiomatize or list the true sentences (Theorem 23.7, Exercise 23.5). But might we be able to define in the formal language the set of true sentences?

If we could define truth, we could also define falsity. Then via the self-reference available to us from our Gödel numberings we could recreate the liar paradox, "This sentence is false."

THEOREM 1 (Gödel, 1934)   The set of sentences true of the natural numbers is not representable in *Arithmetic*.

Proof: Suppose to the contrary that $\{n:$ for some **A**, $n = [\![\mathbf{A}]\!]$ and **A** is true of the natural numbers$\}$ is representable in *Arithmetic*. Then so too is the set

$$F = \{m: \text{ for some } \mathbf{A} \text{ with exactly one free variable } \mathbf{x_1}, \ m = [\![A]\!]$$
$$\text{and } \mathbf{A}(m) \text{ is false}\}$$

because the rest of the definition of F is recursive (we leave the details to you since we'll do a similar proof for Theorem 2). Suppose that $\mathbf{F}(\mathbf{x_1})$ represents F (see Chapter 22.D) and $[\![\mathbf{F}(\mathbf{x_1})]\!] = a$. Then $\mathbf{F}(a) \in$ *Arithmetic* (i.e., $\mathbf{F}(a)$ is true) iff $a \in F$. But $a \in F$ iff $\mathbf{F}(a)$ is false, a contradiction. So F and hence the sentences true of the natural numbers cannot be represented in *Arithmetic* .   ∎

Theorem 1 is sometimes colloquially stated as: *Arithmetical truth is not definable in arithmetic.*

Gödel's insight in 1931 was that "true" need not be the same as "provable". By replacing "true" in the liar paradox by "provable", we get not a paradox but a sentence that expresses its own unprovability (from our vantage point outside the system) and that we can show is indeed unprovable.

THEOREM 2   If Q is consistent, then there is a wff U (which intuitively expresses its own unprovability) such that $\nvdash_Q U$.

Proof: Define a predicate W by

$$W(n, x) \equiv_{\text{Def}} \text{ some } A \text{ with exactly one free variable } x_1,\ n = [\![A]\!]$$
$$\text{and } Prf(x, [\![A(n)]\!])$$

This is recursive [it's $W(n, n, x)$ from the proof of Theorem 23.1]. Hence by Corollary 22.23 it is representable in Q, say by $W(x_1, x_2)$. Consider the wff with one free variable

$$\forall x_2 \, \neg W(x_1, x_2)$$

and then for some a, $[\![\forall x_2 \neg W(x_1, x_2)]\!] = a$. Define

$$U \equiv_{\text{Def}} \forall x_2 \neg W(a, x_2)$$

(intuitively U expresses that it itself is unprovable in Q). Then

$$W(a, x) \text{ iff } x \text{ is the Gödel number of a proof in } Q \text{ of U}$$

Suppose that U is provable in Q. Then there is a number n such that $W(a, n)$. Hence $\vdash_Q W(a, n)$, so by $\exists$-introduction $\vdash_Q \exists x_2 W(a, x_2)$. But that is just $\vdash_Q \neg U$, which contradicts the consistency of Q. Hence $\nvdash_Q U$.   ■

We have not yet produced a wff that is formally undecidable relative to Q. We could claim that U will do if we suppose that all the theorems of Q are true, but we have scrupulously avoided questions of truth and meaning in Theorem 2 except as motivation. In any case, a much weaker syntactic assumption will do. We say that a theory T is ω-*consistent* if for every wff **B**, whenever $\vdash_T B(n)$ for every n, then $\nvdash_T \neg \forall x\, B(x)$. We leave to you the proof that any ω-consistent theory is consistent as well as the following corollary (Exercise 1).

COROLLARY 3   If Q is ω-consistent, then $\nvdash_Q \neg U$, and hence U is formally undecidable relative to Q.

Using only syntactic assumptions we have shown that there must be a wff which is true but not provable in Q, since U is a closed wff and thus one of U or $\neg U$ must be true.

Rosser, 1936, showed how to complicate the definition of U to produce a wff V that requires only the assumption that Q is consistent in order to prove that it is formally undecidable relative to Q (Exercise 2).

Theorem 2 and Corollary 3 generalize to any consistent axiomatizable extension of Q. If T is an axiomatizable theory that extends Q, then proving is a computable procedure in T, too. That is, using your Gödel numbering of the language, the set of Gödel numbers of axioms is recursive and the predicate Prf_T defined by

$$Prf_T(x,y) \equiv_{Def} x \text{ codes a proof sequence in } T \text{ which proves}$$
$$\text{the wff whose Gödel number is } y$$

is recursive. So we can define a wff $\mathbf{U}_T$ exactly as we did $\mathbf{U}$, replacing Prf by Prf_T everywhere, such that $\mathbf{U}_T$ intuitively expresses its own unprovability in T. The proof of the following is then the same as for Theorem 2 and Corollary 3.

THEOREM 4 (Gödel's First Incompleteness Theorem, 1931)
If T is a consistent axiomatizable extension of Q, then $\not\vdash_T \mathbf{U}_T$.
If T is also ω-consistent, then $\not\vdash_T \neg\,\mathbf{U}_T$.

The proofs of Theorem 2 and Theorem 4 are completely constructive. Later mathematicians have produced "intuitive" theorems of arithmetic that can't be proved in *PA* (see *Further Reading* below).

B. The Unprovability of Consistency

What does it mean to say that we cannot prove *PA* consistent within *PA*? We need some way to talk about the consistency of *PA* within *PA*.

Define the predicate *Neg* by

$$Neg(x, y) \equiv_{De} x \text{ is the Gödel number of a wff, and}$$
$$y \text{ is the Gödel number of its negation}$$

This is recursive, as you can show using your Gödel numbering. Then

$$PA \text{ is consistent iff there is no } \mathbf{A} \text{ such that } \vdash_{PA} \mathbf{A} \text{ and } \vdash_{PA} \neg\mathbf{A}$$
$$\text{iff for every } x, y, z, w,$$
$$\text{not } [\, Neg(x,y) \wedge Prf_{PA}(z,x) \wedge Prf_{PA}(w,y)\,]$$

Call the part in brackets $C(x, y, z, w)$. It is recursive and hence is representable in Q, say by $\mathbf{C}$, which also represents it in *PA*. Define

$$\mathbf{Consis}_{PA} \equiv_{Def} \forall x_1 \, \forall x_2 \, \forall x_3 \, \forall x_4 \, \neg\, C(x_1, x_2, x_3, x_4)$$

Viewed from outside the system, $\mathbf{Consis}_{PA}$ expresses that *PA* is consistent.

THEOREM 5 (Gödel's Second Incompleteness Theorem, 1931)
If *PA* is consistent, then $\not\vdash_{PA} \mathbf{Consis}_{PA}$.

Proof: Recall that the proof of Theorem 4 for *PA* is the proof of Theorem 2 with Prf_{PA} in place of Prf. If we choose, we may make that entire proof part of ordinary arithmetic by talking of indices of wffs rather than wffs and interpreting all the discussion of the formal system in terms of predicates on those indices. The interpretation of the formal system does not figure into the proof of:

$$\text{if } PA \text{ is consistent, then } \not\vdash \mathbf{U}_{PA}$$

Moreover, we know that U_{PA} formalizes the proposition that U_{PA} is not provable. So we can establish as an informal theorem of arithmetic:

if $\mathbf{Consis}_{PA}$ then U_{PA}

The entire proof of this arithmetic statement is finitary and, we claim, can be formalized within *PA* (compare how we proved the associative law of arithmetic in *PA* by formalizing the usual informal proof, p. 206). You can imagine how formalizing that informal proof would be tedious and extremely long, and therefore we will not include it here. (You can find it proved in detail in a slightly different version in Shoenfield, pp. 211–213, where his theory *N*, p. 22, is our *Q*, and his theory *P*, p. 204, is our *PA*.) Thus, we have:

$$\vdash_{PA} \mathbf{Consis}_{PA} \rightarrow U_{PA}$$

Since we also have $\nvdash_{PA} U_{PA}$, we must have $\nvdash_{PA} \mathbf{Consis}_{PA}$.   ∎

We can also ask whether Gödel's Second Incompleteness Theorem applies to consistent axiomatizable extensions of *PA* or even *Q*. The answer is "yes" if we are careful about how we interpret consistency as an arithmetic predicate and how the theory is presented (see, e.g., Boolos and Jeffrey, Chapter 16, as well as Feferman, and Bezboruah and Shepherdson). It is possible, however, to give a finitary consistency proof within *PA* of the weak theory *Q*, though not within *Q* itself (see Shoenfield: the finitary consistency proof is on p. 51, and the description of how to convert it to one in *PA* is on p. 214).

Generally, then, we conclude that if an axiom system is consistent and contains as much number theory as *PA*, then we cannot prove the consistency of that system within the system itself.

What is the significance of Gödel's Second Incompleteness Theorem for Hilbert's program? If we are intent on proving finitistically that infinitistic methods are acceptable in mathematics, then at the very least we should be able to prove that *PA* is consistent. There are several possibilities.

1. All finitary methods of proof can be formalized within *PA*. There is good evidence for this: In 1937 Ackermann showed that in essence *PA* is equivalent to set theory without infinite sets (see Moore, p. 279). (Note that we don't claim that *PA* is itself a finitary theory of arithmetic.)

In this case Gödel's theorem demonstrates that Hilbert's program cannot be accomplished. Here is what Shoenfield says (where we've relabeled the theories with our names):

> The theorem on consistency proofs is a limitation on the type of consistency proof which we can give for *PA* . For this to be of any significance, we must know that some types of consistency proofs can be formalized in *PA* . Now it is reasonable to suggest that every finitary consistency proof can be formalized in *PA* (or equivalently, in a recursive[ly axiomatizable] extension of *PA*).

First, a finitary proof deals only with concrete objects, and these may be replaced by natural numbers by assigning such a number to each object (as we have done for expressions [wffs]). Second, the proof deals with these objects in a constructive fashion; so we can expect the functions and predicates which arise to be introducible in recursive extensions of *PA*.

An examination of specific finitary consistency proofs confirms this suggestion. For example, the consistency proof for *Q* given in Chapter 4 can be formalized in *PA*. It is a tedious but elementary exercise to formalize the proof of the consistency theorem. We then have to check that the set of expression numbers of true variable-free formulas of *Q* can be introduced in a recursive extension of *PA*; and this is also straightforward.

We cannot, of course, state with assurance that every finitary consistency proof can be formalized in *PA*, since we have not specified exactly what methods are finitary. ...

Investigations by Kreisel have shown that a consistency proof which could not be formalized in *PA* would have to use some quite different principles from those used in known finitary proofs.

We conclude that it is reasonable to give up hope of finding a finitary consistency proof for *PA*.

Shoenfield, p. 214

2. There are further finitary principles, but we can axiomatize them all in some extension of *PA*. But then our remark that Gödel's theorem can be extended to cover such extensions has the same effect as Theorem 5 in the first case.

3. There are further finitary principles and they can all be expressed in our formal language, but they cannot be axiomatized. This, too, would put an end to Hilbert's hopes.

For all formal systems for which the existence of undecidable arithmetical propositions was asserted above, the assertion of the consistency of the system in question itself belongs to the propositions undecidable in that system. That is, a consistency proof for one of these systems *G* can be carried out only by means of methods of inference that are not formalized in *G* itself. For a system in which all finitary (that is, intuitionistically unobjectionable [see Chapter 25.A]) forms of proof are formalized, a finitary consistency proof, such as the formalists seek, would thus be altogether impossible. However, it seems questionable whether one of the systems set up, say *Principia Mathematica*, is so all-embracing (or whether there is a system so all-embracing at all).

Gödel, 1931a, p. 205

4. There are further finitary principles that may suffice to prove the consistency of *PA* but which cannot be formalized in our first-order language.

I wish to note expressly that [Theorem 5 for the system *P* of *Principia Mathematica*] (and the corresponding results for *M* [formal set theory] and *A* [classical mathematics]) do not contradict Hilbert's formalistic viewpoint. For this viewpoint presupposes only the existence of a consistency proof in which nothing but finitary means of proof is used, and it is conceivable that there exist finitary proofs that *cannot* be expressed in the formalism of *P* (or of *M* or *A*).

Gödel, 1931, p. 195

Indeed, a consistency proof for *PA* was given by Gentzen in 1936 a short time after Gödel's work appeared (see Kleene, 1952, §79 or the Appendix of Mendelson, 1964). Briefly, a binary relation on the natural numbers is a *linear ordering* if it is transitive, antisymmetric, antireflexive, and total. It is a *well-ordering* if in addition every nonempty set of natural numbers has a least element in the ordering. Note that this latter is not a first-order assumption. We gave examples of such orderings in Chapter 13: ω^2, ω^3, ... , ω^n, ... , and argued that there is a valid principle of induction for each. Similarly (see, e.g., Péter, 1967) there are orderings ω^ω, ω^{ω^ω}, ... for each of which there is a principle of induction that can be reduced to ordinary induction. Gentzen showed that we can give a consistency proof of, for example, *PA* if we assume not just one of these principles of induction but all of them at once. (That is, we allow induction on any ordinal less than ε_0, which is the least fixed point of $\lambda x (\omega^x)$.) But if the consistency of *PA* is in question, how much more so the theory of *PA* plus the formalization of Gentzen's induction principle as a scheme of induction in our formal language.

If we hold to finitary or even classical methods of proof, faith cannot be banished from mathematics: we simply have to believe that *PA* is consistent, since any proof which we could formalize will use methods or principles that are more questionable than those we use in the system itself.

Perhaps then we should look at arithmetic from a more constructive viewpoint than classical mathematics, as we will do in Chapter 26.

But before we leave this topic, here is what Gödel, much later, thought of the relation of his theorems to Hilbert's program.

What has been proved is only that the *specific epistemological* objective which Hilbert had in mind cannot be obtained. This objective was to prove the consistency of the axioms of classical mathematics on the basis of evidence just as concrete and immediately convincing as elementary arithmetic.

However, viewing the situation from a purely *mathematical* point of view, consistency proofs on the basis of suitably chosen stronger metamathematical presuppositions (as have been given by Gentzen and others) are just as interesting, and they lead to highly important insights into the proof theoretic structure of mathematics. Moreover, the question remains open whether, or to what extent, it is possible, on the basis of the formalistic approach, to prove 'constructively' the consistency of classical mathematics, i.e., to replace its axioms about abstract entities of an objective Platonic realm by insights about the given operations of our mind.

Gödel, quoted in Reid, 1970, pp. 217–218

C. Historical Remarks

By 1930 research on Hilbert's program was in full swing: In 1929 Presburger had shown that arithmetic without multiplication is decidable, and in 1931 Skolem did the same for arithmetic without addition and successor. Finitary consistency proofs had been given for some restricted but interesting fragments of arithmetic, for example, by Herbrand in 1931. There seemed good reason to believe that a finitary consistency proof could be given for formalized arithmetic.

In 1930 Gödel proved the completeness theorem for first-order logic (see p. 180 above), which justified that the logical axioms and rules we have adopted are sufficient. In 1931 he introduced the idea of numbering a formal system so that theorems about the system could be translated into theorems about natural numbers. To be able to use that idea to talk about a formal theory within the theory itself, he needed to show that various predicates were representable: "is the Gödel number of a wff", "is the Gödel number of an axiom", *Prf*, and so on. To do so he showed that the primitive recursive functions (which he called "recursive") are representable in the first-order part of the formal system of *Principia Mathematica* of Whitehead and Russell. He then observed, as we have in Section A, that arithmetic truth could not be defined in that theory by showing that the liar paradox would result (see Tarski, 1933, p. 247 and pp. 277–78, for historical comments on that). Replacing "true" by "provable" in that example, he then constructed in that theory a wff which intuitively expresses that it itself is unprovable and showed that on the assumption of ω-consistency neither the wff nor its negation can be provable and hence that there must be a wff which is true but unprovable (he used the term "undecidable" for what we call "formally undecidable"). Then, following just as we did here, he showed that the consistency of the theory could not be proved within the theory itself (assuming it is consistent).

The system of *Principia Mathematica* was framed in a language that is much more extensive than the first-order language of arithmetic we have used. Gödel was clearly concerned about the status of the class of functions he had shown to be representable in the system, the primitive recursive ones, and devoted a section of his paper to showing that they could be represented in our formal language of arithmetic. It was for that purpose he devised his β-function (Chapter 22.A). Only later were the undecidability results formulated for simpler theories such as Q (see Tarski, Mostowski, and Robinson, p. 39).

It was not until 1936 that Rosser showed that the assumption of consistency, rather than ω-consistency, was sufficient to establish a formally undecidable sentence. Compare that to how we proved the existence of formally undecidable sentences in Chapter 23: any theory in which the recursive functions are representable is undecidable, in particular *Arithmetic* is undecidable, and hence is not axiomatizable. Gödel did not proceed along those lines because it was not clear to him that the primitive recursive functions were all the computable ones, and without a precise notion of computability there was no precise notion of an axiomatizable theory. Here is how he grappled with that problem in his lectures in 1934:

> A *formal mathematical system* is a system of symbols together with rules for employing them. The individual symbols are called *undefined terms*. *Formulas* are finite sequences of the undefined terms. There shall be defined a class of formulas called *meaningful formulas*, and a class of meaningful formulas called *axioms*. There may be a finite or infinite number of axioms. Further, there shall be specified a list of rules, called *rules of inference*; if such a rule be called R, it defines the relation of *immediate consequence by R* between a set of meaningful formulas $M_1, \dots, M_k$ called the *premises*, and a meaningful formula N, called the *conclusion* (ordinarily $k = 1$ or 2). We require that the rules of inference, and the definitions of meaningful formulas and axioms, be constructive; that is, for each rule of inference there shall be a finite procedure for determining whether a given formula B is an immediate consequence (by that rule) of given formulas $A_1, \dots, A_n$, and there shall be a finite procedure for determining whether a given formula A is a meaningful formula or an axiom.
>
> Gödel, 1934, p. 41

In 1964, in an addendum to his paper of 1934, Gödel commented on how the difficulty of characterizing formal systems was overcome:

> In consequence of later advances, in particular of the fact that, due to A.M. Turing's work, a precise and unquestionably adequate definition of the general concept of formal system can now be given, the existence of undecidable arithmetical propositions and the non-demonstrability of the consistency of a system in the same system can now be proved rigorously for *every* consistent formal system containing a certain amount of finitary number theory. Turing's work gives an analysis of the concept of "mechanical procedure" (alias "algorithm" or "computation procedure" or "finite combinatorial procedure"). This concept is shown to be equivalent with that of a "Turing machine". A formal system can simply be defined to be any mechanical procedure for producing formulas, called provable formulas. For any formal system in this sense there exists one in the sense of [the quotation immediately above] that has the same provable formulas (and likewise vice versa), provided the term "finite procedure" occurring [in the quotation immediately above] is understood to mean "mechanical procedure". This meaning, however, is required by the concept of formal system, whose essence it is that reasoning is completely replaced by mechanical operations on formulas. (Note that the question of whether there exist finite *non-mechanical* procedures not equivalent with any algorithm, has nothing whatsoever to do with the adequacy of the definition of "formal system" and of "mechanical procedure".)
>
> Gödel, 1934, pp. 71–72

Finally, we'll let Gödel tell you how (in retrospect) he believes his work depended on his philosophical views:

A similar remark applies to the concept of mathematical truth, where formalists considered formal demonstrability to be an *analysis* of the concept of mathematical truth and, therefore, were of course not in a position to *distinguish* the two.

I would like to add that there was another reason which hampered logicians in the application to metamathematics, not only of transfinite reasoning, but of mathematical reasoning in general and, most of all, in expressing metamathematics in mathematics itself. It consists in the fact that, largely, metamathematics was not considered as a science describing objective mathematical states of affairs, but rather as a theory of the human activity of handling symbols.

Gödel, quoted in Wang, p. 10

Exercises

1. a. Prove that if a theory T extends Q and is ω-consistent, then it is consistent.
   b. Prove that if Q is consistent and ω-consistent, then $\nvdash_Q \neg U$.
   c. Let T be the theory of the axioms of Q plus $\neg \forall x\,(0 + x = x)$.
      Prove (using Theorem 21.3)
      i. If Q is consistent, then so is T.
      ii. T is ω-inconsistent.

†2. (*Rosser's Theorem*, 1936)
    Define
    $$W^*(n,y) \equiv_{\text{Def}} \text{ some } \mathbf{A} \text{ with exactly one free variable } \mathbf{x}_1 ,$$
    $$n = [\![\mathbf{A}]\!] \text{ and } Prf(y, [\![\neg \mathbf{A}(n)]\!])$$

This is recursive and hence representable in Q, say by $\mathbf{W}^*$. Now consider the wff with one free variable $\mathbf{x}_1$:

$$\forall \mathbf{x}_2 [\mathbf{W}(\mathbf{x}_1, \mathbf{x}_2) \rightarrow \exists \mathbf{x}_3\,(\mathbf{x}_3 < \mathbf{x}_2 \wedge \mathbf{W}^*(\mathbf{x}_1, \mathbf{x}_3))]$$

with Gödel number m. Define

$$\mathbf{V} \equiv_{\text{Def}} \forall \mathbf{x}_2 [\mathbf{W}(m, \mathbf{x}_2) \rightarrow \exists \mathbf{x}_3\,(\mathbf{x}_3 < \mathbf{x}_2 \wedge \mathbf{W}^*(m, \mathbf{x}_3))]$$

1. $W(m, n)$ holds iff n is a Gödel number of a proof in Q of $\mathbf{V}$.
2. $W^*(m, n)$ holds iff n is a Gödel number of a proof in Q of $\neg \mathbf{V}$.

In terms of the standard interpretation, $\mathbf{V}$ intuitively expresses that if it has a proof which is coded by k, then there is also a proof of its negation which is coded by some number less than k.

Prove: if Q is consistent, then $\mathbf{V}$ is formally undecidable relative to Q. (*Hint:* You will need Lemmas 22.15 and 17.)

†3. (*Second-Order Peano Arithmetic*)
    a. Describe what changes we would have to make to the first-order language of arithmetic to express the *full* (*complete*) *induction* principle:

       For every set of natural numbers X: if $0 \in X$, and if for every n, if $n \in X$ then $n + 1 \in X$, then every natural number is in X.

Formalize that principle as a scheme and define *Full (Second-Order) Peano Arithmetic* (**FPA**) in that new language to be **PA** plus that axiom scheme.

b. Show that any model of **FPA** is isomorphic to the natural numbers.

c. Conclude that a wff in the extended language (which includes the old language) is true of the natural numbers iff it is true in every model of **FPA** .

d. Conclude that there can be no axiomatic proof theory for this language that would allow us to deduce formally all consequences of the new axiom.

††4. Give a finitistic consistency proof of **PA** .

Be sure to send us a copy.

Further Reading

Joel Spencer in "Large numbers and unprovable theorems" gives a clear discussion of unprovability in **PA** in terms of functions that grow too fast. He also gives an example of a mathematical statement that is formally undecidable in **PA** which is not a translation via codings of a metamathematical assertion. The first example of a clear combinatorial principle that is true but not provable in *PA* was given in 1977 by J. Paris and L. Harrington in "A mathematical incompleteness in Peano arithmetic".

In "The present state of research into the foundations of mathematics", Gentzen, 1938, gives a very clear discussion of consistency proofs and their relation to Gödel's theorems.

See Epstein's *Classical Mathematical Logic* for a presentation of second-order arithmetic with full induction.

IV

**CHURCH'S THESIS,
CONSTRUCTIVE MATHEMATICS,
and
MATHEMATICS as MODELING**

25 Church's Thesis

In Chapter 10 we first looked at the identification of the notion of an effectively computable function with that of a total recursive function; we called that Church's Thesis. Now we've seen some of the equivalences which establish the Most Amazing Fact (Chapters 18 and 22), and we've seen applications of Church's Thesis in undecidability and incompleteness theorems about formal systems of arithmetic (Chapters 23 and 24). So we can begin to evaluate the significance and nature of that thesis.

A. History

The first statement of what has come to be called Church's Thesis was in an abstract which Church gave for his 1936 paper.

> In this paper a definition of *recursive function of positive integers* which is essentially Gödel's is adopted. And it is maintained that the notion of an effectively calculable function of positive integers should be identified with that of recursive function, since other plausible definitions of effective calculability turn out to yield notions which are either equivalent to or weaker than recursiveness.
>
> Church, 1935

In his paper Church said the following:

> We now define the notion, already discussed, of an *effectively calculable* function of positive integers by identifying it with the notion of a recursive function of positive integers (or of a λ-definable function of positive integers). This definition is thought to be justified by the considerations which follow, so far as positive justification can ever be obtained for the selection of a formal definition to correspond to an intuitive notion.
>
> Church, 1936, p. 100

Thus, Church took the identification to be a *definition*.

It was Post, 1936, who took it to be an hypothesis. To repeat from his paper, which we looked at in Chapter 10:

> The writer expects the present formulation to turn out to be logically equivalent to recursiveness in the sense of the Gödel–Church development. Its purpose, however, is not only to present a system of a certain logical potency but also, in its restricted field, of psychological fidelity. In the latter sense wider and wider formulations are contemplated. On the other hand, our aim will be to show that all such are logically reducible to formulation 1. We offer this conclusion at the present moment as a *working hypothesis*. And to our mind such is Church's identification of effective calculability with recursiveness.* Out of this hypothesis, and because of its apparent contradiction to all mathematical development starting with Cantor's proof of the non-enumerability of the points of a line, independently flows a Gödel–Church development. The success of the above program would, for us, change this hypothesis not so much to a definition or to an axiom but to a *natural law*.
>
> *Actually the work done by Church and others carries this identification considerably beyond the working hypothesis stage. But to mask this identification under a definition hides the fact that a fundamental discovery in the limitations of the mathematicizing power of Homo Sapiens has been made and blinds us to the need of its continual verification.
>
> Post, 1936, p. 291

Post objected to Church's calling the identification a definition. Post was quite explicit: He was attempting to abstract and characterize something about human capabilities. The informal notion of computability or effectiveness was about what people can do.

Turing had the same goal and, independently of Church and Post, attempted to give a formal analogue of the notion of computability, as we saw in Chapter 9. Although his machines are now seen as the most convincing analysis, his statement of his goals was less precise and certainly less memorable than Church's:

> The computable numbers may be described as the real numbers whose expressions as decimals are calculable by finite means. . . . According to my definition, a number is computable if its decimal can be written down by a machine.
>
> Turing, 1936, p. 116

It appears that Turing means by "machine" the kind of machines defined in his paper, but that is not clear. Moreover, he seems to take for granted that a function which can be computed by any one of his machines is computable, and is concerned only with the converse (see Chapter 9.A).

In 1937 Church wrote reviews of both Post's and Turing's papers. There, and in 1938, he continued to maintain that he was making a definition.

> [Post] does not, however, regard his formulation as certainly to be identified with effectiveness in the ordinary sense, but takes this identification as a "working

hypothesis" in need of continual verification. To this the reviewer would object that effectiveness in the ordinary sense has not been given an exact def-inition, and hence the working hypothesis in question has not an exact meaning. To define effectiveness as computability by an arbitrary machine, subject to restrictions of finiteness, would seem to be an adequate representation of the ordinary notion, and if this is done the need for a working hypothesis disappears.

Church, 1937a

This notion of an effective process occurs frequently in connection with mathematical problems, where it is apparently felt to have a clear meaning, but this meaning is commonly taken for granted without explanation. For our present purposes it is desirable to give an explicit definition.

Church, 1938, p. 226

The idea that it was a thesis predominated, however, as Kleene wrote in 1943:

Now, the recognition that we are dealing with a well-defined process which for each set of values of the independent variables surely terminates so as to afford a definite answer, "Yes" or "No," to a certain question about the management of termination, in other words, the recognition of effective decidability in a predi-cate, is a subjective affair. Likewise, the recognition of what may be called *effective calculability* in a function. We may assume, to begin with, an intuitive ability to recognize various individual instances of these notions. In particular, we do recognize the general recursive functions as being effectively calculable, and hence recognize the general recursive predicates as being effectively decidable.

Conversely, as a heuristic principle, such functions (predicates) as have been recognized as being effectively calculable (effectively decidable), and for which the question has been investigated, have turned out always to be general recursive, or in the intensional language, equivalent to general recursive functions (general recursive predicates). This heuristic fact, as well as certain reflections on the nature of symbolic algorithmic processes, led Church to state the following thesis. The same thesis is implicit in Turing's description of computing machines.

Thesis I: *Every effectively calculable function (effectively decidable predicate) is general recursive.*

Since a precise mathematical definition of the term effectively calculable (effectively decidable) has been wanting, we can take the thesis, together with the principle already accepted to which it is the converse, as a definition of it for the purpose of developing a mathematical theory about the term. To the extent that we have already an intuitive notion of effective calculability (effective decidability), the thesis has the character of an hypothesis—a point emphasized by Post and Church. If we consider the thesis and converse as definition, then the hypothesis is an hypothesis about the application of the mathematical theory developed from the definition. For the acceptance of the hypothesis, there are, as we have suggested, quite compelling grounds.

Kleene, 1943, p. 274

There is a certain hedging here that is not unusual: many would have liked to take it as simply an heuristic principle. Gödel originally took it so in 1934.

> Recursive functions [what we now call primitive recursive functions] have the important property that, for each given set of values of the arguments, the value of the function can be computed by a finite procedure.*
>
> * The converse seems to be true, if, besides recursions according to the scheme (2), recursions of other forms (e.g., with respect to two variables simultaneously) are admitted. This cannot be proved, since the notion of finite computation is not defined, but it serves as a heuristic principle.

Gödel, 1934, pp. 43–44

B. A Definition or a Thesis?

1. On definitions

There are two kinds of definitions. One is the kind that mathematicians normally use and is called a *nominal definition*. Included in this type are purely conventional definitions that amount to using one word or symbol in place of others. For instance, in group theory we define the symbol " α^{-1} " to stand for "the β such that $\alpha \cdot \beta = 1$ ". Or an author of a book on group theory will define the word "group" to mean any object that satisfies her axioms.

Also included in this type are definitions that are proposed as formalizations or "rational reconstructions" of imprecise or vague intuitive notions. This is the type of definition Church apparently intended.

Nominal definitions are a matter of convenience, for instance, when a long string of symbols is replaced by a single one, or they are a way to point our attention to a particular object or concept, or an attempt to replace a vague intuitive notion by a precise formal one. They are to be judged by their utility, their fruitfulness, and their aptness.

The other type of definition, more common in philosophy, is called an *absolute*, or *real definition*: a characterization of a concept or object is given by listing its essential features. For example, "human" was defined by Aristotle as "rational animal", for whatever is a human is necessarily a rational animal, and whatever is a rational animal is necessarily a human.

Absolute definitions are not about what words *mean* but about what things *are*. They are assertions in disguise. They assert that a concept or object for which we already had a word is correctly characterized by certain properties. This assumes that the concept, such as computability, which we had thought to be vague was not, either because it refers to a completely precise (though indistinct to us) platonic abstraction as Gödel holds below or because it refers to something in the world of our senses which we had previously understood poorly as Post apparently believed about the notion of effective calculability.

Whether a particular definition is classified as nominal or absolute may depend on your viewpoint. An example is the definition of "∃x" as "￢∀x￢", which we gave in Chapter 21. We put that forward as a nominal definition, a matter of convenience. A classical mathematician, however, would likely take it as an absolute definition, for he would say that it is a necessary fact that there is something which satisfies a predicate if and only if not everything fails to satisfy that predicate.

In this section we will follow the debate about whether Church's Thesis is a definition, in the sense of a nominal definition, to be judged only by whether it is apt or fruitful, or whether it is a thesis, in the sense of an absolute definition, and hence true or false.

2. Kalmár, from "An argument against the plausibility of Church's Thesis"

In 1957 Kalmár revived the debate about whether the identification should be taken as a thesis or a definition.

> In his famous investigations on unsolvable arithmetical problems, Church used a working hypothesis, viz. the identification of the notion of effectively calculable functions with that of general recursive (or, equivalently, λ-definable) functions. This working hypothesis is known under the name Church's thesis. It has several equivalent forms—e.g. Turing's identification of the notion of effectively calculable functions with that of functions computable by means of a Turing machine, or Markov's principle of the normalizability of algorithms—which are generally accepted in the investigations on unsolvable problems.
>
> In the present contribution, I shall not disprove Church's thesis. Church's thesis is not a mathematical theorem which can be proved or disproved in the exact mathematical sense, for it states the identity of two notions only one of which is mathematically defined while the other is used by mathematicians without exact definition. Of course, Church's thesis can be masked under a definition: we call an arithmetical function effectively calculable if and only if it is general recursive, venturing however that once in the future, somebody will define a function which is on the one hand, not effectively calculable in the sense defined thus, on the other hand, its value obviously can be effectively calculated for any given arguments. Similarly, in defining a problem, containing a parameter which runs through the natural numbers, to be solvable if and only if its characteristic function is general recursive, one takes the risk that somebody in the future will solve a problem which is unsolvable under this definition. For this reason, it seems to me better to regard such statements as Church's thesis, or the identity of solvable problems with those having a general recursive characteristic function as propositions rather than definitions, however, not mathematical but "pre-mathematical" ones. The more than two pages of Church's paper [1936] filled with plausibility (hence pre-mathematical) arguments for his thesis, show that his opinion about this question does not differ much from mine.
>
> Kalmár, 1957, pp. 72–73

3. A platonist perspective: Gödel

For most platonists, Church's identification is a thesis. The class of computable functions exists independently of us and our investigations, and the question is whether that class is the same as the class of total recursive functions. The leading exponent of platonism in mathematics recently has been Gödel; Wang has reported on conversations with Gödel about formalizing intuitive concepts.

Gödel on mechanical procedures and the perception of concepts

"If we begin with a vague intuitive concept, how can we find a sharp concept to correspond to it faithfully?" The answer Gödel gives is that the sharp concept is there all along, only we did not perceive it clearly at first. This is similar to our perception of an animal first far away and then nearby. We had not perceived the sharp concept of mechanical procedures sharply before Turing, who brought us to the right perspective. And then we do perceive clearly the sharp concept. There are more similarities than differences between sense perceptions and the perceptions of concepts. In fact, physical objects are perceived more indirectly than concepts. The analog of perceiving sense objects from different angles is the perception of different logically equivalent concepts. If there is nothing sharp to begin with, it is hard to understand how, in many cases, a vague concept can uniquely determine a sharp one without even the *slightest* freedom of choice. "Trying to see (i.e. understand) a concept more clearly" is the correct way of expressing the phenomenon vaguely described as "examining what we mean by a word." . . .

Gödel mentions that the precise concept meant by the intuitive idea of velocity clearly is *ds/dt*, and the precise concept meant by "size" (as opposed to "shape"), e.g. of a lot, clearly is equivalent with Peano measure in the cases where either concept is applicable. In these cases the solutions are *unquestionably* unique, which here is due to the fact that only they satisfy certain axioms which, on closer inspection, we find to be undeniably implied in the concept we had. For example, congruent figures have the same area, a part has no larger size than the whole, etc.

There are cases where we mix two or more exact concepts in one intuitive concept and then we seem to arrive at paradoxical results. One example is the concept of continuity. Our prior intuition contains an ambiguity between smooth curves and continuous movements. We are not committed to the one or the other in our prior intuition. In the sense of continuous movements a curve remains continuous when it includes vibrations in every interval of time, however small, provided only that their amplitudes tend toward 0 if the time interval does. But such a curve is no longer smooth. The concept of smooth curves is seen sharply through the exact concept of differentiability. We find the example of space-filling continuous curves disturbing because we feel intuitively that a continuous curve, in the sense of being a smooth one, cannot fill the space. When we realize that there are two different sharp concepts mixed together in the intuitive concept, the paradox disappears. Here the analogy with sense perception is close. We cannot distinguish two neighboring stars a long distance away. But by using a telescope we can see that there are indeed two stars.

Wang, 1974, pp. 84–86

And here are Gödel's own words about the significance of Turing's formalization of computable functions.

> Tarski has stressed in his lecture (and I think justly) the great importance of the concept of general recursiveness (or Turing's computability). It seems to me that this importance is largely due to the fact that with this concept one has for the first time succeeded in giving an absolute definition of an interesting epistemological notion, i.e., one not depending on the formalism chosen. In all other cases treated previously, such as demonstrability or definability, one has been able to define them only relative to a given language, and for each individual language it is clear that the one thus obtained is not the one looked for. For the concept of computability however, although it is merely a special kind of demonstrability or decidability the situation is different. By a kind of miracle it is not necessary to distinguish orders, and the diagonal procedure does not lead outside the defined notion.

Gödel, 1946, p. 84

4. Other examples: definitions or theses?

Let us consider the example of a schoolteacher who wishes to teach a child the definition of a circle. The child already knows "what a circle is", and we can take that either in the sense of having learned the notion from experience or having acquired it in the platonic heaven above the heavens. The teacher then tells him that a circle is the locus of all points equidistant from some one point. That sounds pretty odd, and the words themselves may have to be explained to the child. Certainly, the teacher will have to convince him that the definition is right by reflecting on the meaning of the words and showing how a circle can be drawn by tying down one end of a string and attaching a pencil or piece of chalk to the other. The teacher has to show the child that his intuitive concept of circle, which he might only be able to express by "It's round," corresponds to the rigid formal definition.

Now does this "evidence" show that the teacher was really thinking of a "circle thesis", as Kalmár might suggest?

Or consider the example of attempts to characterize continuity. There was, and we have an intuitive notion of a continuous curve, one which can be drawn with no breaks. The "can be drawn" was refined into: at every point the function agrees with its limit at that point. And in the nineteenth century we come to the characterization due to Cauchy and Weierstrass:

> A function is continuous iff for every point x in its domain, given any positive number ε there is a positive number δ such that if $|z - x| < \delta$, then $|f(z) - f(x)| < \varepsilon$.

This is presented nowadays as a definition. But shouldn't we call this the Cauchy–Weierstrass Thesis? What guarantee do we have that this story of δ's and ε's is exactly what our intuitive idea of limit was? That doubt is not absurd when we remember that behind the δ's and ε's are the natural numbers (the archimedean principle that for every $\varepsilon > 0$ there is some natural number n such that $1 < n \cdot \varepsilon$) so that we have reduced the idea of an approximation and a curve with no breaks to the idea of natural numbers and inequalities. Certainly that is something we should justify, especially since a consequence of this "definition" and the characterization of smoothness as differentiability is that there is a curve which has no breaks but is not smooth anywhere (a continuous nowhere differentiable function).

Why should we consider these examples different from what Church did? Oswaldo Chateaubriand has suggested to us that it might be because the preformal notions in these two examples were originally conceived to be part of mathematics, whereas constructivity or computability was all along considered to be a nonmathematical notion. Perhaps, but it seems to us the notion of a circle or an unbroken curve is not inherently mathematical and is learned by children long before they can add.

No, the difference is that these definitions resolved only mathematical queries; they led to certain formal mathematics being applied to problems from our physical experience. Church's thesis/definition, on the other hand, was motivated and adopted to resolve philosophical problems about the meaning and justification of mathematics. In philosophical contexts, agreements are hard to reach and the near unanimity of agreement on Church's thesis/definition, which we will discuss in the next section, has been nothing less than astounding.

We believe that each one of these examples is a definition, or each is a thesis. But which of these we call them depends on what we believe we are doing when we do mathematics: abstracting from experience, or coming to see clearly platonic abstract concepts.

5. On the use of Church's Thesis

All the authors we have quoted above are in agreement on one thing: the thesis/definition is not part of mathematics, it is not part of the theory of recursive functions or Turing machines. Those theories are interesting in their own right even if Church's thesis/definition were to be abandoned. Whatever else the Most Amazing Fact establishes, it shows that the notion which is stable under so many different formulations must be fundamental. Thus, it is simply a confusion when a modern practitioner of recursion theory writes the following:

> Church's Thesis is more than a philosophical statement about the nature of computability. It is a useful tool in proofs. We shall often find, in the following chapters, that it is easy to give an informal description of a function from which it appears that the function is computable. Again, we may give an informal description of how to decide whether or not an element is in a given set. To show

that the function is partial recursive, or that the set is recursive, might involve some long and messy calculations. Such calculations, which are not likely to give any insight as to what is happening, are usually replaced by an appeal to Church's Thesis. That is, since we have given an intuitive argument that the function is computable (or that the set is decidable), we then claim that Church's Thesis tells us that the function is partial recursive. This saves tedious calculations; readers should convince themselves, however, that any time Church's Thesis is used, a formal proof can be made by anyone who is sufficiently industrious.

D. Cohen, 1987, p. 104

To invoke Church's thesis when "the proof is left to the reader" is meant amounts to giving a fancy name to a routine piece of mathematics while at the same time denigrating the actual mathematics.

Rather, Church's thesis/definition is about the applicability of the formal mathematical theory. It is a bridge between the mathematics and the philosophical problems that generated the mathematics. Whether it is the right bridge is the question, and this can be asked whether we take it as a definition (is the definition apt?) or as a thesis (is the thesis true?).

Why then do people believe it is the right bridge?

C. Arguments For and Against

1. For

Kleene has stated the case for Church's thesis/definition. He takes it as evident that every recursive function is computable and discusses only whether every computable function is recursive.

(A) Heuristic evidence

(A1) Every particular effectively calculable function, and every operation for defining a function effectively from other functions, for which the question has been investigated, has proved to be general recursive. A great variety of effectively calculable functions, of classes of effectively calculable functions, and of operations for defining functions effectively from other functions, selected with the intention of exhausting known types, have been investigated.

(A2) The methods for showing effectively calculable functions to be general recursive have been developed to a degree which virtually excludes doubt that one could describe an effective process for determining the values of the function which would not be transformed by these methods into a general recursive definition of the function.

(A3) The exploration of various methods which might be expected to lead to a function outside the class of the general recursive functions has in every case shown either that the method does not actually lead outside or that the new function obtained cannot be considered as effectively defined, i.e. its definition provides no effective process of calculation. In particular, the latter is the case for the Cantor diagonal method. . . .

(B) Equivalence of diverse formulations.
[Kleene discusses what we have called "The Most Amazing Fact".]

(C) Turing's concept of a computing machine.
Turing's computable functions (1936–7) are those which can be computed by a machine of a kind which is designed, according to his analysis, to reproduce all the sorts of operations which a human computer could perform, working according to preassigned instructions. Turing's notion is thus the result of a direct attempt to formulate mathematically the notion of effective calculability, while the other notions [e.g. λ-definability, recursiveness] arose differently and were afterwards identified with effective calculability. Turing's formulation hence constitutes an independent statement of Church's thesis (in equivalent terms). Post 1936 gave a similar formulation. . . .

(D) Symbolic logics and symbolic algorithms.
In brief, . . . if the individual operations or rules of a formal system or symbolic algorithm used to define a function are general recursive, then the whole is general recursive [see Chapter 19.F, Chapter 21.F, and p. 217]. So we could include [these] as particular examples of operations or methods of definition under (*A1*).

Kleene, 1952, pp. 319–323

In Section D we will evaluate the significance of this evidence. But first let's turn to the arguments against Church's thesis/definition.

2. Not every recursive function is computable: theoretical vs. actual computability

A major criticism of the thesis/definition is that there are many recursive functions (or, nonextensionally, programs) for which too much time or material would be required to perform the calculations—not just now, but ever. Recall, for instance, the Ackermann-Péter function ψ. Try to calculate $\psi(47, 14)$; or if you have a year or two available on a supercomputer, start the calculation of $\psi(8489727, 12)$. We know that ψ dominates all primitive recursive functions, and so by suitably formalizing the notion of number of steps in a computation, say by using the universal computation predicate, we know that for all but a finite and actually rather small number of m, the calculation of $\psi(m, n)$ takes more steps than, say:

$$\left.\begin{array}{l} 100^{\displaystyle 100^{\displaystyle \cdot^{\displaystyle \cdot^{\displaystyle \cdot^{100}}}}} \\ 100 \end{array}\right\} n \text{ times}$$

So we could not actually calculate such values. And this doesn't depend on the value of $\psi(m, n)$ being very large, since the same applies to $Ev \circ \psi$. Worse, we know that ψ is rather low on the hierarchy of complexity of computation (see Chapter 13).

Further, we can produce a recursive function f that by any even generous plausible measure of the amount of material needed to do the calculations (say one atom for each decimal digit in the calculation) would require more matter than there is in the solar system to calculate $f(0)$ or $f(x)$ for any x. Are we justified in calling such a function computable? The usual response is of the sort given by Mendelson:

> Human computability is not the same as effective computability. A function is considered effectively computable if its value can be computed in an effective way in a finite number of steps, but there is no bound on the number of steps required for any given computation. Thus, the fact that there are effectively computable functions which may not be humanly computable has nothing to do with Church's Thesis.
>
> Mendelson, 1963, p. 202

This is too cavalier. Certainly what Turing and Post were doing was analyzing what a person or machine could do, and therein lies the power of their analyses to convince. The question is what is meant by "could". The same problem arises when we say that a natural number is any number we can reach by successively adding 1, starting at 0. If we say there are infinitely many natural numbers, then we must be interpreting the word "can" quite generously, since there is no way we or anything else on earth can add 1 more times than there are atoms on the earth to represent addition (or if you don't want to represent them, consider how long it would take to add in your mind). Theoretical versus actual computation is analogous to theoretical versus actual finiteness, a topic taken up in the paper by van Dantzig in the next chapter.

Because of the evidence for it, Church's thesis/definition is useful as a bound, a limitation on what we would willingly call computable. Perhaps a function being recursive does not mean that we could compute its values in our lifetime for even small inputs; but if a function can be shown not to be recursive, then we can feel confident that no one would be justified in calling it computable. That is, Church's thesis/definition is useful primarily in establishing negative results about computability (see Goodstein, 1951a). And this was what we were interested in when we began our studies of computability: Is there an effective procedure for deciding the truth-value of each statement in arithmetic? Is there an effective procedure for establishing the consistency of infinitistic mathematics? With Church's thesis/definition providing an upper bound for computability, we can answer these queries in the negative. On the other hand, in those cases such as arithmetic with addition but not multiplication where a recursive procedure has been found for deciding the truth-value of statements in the theory, the question immediately arises: Can it be done in a polynomial number of steps depending on the length of the wff? That is, is the decision procedure actually (versus theoretically) computable?

Note the similarity to the case of continuity described earlier: the example of a continuous nowhere differentiable function is likely to make us think of the definition of continuity as an upper bound on what we would be willing to call continuous.

3. Interpretation of the quantifiers in the thesis/definition

Let's state Church's thesis/definition in a precise form for recursive functions using our notation.

> A function f is computable iff *there exists* an index e such that
> *for all* $\vec{x}$, *there exists* a computation (coded by some q) such that
> $\varphi_e(\vec{x})\downarrow = f(\vec{x})$, that is, $C(e, \langle \vec{x} \rangle, (q)_0, q)$ and $(q)_0 = f(\vec{x})$.

There are two existential quantifiers. Péter, 1957, and Heyting, 1962, have argued that both of these must be interpreted effectively. That is, given a function which is claimed to be computable, an index should be effectively produced for it and there should be an effective proof that for each $\vec{x}$ the computation halts.

Church had already anticipated this objection in 1936 (for his definition, a function is recursive if equations of a certain type can be found for it).

> The reader may object that this algorithm cannot be held to provide an effective calculation of the required particular value ... unless the proof is constructive that the required equation ... will ultimately be found. But if so this merely means that he should take the existential quantifier which appears in our definition of a set of recursion equations in a constructive sense. What the criterion of constructiveness shall be is left to the reader.
>
> Church, 1936, p. 95 n

But here is what Heyting has to say:

> The notion of a recursive function, which had been invented in order to make that of a calculable function more precise, is interpreted by many mathematicians in such a way, that it loses every connection with calculability, because they interpret non-constructively the existential quantifier which occurs in the definition.
>
> Of course every finite set is primitive recursive. But is every subset of a finite set recursive? Who can calculate the Gödel number of the characteristic function of the set of all non-Fermat exponents less than 10^{10} [those numbers n for which there are x, y, z such that $x^n + y^n = z^n$] or of the set
> $$P_n = \{ x \mid x < n \ \& \ (\mathrm{E}y) T_1(x,x,y) \} \quad [\text{i.e. } \{ x : x < n \ \& \ \varphi_x(x)\downarrow \}],$$
> where n is a given natural number? The answer depends upon the logic which is adopted. If recursiveness is interpreted non-constructively, then P_n constitutes a counter-example to the converse of Church's thesis.
>
> Heyting, 1962, pp. 195–196

That is, the existential quantifier must be constructive (non-extensional) or it has no force. But, as Péter argues, if we interpret it constructively then we have a vicious circle, for we are defining constructivity in terms of itself.

Platonistically there is no problem: the quantifiers need not be interpreted constructively, for such sets as Heyting describes either are or are not recursive. It is a different problem whether we can prove that they are recursive, or whether we can exhibit the index of a machine that computes their characteristic function. Mendelson states it rather baldly for the characterization of computable functions in terms of equations.

> In addition, for a function to be computable by a system of equations it is not necessary that human beings ever know this fact, just as it is not necessary for human beings to prove a given function continuous in order that the function be continuous.
>
> Mendelson, 1963, p. 202

This is no argument; it is only a viewpoint. In the next chapter we will look at the brand of constructivism of which Heyting is a spokesman, called intuitionism.

The debate about the second existential quantifier is about whether we should demand that we know in advance that all computations halt, as we do with the primitive recursive functions. We know from Chapter 15 that we cannot in general predict recursively which computations halt, for if we could we would have a contradiction by diagonalizing.

But the requirement that we must know in advance that the computations halt for the function to be deemed computable has no force against the characterization of effective processes (as opposed to functions) as partial recursive programs. The nature of effectiveness is such that we cannot expect to be rewarded like good little children who play by the rules. The significance of the unsolvability of the halting problem is that the fundamental notion is that of an effective procedure, not an effective function: What is effective step by step may lead us nowhere (cf. the remarks by Gödel in Chapter 15.D, pp. 126–127). This, as our inability to reduce the infinite to the finite, seems to be another example that there is no certainty in an uncertain world.

4. A paradoxical consequence?

After what we've discussed so far you might find it surprising that Kalmár, 1957, has argued that Church's thesis/definition gives too narrow a characterization of effectivity. Kalmár doesn't actually produce a function that he claims to be computable but not recursive. Rather he shows what he considers to be a paradoxical consequence of the thesis/definition. Here is his argument.

Take a recursively enumerable set that is not recursive, such as K. Consider the following procedure:

Given a number p, to decide "$p \in K$", simultaneously:
1. Generate the elements of K, and
2. Look for a proof "not in the frame of some fixed postulate system but by means of arbitrary—of course, correct—arguments" that $p \notin K$.

Since K is not recursive, by Church's thesis/definition this procedure is not effective. But then there must be some p for which the proposition "$p \in K$" is false, that is $p \notin K$ yet that fact cannot be proved by any correct means.

> The proposition stating that, for this p, ["$p \in K$"] would be undecidable, with other words, the problem if this proposition holds or not, would be unsolvable, not in Gödel's sense of a proposition neither provable nor disprovable in the frame of a fixed postulate system, nor in Church's sense of a problem with a parameter for which no general recursive method exists to decide, for any given value of the parameter in a finite number of steps, which is the correct answer to the corresponding particular case of the problem, "yes" or "no". As a matter of fact, the problem, if the proposition in question holds or not, does not contain any parameter and, supposing Church's thesis, *the proposition itself can be neither proved nor disproved*, not only in the frame of a fixed postulate system, but *even admitting any correct means*. It cannot be proved for it is false and it cannot be disproved for its negation cannot be proved. According to my knowledge, this consequence of Church's thesis, viz. the existence of a proposition (without parameter) which is undecidable in this, *really absolute* sense, has not been remarked so far.
>
> However, this "absolutely undecidable proposition" has a defect of beauty: we can decide it, for we know, it is false. Hence, *Church's thesis implies the existence of an absolutely undecidable proposition which can be decided viz., it is false*, or, in another formulation, *the existence of an absolutely unsolvable problem with a known definite solution*, a very strange consequence indeed.
>
> Kalmár, 1957, p. 75

Kalmár's argument is flawed. Even granting that the description actually corresponds to a function (which it certainly does by intuitionist standards, cf. Chapter 26.A.2), there is a problem. Kalmár refers to "this proposition"—in our formulation "$p \in K$"—as if we indeed had such a number p. Yet we cannot produce such a proposition, for if we could we would have a contradiction: if we can prove that the proposition is "$p \in K$" then that must be because $p \notin K$, and hence we have a proof that $p \notin K$ *and* that there is no proof that $p \notin K$. What we can do is prove that there must exist some $p \notin K$ for which there is no proof that $p \notin K$, though we cannot say which p this is. At least in classical mathematics that is quite sufficient. But for a constructivist it seems more than a little odd that not only do we have a nonconstructive existence proof, but a proof that it cannot be made constructive.

D. Interpreting the Evidence

The most significant pieces of evidence for Church's thesis/definition, as discussed above, are (i) the fact that despite many concerted efforts no one has been able to produce a function that is clearly computable and not recursive, (ii) Turing's analysis that the recursive functions are computable, and (iii) the Most Amazing Fact that the various attempts to formalize the notion of computability by what appeared to be radically different means have all been shown by effective translations to be equivalent. How are we to interpret these?

The platonist can explain the evidence easily: the class of computable functions was there all along and we have finally managed to "see" it. Indeed, the Most Amazing Fact is taken as good evidence that there is a platonic reality of abstract objects independent of us.

But we can argue along with Post that Church's thesis/definition is about human capabilities. Perhaps the Most Amazing Fact is a consequence or reflection of the structure of our bodies and brains.

Or perhaps it is a cultural artifact, a product of our Western mathematical–scientific culture and age. That no one has yet produced a computable function which is not recursive establishes nothing: for millennia logicians thought that all forms of correct reasoning which could be codified were in Aristotle's syllogistic, whereas now many other formal methods are taken (seen?) to be correct (see, e.g., Epstein, 1994). Moreover, in a different age the ancient Greeks took the notion of constructivity, for both numbers and geometric figures, to be quite different: constructible by straightedge and compass.

Or finally, consider what Kreisel has to say:

> The support for Church's Thesis ... consists above all in the analysis of
> machine-like behavior and in a number of closure conditions, for example
> diagonalization. ... It certainly does not consist in the so-called empirical
> support; namely the equivalence of different characterizations: what excludes
> the case of a *systematic* error? (Cf. the overwhelming empirical support from
> ordinary mathematics for: if an arithmetic identity is provable at all, it is
> provable in classical first-order arithmetic; they all overlook the principle
> involved in, for example, consistency proofs.)
>
> Kreisel, 1965, p. 144

And whether it is an error, or whether there is any sense in which we are "right" and if so why, we are incapable of verifying. These various interpretations do not correspond to various different assertions whose truth or falsity we can investigate, for what "evidence" could there ever be for Plato's heaven above the heavens? These various interpretations correspond to different ways of understanding the nature of mathematics, which we will explore in the next two chapters. Amongst these, there appears to be no definitive choice.

Exercises

1. Is it a thesis or a definition? Rewrite the quotation from Kalmár in Section B.2 up to the end of the second sentence in the second paragraph replacing "Church's thesis" everywhere by "the Cauchy–Weierstrass continuity thesis" and making the other necessary changes. Is the analogy apt? Is our analogy of Church's thesis/ definition with the definition of a circle apt?

2. Are we justified in using Church's thesis within the mathematical development of recursive function theory? Rewrite the quotation from Cohen above in Section B.5 replacing "Church's thesis" everywhere by "the Cauchy–Weierstrass continuity thesis" and making the other necessary changes. Would that paragraph be acceptable in a freshman calculus course? In a graduate level research text (cf. Lerman, p. 9) ?

3. Compare Wang's report on Gödel's views contained in Section B.3 with the quotations from Plato in Chapter 2. In what way could the evidence for Church's thesis/definition be construed as evidence for platonism?

4. In Section C.2 we described certain functions as being recursive but not "actually computable". Are we justified in calling such a function "effective" or "computable"? Can you make precise the distinction between actually computable and theoretically computable?

5. Are we justified in making the distinction between a process which is effective and which might not halt, and a function which is computable and must give an output for every natural number (Section C.3)?

6. Argue against our conclusion that there is no possible way to resolve which of the various interpretations of the evidence for Church's thesis/definition is correct.

Further Reading

Odifreddi in *Classical Recursion Theory* presents a thorough discussion of Church's thesis/definition, which we highly recommend. He analyses it in its various guises: as a thesis about mechanism and hence about physics and computers, as a thesis about computers and thought, as a thesis about the nature of the brain, and as a thesis about constructivism in mathematics.

Our discussion of the history of Church's Thesis has been confined to the published sources from that period. Several recent articles draw on personal recollections, in particular an interesting taped oral discussion edited by Crossley, 1975, and Kleene's "Origins of recursive function theory" (a very difficult paper). Several articles in *The Universal Turing Machine, A Half-Century Survey*, edited by Rolf Herken, discuss the history and significance of Church's Thesis.

For a survey and history of various notions of definition see Abelson's entry, "Definitions" in *The Encyclopedia of Philosophy*.

26 Constructivist Views of Mathematics

In this chapter we will look at several views of the nature of mathematics which to one degree or another reject infinitistic assumptions.

Kronecker had been an early and powerful opponent of the introduction of infinite sets and nonconstructive existence proofs into mathematics even before the crises of the set-theoretic paradoxes. As Professor of Mathematics in Berlin he had considerable influence, making it difficult for Cantor to publish anything and impossible for Cantor to be promoted. Hilbert, too, had to take into account Kronecker's views in the early part of his career, and his paper "On the infinite" (Chapter 7) should be seen in part as the culmination of his concern that Kronecker's views would predominate.

However, the first fully developed alternative to what is now called *classical mathematics*, that is, mathematics based on the use of infinite sets and classical reasoning (Chapter 21.B), was the movement called intuitionism initiated by Brouwer. His paper in Section A is one of the first on the subject, and it is important to keep in mind that it was published in 1913, when axiomatic set theory was still in its infancy and long before Hilbert's paper "On the infinite" (see p. 57).

Recursive analysis, which we consider in Section B, is quite different from Brouwer's intuitionism. Both Goodstein and a Russian school develop the theory of real numbers by embracing Church's Thesis: a real number is, in essence, a decimal whose digits are the output of a recursive function.

Bishop, in Section C, criticizes Brouwer's work as too imprecise and infinitistic, and recursive analysis as too formal and limited. His goal is to replace classical mathematics with a body of constructive results; negative theorems, which would state that something is not constructive or does not exist, are of no interest to him since they would involve precisely delimiting the boundaries of constructivity. He takes the notion of a constructive function as primitive and, along with Brouwer, refuses to identify it with any formal notion.

But Nicolas Goodman, in Section D, criticizes Bishop's work in turn by pointing out that a theoretical construction which could not be realized in practice or a natural number we "could" write but never will because there are not enough physical resources on earth to do so, are abstract, not finitistic, and hence we are justified in using classical methods in reasoning about them. In this sense, he says, constructive mathematics is a partner with classical mathematics, its interest being directed to different results.

Van Dantzig and Isles make the same criticisms in Section E but conclude instead that a more finitistic approach is called for. We do not "have" the natural numbers as a well-defined unique sequence but only different notations for specific natural numbers, notations which may be incomparable. The status of induction is then in question, and we conclude our readings with Isle's reinterpretation of the Halting Problem for Turing machines.

A. Intuitionism

1. L. E. J. Brouwer, from "Intuitionism and formalism", 1913

The subject for which I am asking your attention deals with the foundations of mathematics. To understand the development of the opposing theories existing in this field one must first gain a clear understanding of the concept "science"; for it is as a part of science that mathematics originally took its place in human thought.

By science we mean the systematic cataloguing by means of laws of nature of causal sequences of phenomena, i.e., sequences of phenomena which for individual or social purposes it is convenient to consider as repeating themselves identically,—and more particularly of such causal sequences as are of importance in social relations.

That science lends such great power to man in his action upon nature is due to the fact that the steadily improving cataloguing of ever more causal sequences of phenomena gives greater and greater possibility of bringing about desired phenomena, difficult or impossible to evoke directly, by evoking other phenomena connected with the first by causal sequences. And that man always and everywhere creates order in nature is due to the fact that he not only isolates the causal sequences of phenomena (i.e., he strives to keep them free from disturbing secondary phenomena) but also supplements them with phenomena caused by his own activity, thus making them of wider applicability. Among the latter phenomena the results of counting and measuring take so important a place, that a large number of the natural laws introduced by science treat only of the mutual relations between the results of counting and measuring. It is well to notice in this connection that a natural law in the statement of which measurable magnitudes occur can only be understood to hold in nature with a certain degree of approximation; indeed natural laws as a rule are not proof against sufficient refinement of the measuring tools.

The exceptions to this rule have from ancient times been practical arithmetic and geometry on the one hand, and the dynamics of rigid bodies and celestial mechanics on the other hand. Both these groups have so far resisted all improvements in the tools of observation. But while this has usually been looked upon as something accidental and temporal for the latter group, and while one has always been prepared to see these sciences descend to the rank of approximate theories, until comparatively recent times there has been absolute confidence that no experiment could ever disturb the exactness of the laws of arithmetic and geometry; this confidence is expressed in the statement that mathematics is "the" exact science.

On what grounds the conviction of the unassailable exactness of mathematical laws is based has for centuries been an object of philosophical research, and two points of view may here be distinguished, *intuitionism* (largely French) and *formalism* (largely German). In many respects these two viewpoints have become more and more definitely opposed to each other; but during recent years they have reached agreement as to this, that the exact validity of mathematical laws as laws of nature is out of the question. The question where mathematical exactness does exist, is answered differently by the two sides; the intuitionist says: in the human intellect, the formalist says: on paper.

In Kant we find an old form of intuitionism, now almost completely abandoned, in which time and space are taken to be forms of conception inherent in human reason. For Kant the axioms of arithmetic and geometry were synthetic a priori judgments, i.e., judgments independent of experience and not capable of analytical demonstration; and this explained their apodictic [necessarily true] exactness in the world of experience as well as in abstracto. For Kant, therefore, the possibility of disproving arithmetical and geometrical laws experimentally was not only excluded by a firm belief, but it was entirely unthinkable.

Diametrically opposed to this is the view of formalism, which maintains that human reason does not have at its disposal exact images either of straight lines or of numbers larger than ten, for example, and that therefore these mathematical entities do not have existence in our conception of nature any more than in nature itself. It is true that from certain relations among mathematical entities, which we assume as axioms, we deduce other relations according to fixed laws, in the conviction that in this way we derive truths from truths by logical reasoning, but this non-mathematical conviction of truth or legitimacy has no exactness whatever and is nothing but a vague sensation of delight arising from the knowledge of the efficacy of the projection into nature of these relations and laws of reasoning. For the formalist therefore mathematical exactness consists merely in the method of developing the series of relations, and is independent of the significance one might want to give to the relations or the entities which they relate. And for the consistent formalist these meaningless series of relations to which mathematics is reduced have mathematical existence only when they have been represented in spoken or written language together with the mathematical-logical laws upon which their development depends, thus forming what is called symbolic logic.

Because the usual spoken or written languages do not in the least satisfy the requirements of consistency demanded of this symbolic logic, formalists try to avoid the use of ordinary language in mathematics. How far this may be carried is shown by the modern Italian school of formalists, whose leader, Peano, published one of his most important discoveries concerning the existence of integrals of real differential equations in the *Mathematische Annalen* in the language of symbolic logic; the result was that it could only be read by a few of the initiated and that it did not become generally available until one of these had translated the article into German.

The viewpoint of the formalist must lead to the conviction that if other symbolic formulas should be substituted for the ones that now represent the fundamental mathematical relations and the mathematical-logical laws, the absence of the sensation of delight, called "consciousness of legitimacy," which might be the result of such substitution would not in the least invalidate its mathematical exactness. To the philosopher or to the anthropologist, but not to the mathematician, belongs the task of investigating why certain systems of symbolic logic rather than others may be effectively projected upon nature. Not to the mathematician, but to the psychologist, belongs the task of explaining why we believe in certain systems of symbolic logic and not in others, in particular why we are averse to the so-called contradictory systems in which the negative as well as the positive of certain propositions are valid.*

As long as the intuitionists adhered to the theory of Kant it seemed that the development of mathematics in the nineteenth century put them in an ever weaker position with regard to the formalists. For in the first place this development showed repeatedly how complete theories could be carried over from one domain of mathematics to another: projective geometry, for example, remained unchanged under the interchange of the roles of point and straight line, an important part of the arithmetic of real numbers remained valid for various complex number fields and nearly all the theorems of elementary geometry remained true for non-archimedian geometry, in which there exists for every straight line segment another such segment infinitesimal with respect to the first. These discoveries seemed to indicate indeed that of a mathematical theory only the logical form was of importance and that one need no more be concerned with the material than it is necessary to think of the significance of the digit groups with which one operates, for the correct solution of a problem in arithmetic.

But the most serious blow for the Kantian theory was the discovery of non-euclidean geometry, a consistent theory developed from a set of axioms differing from that of elementary geometry only in this respect that the parallel axiom was replaced by its negative. For this showed that the phenomena usually described in the language of elementary geometry may be described with equal exactness, though frequently less compactly in the language of non-euclidean geometry; hence it is not only impossible to hold that the space of our

* See Mannoury, *Methodologisches und Philosophisches zur Elementarmathematik*, pp. 149–154.

experience has the properties of elementary geometry but it has no significance to ask for *the* geometry which would be true for the space of our experience. It is true that elementary geometry is better suited than any other to the description of the laws of kinematics of rigid bodies and hence of a large number of natural phenomena, but with some patience it would be possible to make objects for which the kinematics would be more easily interpretable in terms of non-euclidean than in terms of euclidean geometry.*

However weak the position of intuitionism seemed to be after this period of mathematical development, it has recovered by abandoning Kant's apriority of space but adhering the more resolutely to the apriority of time. This neo-intuitionism considers the falling apart of the moments of life into qualitatively different parts, to be reunited only while remaining separated by time, as the fundamental phenomenon of the human intellect, passing by abstracting from its emotional content into the fundamental phenomenon of mathematical thinking, the intuition of the bare two-oneness. This intuition of two-oneness, the basal intuition of mathematics, creates not only the numbers one and two, but also all finite ordinal numbers, inasmuch as one of the elements of the two-oneness may be thought of as a new two-oneness, which process may be repeated indefinitely; this gives rise still further to the smallest infinite ordinal number ω. Finally this basal intuition of mathematics, in which the connected and the separate, the continuous and the discrete are united, gives rise immediately to the intuition of the linear continuum, i.e., of the "between," which is not exhaustible by the interposition of new units and which therefore can never be thought of as a mere collection of units.

In this way the apriority of time does not only qualify the properties of arithmetic as synthetic a priori judgments, but it does the same for those of geometry, and not only for elementary two- and three-dimensional geometry, but for non-euclidean and n-dimensional geometries as well. For since Descartes we have learned to reduce all these geometries to arithmetic by means of the calculus of coordinates.

From the present point of view of intuitionism therefore all mathematical sets of units which are entitled to that name can be developed out of the basal intuition, and this can only be done by combining a finite number of times the two operations: "to create a finite ordinal number" and "to create the infinite ordinal number ω"; here it is to be understood that for the latter purpose any previously constructed set or any previously performed constructive operation may be taken as a unit. Consequently the intuitionist recognizes only the existence of denumerable sets, i.e., sets whose elements may be brought into one-to-one correspondence either with the elements of a finite ordinal number or with those of the infinite ordinal number ω. And in the construction of these sets neither the ordinary language nor any symbolic language can have any other role than that of serving as a non-mathematical auxiliary, to assist the mathematical memory or to enable different individuals to build up the same set.

For this reason the intuitionist can never feel assured of the exactness of a

* See Poincaré, *Science and Hypothesis*, [Dover, 1952] p. 104

mathematical theory by such guarantees as the proof of its being non-contradictory, the possibility of defining its concepts by a finite number of words, or the practical certainty that it will never lead to a misunderstanding in human relations.

As has been stated above, the formalist wishes to leave to the psychologist the task of selecting the "truly-mathematical" language from among the many symbolic languages that may be consistently developed. Inasmuch as psychology has not yet begun in this task, formalism is compelled to mark off, at least temporarily, the domain that it wishes to consider as "true mathematics" and to lay down for that purpose a definite system of axioms and laws of reasoning if it does not wish to see its work doomed to sterility. The various ways in which this attempt has actually been made all follow the same leading idea, viz., the presupposition of the existence of a world of mathematical objects, a world independent of the thinking individual, obeying the laws of classical logic and whose objects may possess with respect to each other the "relation of a set to its elements." With reference to this relation various axioms are postulated, suggested by the practice with natural finite sets; the principal of these are: "*a set is determined by its elements*"; "*for any two mathematical objects it is decided whether or not one of them is contained in the other one as an element*"; "*to every set belongs another set containing as its elements nothing but the subsets of the given set*"; the axiom of selection: "*a set which is split into subsets contains at least one subset which contains one and not more than one element of each of the first subsets*"; the axiom of inclusion: "*if for any mathematical object it is decided whether a certain property is valid for it or not, then there exists a set containing nothing but those objects for which the property does hold*"; the axiom of composition: "*the elements of all sets that belong to a set of sets form a new set.*"

On the basis of such a set of axioms the formalist develops now in the first place the theory of "finite sets." A set is called finite if its elements can not be brought into one-to-one correspondence with the elements of one of its subsets; by means of relatively complicated reasoning the principle of complete induction is proved to be a fundamental property of these sets; this principle states that a property will be true for all finite sets if, first, it is true for all sets containing a single element, and second, its validity for an arbitrary finite set follows from its validity for this same set reduced by a single one of its elements. That the formalist must give an explicit proof of this principle, which is self-evident for the finite numbers of the intuitionist on account of their construction, shows at the same time that the former will never be able to justify his choice of axioms by replacing the unsatisfactory appeal to inexact practice or to intuition equally inexact for him by a proof of the non-contradictoriness of his theory. For in order to prove that a contradiction can never arise among the infinitude of conclusions that can be drawn from the axioms he is using, he would first have to show that if no contradiction had as yet arisen with the nth conclusion then none could arise with the $(n+1)$th conclusion, and secondly, he would have to apply the principle of complete induction intuitively. But it is this last step which the formalist may not take, even though he should have proved the principle of complete induction; for this would require mathematical certainty that the set of properties obtained after the nth conclusion had been reached, would satisfy for an arbitrary n his definition for finite sets, and in order to secure this certainty he would have to have recourse not only to the unpermissible application of a symbolic criterion to a concrete example but also to another intuitive application of the principle of complete induction; this would lead him to a vicious circle reasoning.

In the domain of finite sets in which the formalistic axioms have an interpretation perfectly clear to the intuitionists, unreservedly agreed to by them, the two tendencies differ solely in their method, not in their results; this becomes quite different however in the domain of infinite or transfinite sets, where, mainly by the application of the axiom of inclusion, quoted above, the formalist introduces various concepts, entirely meaningless to the intuitionist, such as for instance *"the set whose elements are the points of space," "the set whose elements are the continuous functions of a variable," "the set whose elements are the discontinuous functions of a variable,"* and so forth. In the course of these formalistic developments it turns out that the consistent application of the axiom of inclusion leads inevitably to contradictions. [Here he describes the Burali–Forti paradox, which is a variation on Cantor's antinomy of the set of all sets presented in Chapter 6.D]. . . .

Although the formalists must admit contradictory results as mathematical if they want to be consistent, there is something disagreeable for them in a paradox like that of Burali–Forti because at the same time the progress of their arguments is guided by the principium contradictionis, i.e., by the rejection of the simultaneous validity of two contradictory properties. For this reason the axiom of inclusion has been modified to read as follows: *"If for all elements of a set it is decided whether a certain property is valid for them or not, then the set contains a subset containing nothing but those elements for which the property does hold."*

In this form the axiom permits only the introduction of such sets as are subsets of sets previously introduced; if one wishes to operate with other sets, their existence must be explicitly postulated. Since however in order to accomplish anything at all the existence of a certain collection of sets will have to be postulated at the outset, the only valid argument that can be brought against the introduction of a new set is that it leads to contradictions; indeed the only modifications that the discovery of paradoxes has brought about in the practice of formalism has been the abolition of those sets that had given rise to these paradoxes. One continues to operate without hesitation with other sets introduced on the basis of the old axiom of inclusion; the result of this is that extended fields of research, which are without significance for the intuitionist are still of considerable interest to the formalist.

[The rest of the paper is devoted to a critique of transfinite set theory.]

Brouwer, pp. 77–84

2. Modern intuitionism

Intuitionism as developed by Brouwer, Heyting, and others who followed has become a distinct alternative to what is now known as classical mathematics. In the theory of finite mathematical objects they agree with the classical results; they disagree with the classical attempt to generalize proof procedures from the domain of finite mathematical objects to infinite ones. In particular they require that to establish an existence statement one must exhibit an object satisfying the desired property. Facts of existence must be justified by construction.

For a classical mathematician there are only two possibilities for any mathematical problem: true or false (although we may not know which it is). But for an intuitionist this is not the case. Here is an example. Consider the decimal $a = .a_0 a_1 \cdots a_n \cdots$ where

$$a_n = \begin{cases} 3 & \text{if no string of seven consecutive 7's appears before the } n^{\text{th}} \\ & \text{decimal place in the decimal expansion of } \pi \\ 0 & \text{otherwise} \end{cases}$$

We may prove, the intuitionist says, that $\neg\neg(a \text{ is rational})$ by showing that $\neg(a \text{ is rational})$ leads to a contradiction. For if a were not rational, then it could not be a finite string of 3's, $.333\cdots3$. So it would have to be $1/3$, which is a contradiction. But it is, at present, not correct to assert that a is rational, for no method is known to compute numbers p and q such that $a = p/q$. The intuitionist does not accept that from $\neg\neg A$ we can conclude A. Alternatively, the example can be viewed as a rejection of the law of excluded middle, $A \vee \neg A$, since we have shown that "$\neg(a \text{ is rational})$" is false.

> The solution is to abandon the principle of bivalence, and suppose our statements to be true just in case we have established that they are, i.e. if mathematical statements are in question, when we at least have an effective method of obtaining a proof of them.
>
> Dummett and Minio, p. 375

An explanation of proof is then needed. The simplest (atomic) arithmetic statements are equalities of terms, for example, $47 + 82 = 118$, $10^{10} = 100$; and these we can prove or disprove by a computation procedure. For compound statements, Dummett and Minio explain:

> The logical constants fall into two groups. A proof of $A \wedge B$ is anything that is a proof of A and of B. A proof of $A \vee B$ is anything that is a proof either of A or of B. A proof of $\exists x\, A(x)$ is anything that is a proof, for some n, of the statement $A(n)$. Note that any proof of any sentence containing only the constants $\wedge$, $\vee$, and $\exists$ is a computation or a finite set of computations.
>
> The second group is composed of $\forall$, $\rightarrow$, and $\neg$. A proof of $\forall x\, A(x)$

is a construction of which we can recognize that, when applied to any number n, it yields a proof of $A(n)$. Such a proof is therefore an *operation* that carries natural numbers into proofs. A proof of $A \to B$ is a construction of which we can recognize that, applied to any proof of A, it yields a proof of B. Such a proof is therefore an operation carrying proofs into proofs. Note that it would be incorrect to characterize a proof of $\forall x \, A(x)$ merely as "a construction which, when applied to any number n, yields a proof of $A(n)$", or a proof of $A \to B$ as "a construction which transforms every proof of A into a proof of B", since we should then have no right to suppose that we could effectively recognize a proof whenever we were presented with one. . . .

A proof of $\neg A$ is usually characterized as a construction of which we can recognize that, applied to any proof of A, it will yield a contradiction. This is unsatisfactory because "a contradiction" is naturally understood to be a statement $B \wedge \neg B$, so that it seems we are defining $\neg$ in terms of itself. We can avoid this in either of two ways. We can choose some one absurd statement, say $0 = 1$, and say that a proof of $\neg A$ is a proof of $A \to 0 = 1$ . . . . Alternatively, we may regard the sense of $\neg$, when applied to atomic statements, as being given by the computational procedure which decides those statements as true or false, and then define a proof of $\neg A$, for any non-atomic statement A, as being a proof of $A \to B \wedge \neg B$, where B is an atomic statement.

Dummett and Minio, pp. 12–14

Dummett and Minio's discussion of the nature of proof is intended to be more than a suggestion to aid our understanding in the spirit of Brouwer. Since Heyting first tried to codify some of the laws of reasoning acceptable to intuitionists in 1930, a formal logic of intuitionism has been a major concern of *intuitionists*, as the followers of Brouwer are now called, and it is that which they are trying to explain (see Epstein 1990, Chapter VII).

The main area of research for intuitionists, however, has been an alternative conception of real numbers based on the idea of a "free-choice sequence", a notion which they use in their version of real analysis (the theory of functions of a real variable that underlies the calculus). For Brouwer a sequence is something which is freely constructed. Time is divided into discrete stages, and at any moment n we can tell whether, say, we have a proof of Goldbach's Conjecture (see p. 122). So we may define a real number $x = (x_n)$ by taking

$$x_{2n} = \begin{cases} 1 & \text{if there is a } w \text{ and no primes } p, q \text{ such that} \\ & p, q, w \leq n \text{ and } p + q = 2w \\ 0 & \text{otherwise} \end{cases}$$

$$x_{2n+1} = \begin{cases} 1 & \text{if a proof of Goldbach's Conjecture has been obtained} \\ & \text{by stage } n \\ 0 & \text{otherwise} \end{cases}$$

This is an example of a choice sequence for which there is no determinate procedure for calculating x_n, because "proof" is not to be understood as a proof in some specific formal system, but any arbitrary correct proof.

We may prove intuitionistically that $x \neq 0$. Suppose to the contrary that we had a proof that $x = 0$. Then

i. We would have a proof that for all n, $x_{2n} = 0$, that is, a proof that there is no counterexample to Goldbach's Conjecture.

and

ii. We would have a proof that for all n, $x_{2n+1} = 0$, that is, a proof that we shall never obtain a proof of Goldbach's Conjecture.

But (ii) contradicts (i). So from the assumption that we have a proof of $x = 0$ we get a contradiction, and hence $x \neq 0$.

Nonetheless we cannot produce a number q such that $x > 1/q$.

But the notion of a free-choice sequence, the acceptance of completed infinities (even though denumerable), and the emphasis on formal logic have been unacceptable to many who are concerned with constructivity in mathematics.

B. Recursive Analysis

Turing's paper in Chapter 9 was called "On computable numbers". His definition of computability via machines was intended to be a basis for a computable version of real analysis:

> The "computable" numbers may be described briefly as the real numbers whose expressions as a decimal are calculable by finite means. . . . According to my definition, a number is computable if its decimal can be written down by a machine.

> We shall say that a sequence β_n of computable numbers *converges computably* if there is a computable integral valued function $N(\varepsilon)$ of the computable variable ε, such that we can show that, if $\varepsilon > 0$ and $n > N(\varepsilon)$ and $m > N(\varepsilon)$, then $|\beta_n - \beta_m| < \varepsilon$.

> We can then show that

> vii. A power series whose coefficients form a computable sequence of computable numbers is computably convergent at all computable points in the interior of its interval of convergence.

> viii. The limit of a computably convergent sequence is computable.
>   And with the obvious definition of "uniformly computably convergent":

> ix. The limit of a uniformly computably convergent computable sequence of computable functions is a computable function. Hence

> x. The sum of a power series whose coefficients form a computable sequence is a computable function in the interior of its interval of convergence.
>   From (viii) and $\pi = 4(1 - \frac{1}{3} + \frac{1}{5} - \cdots)$ we deduce that π is computable.
>   From $e = 1 + 1 + \frac{1}{2!} + \frac{1}{3!} + \cdots$ we deduce that e is computable.

Turing, p. 116 and p. 142

Turing's ideas have been developed by a Russian school of mathematicians based on the acceptance of Church's Thesis (see Bridges and Richman, or Troelstra and van Dalen).

Goodstein, whose constructivist views we studied in Chapter 2.B and Chapter 5.G, was one of the first proponents of recursive analysis (see Goodstein, 1951a and 1961). He, however, argues that computable functions must halt, and since we cannot predict if a general recursive function will halt, he uses only primitive recursive functions.

Computable analysis in Turing's or even Goodstein's sense differs from classical numerical analysis (the study of numerical solutions to equations involving functions of real variables) as performed on a computer (with no limitations of time or memory) only to the extent that the reasoning involved is constructive.

> It has been contended, by Rudolf Carnap and others, that since we are unable to find in application an absolute standard by which the validity of a formal system may be tested we are free to choose what formalisation of mathematics we please, technical considerations alone leading us to prefer one system to another. If we accept this standpoint then the distinction between constructive and non-constructive systems is a distinction without a difference and the constructive system becomes little more than a poor relation of the non-constructive.
>
> I consider this view to be wholly mistaken. Even if we leave out of account the question of demonstrable freedom from contradiction, the *Principia* [*Mathematica* of Whitehead and Russell] and the *Grundlagen* [*der Mathematik* of Hilbert and Bernays] must be rejected as formalisations of mathematics for their failure to express adequately the concepts of universality and existence. Even though we do not discover a contradiction in a formal system by showing that the existential quantifier fails to express the notion of existence, for we have no right to pre-judge the meaning of the signs of the system—and to this extent Carnap is right —none-the-less when a mathematician seeks to establish the existence of a number with a certain property he will not, and should not, be satisfied to find that all he has proved is a formula in some formal system, which whatever it may affirm assuredly does not say that a number exists with the desired property.
>
> Goodstein, 1951a, p. 24

C. Bishop's Constructivism

1. Errett Bishop, from *Foundations of Constructive Analysis*

Preface

If every mathematician occasionally, perhaps only for an instant, feels an urge to move closer to reality, it is not because he believes mathematics is lacking in meaning. He does not believe that mathematics consists in drawing brilliant conclusions from arbitrary axioms, of juggling concepts devoid of pragmatic content, of playing a meaningless game. On the other hand, many mathematical statements have a rather peculiar pragmatic content. Consider the theorem that either every even integer greater than 2 is the sum of two primes, or else there exists an even integer greater than 2 that is not the sum of two primes. The pragmatic content of this theorem is not that if we go to the integers and observe we shall see certain things happening. Rather the pragmatic content of such a theorem, if it exists, lies in the circumstance that we are going to use it to help derive other theorems, themselves of peculiar pragmatic content, which in turn will be the basis for further developments.

It appears then that there are certain mathematical statements that are merely evocative, which make assertions without empirical validity. There are also mathematical statements of immediate empirical validity, which say that certain performable operations will produce certain observable results, for instance, the theorem that every positive integer is the sum of four squares. Mathematics is a mixture of the real and the ideal, sometimes one, sometimes the other, often so presented that it is hard to tell which is which. The realistic component of mathematics – the desire for pragmatic interpretation—supplies the control which determines the course of development and keeps mathematics from lapsing into meaningless formalism. The idealistic component permits simplifications and opens possibilities which would otherwise be closed. The methods of proof and the objects of investigation have been idealized to form a game, but the actual conduct of the game is ultimately motivated by pragmatic considerations. . . .

There have been, however, attempts to constructivize mathematics, to purge it completely of its idealistic content. The most sustained attempt was made by L.E.J. Brouwer, beginning in 1907. The movement he founded has long been dead, killed partly by extraneous peculiarities of Brouwer's system which made it vague and even ridiculous to practicing mathematicians, but chiefly by the failure of Brouwer and his followers to convince the mathematical public that abandonment of the idealistic viewpoint would not sterilize or cripple the development of mathematics. Brouwer and other constructivists were much more successful in their criticisms of classical mathematics than in their efforts to replace it with something better. Many mathematicians familiar with Brouwer's objections to classical mathematics concede their validity but remain unconvinced that there is any satisfactory alternative. . . .

A Constructivist Manifesto

1. The descriptive basis of mathematics

Mathematics is that portion of our intellectual activity which transcends our biology and our environment. The principles of biology as we know them may apply to life forms on other worlds, yet there is no necessity for this to be so. The principles of physics should be more universal, yet it is easy to imagine another universe governed by different physical laws. Mathematics, a creation of mind, is less arbitrary than biology or physics, creations of nature; the creatures we imagine inhabiting another world in another universe, with another biology and another physics, will develop a mathematics which in essence is the same as ours. In believing this we may be falling into a trap: Mathematics being a creation of our mind, it is, of course, difficult to imagine how mathematics could be otherwise without actually making it so, but perhaps we should not presume to predict the course of the mathematical activities of all possible types of intelligence. On the other hand, the pragmatic content of our belief in the transcendence of mathematics has nothing to do with alien forms of life. Rather it serves to give a direction to mathematical investigation, resulting from the insistence that mathematics be born of an inner necessity.

The primary concern of mathematics is number, and this means the positive integers. We feel about number the way Kant felt about space. The positive integers and their arithmetic are presupposed by the very nature of our intelligence and, we are tempted to believe, by the very nature of intelligence in general. The development of the theory of the positive integers from the primitive concept of the unit, the concept of adjoining a unit, and the process of mathematical induction carries complete conviction. In the words of Kronecker, the positive integers were created by God. Kronecker would have expressed it even better if he had said that the positive integers were created by God for the benefit of man (and other finite beings). Mathematics belongs to man, not to God. We are not interested in properties of the positive integers that have no descriptive meaning for finite man. When a man proves a positive integer to exist, he should show how to find it. If God has mathematics of his own that needs to be done, let him do it himself.

Almost equal in importance to number are the constructions by which we ascend from number to the higher levels of mathematical existence. These constructions involve the discovery of relationships among mathematical entities already constructed, in the process of which new mathematical entities are created. The relations which form the point of departure are the order and arithmetical relations of the positive integers. From these we construct various rules for pairing integers with one another, for separating out certain integers from the rest, and for associating one integer to another. Rules of this sort give rise to the notions of sets and functions.

A set is not an entity which has an ideal existence. A set exists only when it has been defined. To define a set we prescribe, at least implicitly, what we (the constructing intelligence) must do in order to construct an element of the set, and what we must do to show that two elements of the set are equal. A similar remark applies to the definition of a function: in order to define a function from a set A to a set B, we prescribe a finite routine which leads from an element of A to an element of B, and show that equal elements of A give rise to equal elements of B.

Building on the positive integers, weaving a web of ever more sets and more functions, we get the basic structures of mathematics: the rational number system, the real number system, the euclidean spaces, the complex number system, the algebraic number fields, Hilbert space, the classical groups, and so forth. Within the framework of these structures most mathematics is done. Everything attaches itself to number, and every mathematical statement ultimately express the fact that if we perform certain computations within the set of positive integers we shall get certain results. . . .

The transcendence of mathematics demands that it should not be confined to computations that I can perform, or you can perform, or 100 men working 100 years with 100 digital computers can perform. Any computation that can be performed by a finite intelligence—any computation that has a finite number of steps—is permissible. This does not mean that no value is to be placed on the efficiency of a computation. An applied mathematician will prize a computation for its efficiency above all else, whereas in formal mathematics much attention is paid to elegance and little to efficiency. Mathematics should and must concern itself with efficiency, perhaps to the detriment of elegance, but these matters will come to the fore only when realism has begun to prevail. Until then our first concern will be to put as much mathematics as possible on a realistic basis without close attention to questions of efficiency.

2. The idealistic component of mathematics

. . . Brouwer fought the advance of formalism and undertook the disengagement of mathematics from logic. He wanted to strengthen mathematics by associating to every theorem and every proof a pragmatically meaningful interpretation. His program failed to gain support. He was an indifferent expositor and an inflexible advocate, contending against the great prestige of Hilbert and the undeniable fact that idealistic mathematics produced the most general results with the least effort. More important, Brouwer's system itself had traces of idealism [the view that ideal (abstract) objects have a real existence] and, worse, of metaphysical speculation. There was a preoccupation with the philosophical aspects of constructivism at the expense of concrete mathematical activity. A calculus of negation was developed which became a crutch to avoid the necessity of getting precise constructive results. It is not surprising that some of Brouwer's precepts were then formalized, giving rise to so-called intuitionistic number theory, and that the formal system so obtained turned out not to be of any constructive value. In fairness to Brouwer it should be said that he did not associate himself with these efforts to formalize reality; it is the fault of the logicians that many mathematicians who think they know something of the constructive point of view have in mind a dinky formal system or, just as bad, confuse constructivism with recursive function theory.

Brouwer became involved in metaphysical speculation by his desire to improve the theory of the continuum. A bugaboo of both Brouwer and the logicians has been compulsive speculation about the nature of the continuum. In the case of the logicians this leads to contortions in which various formal systems, all detached from reality, are interpreted within one another in the hope that the nature of the continuum will somehow emerge. In Brouwer's case there seems to have been a nagging suspicion that unless he personally intervened to prevent it the continuum would turn out to be discrete. He therefore introduced the method of free-choice sequences for constructing the continuum, as a consequence of which the continuum cannot be discrete because it is not well enough defined. This makes mathematics so bizarre it becomes unpalatable to mathematicians, and foredooms the whole of Brouwer's program. This is a pity, because Brouwer had a remarkable insight into the defects of classical mathematics, and he made a heroic attempt to set things right.

3. The constructivization of mathematics

A set is defined by describing exactly what must be done in order to construct an element of the set and what must be done in order to show that two elements are equal. There is no guarantee that the description will be understood; it may be that an author thinks he has described a set with sufficient clarity but a reader does not understand. As an illustration consider the set of all sequences $\{n_k\}$ of integers. To construct such a sequence we must give a rule which associates an integer n_k to each positive integer k in such a way that for each value of k the associated integer n_k can be determined in a finite number of steps by an entirely routine process. Now this definition could perhaps be interpreted to admit sequences $\{n_k\}$ in which n_k is constructed by a search, the proof that the search actually produces a value of n_k after a finite number of steps being given in some formal system. Of course, we do not have this interpretation in mind, but it is impossible to consider every possible interpretation of our definition and say whether that is what we have in mind. There is always ambiguity, but it becomes less and less as the reader continues to read and discovers more and more of the author's intent, modifying his interpretations if necessary to fit the intentions of the author as they continue to unfold. At any stage of the exposition the reader should be content if he can give a reasonable interpretation to account for everything the author has said. The expositor himself can never fully know all the possible ramifications of his definitions, and he is subject to the same necessity of modifying his interpretations, and sometimes his definitions as well, to conform to the dictates of experience.

The constructive interpretations of the mathematical connectives and quantifiers have been established by Brouwer [see pp. 246–247 of this text]. ...

Brouwer's system makes essential use of negation in defining, for instance, inequality and set complementation. Thus two elements of a set A are unequal according to Brouwer if the assumption of their equality somehow allows us to compute that $0 = 1$. It is natural to want to replace this negativistic definition by something more affirmative, phrased as much as possible in terms of specific computations leading to specific results. Brouwer does just this for the real number system, introducing an affirmative and stronger relation of inequality in addition to the negativistic relation already defined. Experience shows that it is not necessary to define inequality in terms of negation. For those cases in which an inequality relation is needed, it is better to introduce it affirmatively. The same remarks apply to set complementation.

Van Dantzig and others have gone as far as to propose that negation could be entirely avoided in constructive mathematics. Experience bears this out. In many cases where we seem to be using negation—for instance, in the assertion that either a given integer is even or it is not—we are really asserting that one of two finitely distinguishable alternatives actually obtains. Without intending to establish a dogma, we may continue to employ the language of negation but reserve it for situations of this sort, at least until experience changes our minds, and for counterexamples and purposes of motivation. This will have the advantage of making mathematics more immediate and in certain situations forcing us to sharpen our results. ...

Constructive existence is much more restrictive than the ideal existence of classical mathematics. The only way to show that an object exists is to give a finite routine for finding it, whereas in classical mathematics other methods can be used. In fact the following principle is valid in classical mathematics: *Either all elements of A have property P or there exists an element of A with property not P.* This principle, which we shall call the *principle of omniscience*, lies at the root of most of the unconstructivities of classical mathematics. This is already true of the principle of omniscience in its simplest form: if $\{n_k\}$ is a sequence of integers, then either $n_k = 0$ for some k or $n_k \neq 0$ for all k. We shall call this the *limited principle of omniscience*. Theorem after theorem of classical mathematics depends in an essential way on the limited principle of omniscience, and is therefore not constructively valid. Some instances of this are the theorem that a continuous real-valued function on a closed bounded interval attains its maximum, the fixed-point theorem for a continuous map of a closed cell into itself, the ergodic theorem, and the Hahn–Banach theorem. Nevertheless these theorems are not lost to constructive mathematics. Each of these theorems P has a constructive substitute Q, which is a constructively valid theorem Q implying P in the classical system by a more or less simple argument based on the limited principle of omniscience. For instance, the statement that every continuous function from a closed cell in euclidean space into itself admits a fixed point finds a constructive substitute in the statement that such a function admits a point which is arbitrarily near to its image. ...

Almost every conceivable type of resistance has been offered to a straightforward realistic treatment of mathematics, even by constructivists. Brouwer, who has done more for constructive mathematics than anyone else, thought it necessary to introduce a revolutionary, semimystical theory of the continuum. Weyl, a great mathematician who in practice suppressed his constructivist convictions, expressed the opinion that idealistic mathematics finds its justification in its applications to physics. Hilbert, who insisted on constructivity in metamathematics but believed the price of a constructive mathematics was too great, was willing to settle for consistency. Brouwer's disciples joined forces with the logicians in attempts to formalize constructive mathematics. Others seek constructive truth in the framework of recursive function theory. Still others look for a short cut to reality, a point of vantage which will suddenly reveal classical mathematics in a constructive light. None of these substitutes for a straightforward realistic approach has worked. It is no exaggeration to say that a straightforward realistic approach to mathematics has yet to be tried. It is time to make the attempt.

Bishop, 1967, pp. viii–ix and pp. 1–10

2. Some definitions from Bishop's program

Bishop followed up on his manifesto by developing a great deal of modern mathematics in a constructive vein (see *Further Readings* below). To give you an idea of his work, we present some of his definitions and basic concepts.

For Bishop the notion of a (constructive) function on the natural numbers is taken as primitive; he does not identify it with the notion of recursive function. A function is total and at every stage n of its calculation it is completely determined how to calculate its value for $n+1$ (cf. the example of a free-choice sequence in Section A.2).

He also takes the integers as primitive and derives from them in the usual way the rationals with the operations of addition, subtraction, multiplication, and division and the relations of equality, inequality, and $<$.

A *sequence* is defined as a function from the positive integers. Then a *real number* is defined as a sequence (x_n) of rationals such that for all m, n, $|x_m - x_n| \le 1/m + 1/n$ (the ordering is on the rationals). That is, a real number is a (constructive) Cauchy sequence of rationals with predetermined rate of convergence.

Two reals $x = (x_n)$ and $y = (y_n)$ are *equal* if for all n, $|x_n - y_n| \le 2/n$. A real number $x = (x_n)$ is *positive* if for some n, $x_n > 1/n$. The ordering and inequality relations are then defined as: $x > y$ iff $x - y$ is positive, and $x \ne y$ iff $x > y$ or $y > x$, where $x - y = (x_{2n} - y_{2n})$. We leave for you to show (Exercise 9) that $x = (x_n)$ is positive iff there are numbers q, m such that for all $n > m$, $x_n > 1/q$ (compare this to the proof that $x \ne 0$ in Section A.2).

D. Criticisms of Intuitionism and Bishop's Constructivism

If intuitionists and constructivists in the line of Bishop feel that classical mathematics allows abstract notions that have no concrete intuitive sense, then there are others who believe the same criticism can be applied to intuitionism and constructivism.

1. Paul Bernays on intuitionism

Intuitionism makes no allowance for the possibility that, for very large numbers, the operations required by the recursive method of constructing numbers can cease to have a concrete meaning. From two integers k, l one passes immediately to k^l; this process leads in a few steps to numbers which are far larger than any occurring in experience, e.g., $67^{(257^{729})}$.

Intuitionism, like ordinary mathematics, claims that this number can be represented by an Arabic numeral. Could not one press further the criticism which intuitionism makes of existential assertions and raise the question: What does it mean to claim the existence of an arabic numeral for the foregoing number, since in practice we are not in a position to obtain it ?

Bernays, 1935, p. 265

2. Nicolas Goodman, from "Reflections on Bishop's philosophy of mathematics"

Nicolas Goodman argues that Brouwer's conception of mathematics is too subjective, for there is nothing in it to preclude accepting a proof of a contradiction. On the other hand, according to Goodman, Bishop's constructivism relies on an unknowable but objective reality just as classical mathematics does, and hence classical logic should be acceptable to it.

> I have tried to describe certain aspects of the experience of doing mathematics by using the metaphor of seeing. As a matter of fact, for me mathematical insight does have a strong specifically visual component. When I do mathematics, I see vague, almost dreamlike, images. Nevertheless, doing mathematics is not like dreaming. It makes sense to say that I have made a mistake. Often, when I am working on a problem, an idea comes to me, I experience that sense of relief that comes from the breaking of the tension, and then, to my dismay, I see that the idea is not correct. In a dream, on the other hand, there are no errors. Everything is arbitrary, and so everything is correct. It is impossible to be mistaken. It is only after waking from the dream that I can criticize the dream.
>
> The essential attribute of a mathematical proof is not that it enables us to visualize or grasp a certain pattern, but that it enables us to recognize that a certain proposition is true. The standard intuitionistic definition of negation as implying a contradiction is a definition of negation only because no one thinks that we will ever prove a contradiction. A correct constructive proof that $0 = 1$ would amount to a certification of the insanity of the human race. From this it follows that truth cannot consist merely in provability. If to assert the truth of a theorem is *only* to assert that one has a proof of the theorem, then it becomes incomprehensible why we should not be able to prove mutually contradictory theorems.
>
> There seems to be nothing in Brouwer's philosophy to prevent mathematicians from proving two mutually contradictory theorems. For Brouwer, doing mathematics is a constantly repeated act of free creation. Mathematical objects have the properties I say they have because they are my creations and I see that they have those properties. I create them so as to have those properties. It may perfectly well happen, on Brouwer's view, that you and I see different things— create differently. Brouwer's mathematics is dreamlike in that, though it may have internal coherence, it has no referent outside of itself. Doing mathematics is a private and subjective experience. In fact, there seems to be nothing to prevent Brouwer from changing his mind. Today he sees one thing and tomorrow he sees another. The dreamer dreams and dreams again. Errett Bishop, on the other hand, is no subjectivist. He insists on the objective character of mathematical knowledge. For example, he asserts that
>
> > Mathematics, a creation of mind, is less arbitrary than biology or physics, creations of nature; the creatures we imagine inhabiting another world in another universe, with another biology and another physics, will develop a mathematics which in essence is the same as ours [pp. 250–251 above].

Such objectivity must somehow be grounded in a mathematical reality.

Let me put this point in terms of the visual metaphor I have been using. The finitary mathematician I considered above will never prove two mutually contradictory theorems because it is impossible that he should clearly visualize a finite pattern and see in that pattern two mutually contradictory facts. If he were to find only one element of order two in the cyclic group of order four yesterday and two such elements today—well, then, we can only conclude that today he is not paying close enough attention. We know this because we know that the pattern formed by the cyclic group of order four is an actuality which cannot display contradictory features. Similarly, if the constructive analyst correctly proves a theorem, then that proof displays an aspect of the actual pattern that he is studying. That pattern also actually exists, and therefore it also cannot display contradictory features. Hence the constructive analyst also cannot correctly prove a theorem today that contradicts a theorem he correctly proved yesterday.

Suppose we say that the infinity dealt with by the constructive analyst is only potential, not actual. Then we must ask whether the analyst is free to actualize that potentiality in a creative and spontaneous and unpredictable way. If so, then we have Brouwerian subjectivism. It is the free choice sequences, objects which are forever indeterminate, that allow Brouwer to refute the law of the excluded middle. To reject classical logic is to affirm that mathematical reality is inherently vague. It is the indeterminacy introduced by the unknown and unknowable future action of Brouwer's subjective will that produces this vagueness of Brouwer's version of mathematical reality. Nothing like this is to be found in Bishop. Of course, Bishop does not claim to be able to refute classical logic. It seems to me, however, that Bishop's insistence on objectivity ought to force him to accept classical logic. In order to make mathematics objective, Bishop must hold that the mathematical potential infinity can be actualized in only one way. In Cantor's language, there is only one road which we travel when we, for example, extend the sequence of natural numbers. But then, as Cantor argued,* that road is actually infinite, not merely potentially infinite. Its character is determined independently of our activity. It is that actual infinity which grounds the objectivity of the mathematician's knowledge. But the actual infinity will plainly also make every well-defined assertion either true or false.

In order to give an account of the objective character of mathematical propositions, we must recognize that the theorems we prove are true about a determinate structure which we do not dream, but which is actual and independent of which of us is studying it. It seems to me irrelevant to this point whether, as Bishop seems to hold, this structure is a "creation of mind," or whether, as I hold, it exists independently of any mind, and would exist even if there were no mind. In either case its properties do not depend on the beholder or on the particular mind that creates it. Mathematical theorems are true in the

* Georg Cantor, *Abhandlungen Mathematischen und Philosophischen Inhalts,* ed. E. Zermelo, Olms, Hildesheim, 1966, pp. 136–7

sense that they correctly describe a structure whose properties are independent of our knowledge.

If this is the case, then there can be no objection in principle to the nonconstructive application of classical logic in mathematical reasoning. The same determinate structure that is needed to ground the objective character of mathematics will also suffice to ground classical logic. The desire for constructive proofs, then, becomes a matter of preference and not a matter of principle. If Kronecker prefers constructive arguments and Noether prefers conceptual arguments, let each go her own way. It is a matter of taste, not a question of foundational correctness.

The theorems that the analyst proves are true of a structure which, in some sense, must be actual. It is not at all clear, however, just what the sense of that actuality is to be. As Bishop might say, it is "out of this world" (see Bishop, p. viii). Specifically, the structure that mathematics is about is infinite and therefore neither surveyable nor physically realizable. It can exist neither in a mind nor in the physical world. Where, then, can it exist? My argument for the existence of such a structure is itself nonconstructive. I have not shown the structure. I have merely argued that without it there can be no mathematical activity.

Of course, Bishop himself is in favor of a certain degree of idealization of mathematical existence. Thus he says that

> The transcendence of mathematics demands that it should not be
> confined to computations that I can perform, or you can perform,
> or 100 men working 100 years with 100 digital computers can
> perform. Any computation that can be performed by a finite
> intelligence—any computation that has a finite number of steps—
> is permissible [p. 252 in this text].

Nevertheless, we must certainly draw the line somewhere. For, to quote Bishop once more,

> We are not interested in properties of the positive integers that have
> no descriptive meaning for finite man. When a man proves a
> positive integer to exist, he should show how to find it. If God has
> mathematics of his own that needs to be done, let him do it himself
> [p. 251 in this text].

It is not obvious, however, just where that line should be drawn. A computation which is not feasible in the actual physical universe is, to paraphrase Bishop, of interest only to God. Indeed, potential computations are of a great many kinds. There are computations that I actually perform and which I can survey, like multiplying two three-digit integers. Then there are computations which I actually perform, but which I cannot survey. Perhaps contemporary mathematicians do not do many such computations, but some mathematicians of previous generations devoted a great deal of their professional time to extensive computations carried out by hand. Kepler, for example, made trigonometric tables by hand. Then there are computations which I could actually perform by hand, but which I have a machine do for me. There are computations performed by machine which it would not be feasible to do by hand. There are computations which I could perform by hand but which, as a matter of fact, I do not bother to perform at all. There are computations which could be performed by a presently existing machine, but which I cannot afford the expense of carrying out. There are computations which are beyond the powers of any machine now on the market, but which could be performed by some physically possible machine. There are computations which, though finite, are not performable on any machine which could actually exist in this physical universe. Then there are computations which are possibly infinite, but which are quite simple. An example is the computation called for by the following instruction: "Systematically search for a counter-example to the Fermat conjecture. If, after checking all possibilities, you have not found a counter-example, certify the conjecture as true." Then there are more complex infinite computations, such as the one called for by the following instruction: "List all and only those Gödel numbers of partial recursive functions which happen to give total functions." . . .

There is nothing very problematical about a computation which I have actually carried out and which I can survey. Such a computation is a completely constructive object. It is, so to speak, the paradigm case of a construction in the sense of Heyting or Brouwer. But already a computation which, although I have carried it out, I cannot survey, has something hypothetical about it. I do not actually see that the outcome of the computation is what I say it is. I may have great faith that I have not made an error, perhaps because I have checked my computation repeatedly, but that faith is still only faith, not mathematical certainty. A computation carried out by machine has the result it has because of the laws of physics, which are presumably only empirical. There is something very nonconstructive about relying on the results of machine computations. After all, the entire computational process is hidden. A computation which no one has carried out is a figment. It does not exist. A computation which, though finite, could not be carried out, therefore could not actually exist. It is not even potentially an ingredient in a mind or in the physical universe. Vague talk about possibility "in principle" should not be allowed to distract us from the clear and fundamental distinction between what we can actually do and what we cannot actually do.

It seems to me, therefore, that if a line must be drawn between actual existence and merely ideal mathematical existence, then that line should be drawn not between the finite and the infinite, as Bishop urges, but between the feasible and the infeasible as, for example, Kino has suggested.* If the line is drawn there, however, then the restriction to actual existence would maim mathematical practice in a fundamental and incurable way.

Finite objects which are so large as not to be mentally or physically realizable have a rather strange status in constructive mathematics. For example, a natural number, no matter how large, is constructively either prime or not prime. It constructively either is or is not a counter-example to the Fermat conjecture. These cases of excluded middle apply to such numbers not

* See Akiko Kino, "How long are we prepared to wait?—A note on constructive mathematics." Preprint.

because of the existence of any actual or physically potential computations which would decide which of the two cases applies, but only by virtue of the logical force of the principle of mathematical induction. The proof that every positive integer is either prime or not prime is based on an application of mathematical induction which presupposes that if a computation taking n steps can actually be carried out, then so can a computation taking $n + 1$ steps. But that is false because our capacities are finite and bounded, not potentially infinite. Thus the principle of mathematical induction commits the constructive mathematician to the mathematical existence of objects which cannot actually exist. But more than that. It commits the constructive mathematician to the use of classical logic for these objects. As a matter of fact, whether a sufficiently large integer is prime can only be settled by a proof. It cannot be settled by a routine computation. Therefore, on the usual constructive grounds, we ought not to assert that such a number is or is not prime.

Classical logic is every bit as misleading when applied to the arbitrarily large but finite as it is when applied to the infinite. For example, the constructive mathematician is committed to holding that there is an integer which is prime if the 10^{100}th digit in the decimal expansion of π is even, and which is composite if that 10^{100}th digit is odd. It seems unlikely that anyone will ever know the value of such an integer. Certainly no one now knows its value. It seems to me every bit as empty to say that the value of such an integer could be computed "in principle" as it is to say that such an integer must exist because of the law of the excluded middle.

Goodman, 1981, pp. 139–144

E. Strict Finitism

Many of Goodman's criticisms of Bishop do not have the same weight if we agree that a computation is indeed something we can do, not theoretically but actually.

1. D. van Dantzig, "Is $10^{10^{10}}$ a finite number?"

1. Unless one is willing to admit fictitious "superior minds" like Laplace's "intelligence", Maxwell's "demon" or Brouwer's "creating subject", it is necessary, in the foundations of mathematics like in other sciences, to take account of the limited possibilities of the human mind and of mechanical devices replacing it.

2. Whether a natural number be defined according to Peano, Whitehead and Russell and Hilbert as a sequence of printed signs (e.g. primes, affixed to a zero) or, according to Brouwer, as a sequence of elementary mental acts, in both cases it is required that each individual sequence can be recognized and two different ones can be distinguished. If—as it is usually done both by formalists, logicists and intuitionists—one assumes that by such a procedure in a limited time arbitrarily large natural numbers could be constructed, this would imply the rejection of at least one of the fundamental statements of modern physics (quantum theory, finiteness of the universe, necessity of at least one quantum jump for every mental act). Modern physics implies an upper limit, by far surpassed by $10^{10^{10}}$ for numbers which actually can be constructed in this way. Weakening the requirement of actual constructibility by demanding only that one can *imagine* that the construction could actually be performed—or, perhaps one should say rather, that one can *imagine* that one *could* imagine it—means imagining that one would live in a different world, with different physical constants, which might replace the above mentioned upper limit by a higher one, without anyhow solving the fundamental difficulty.

3. The result of 2. seems to be contradictory: it is impossible to construct natural numbers as large as $10^{10^{10}}$, but $10^{10^{10}}$ *is* a natural number. The contradiction, however, is apparent only, as one has meanwhile unconsciously changed the meaning of the term "natural number".

4. In fact, assume that natural numbers in the original sense, say 0, $0'$, $0''$, $0'''$, ... *have been* constructed up to a definite one which we abbreviate by n_1, and let S_1 be the set they form. The definition by complete induction of the sums:

$$x + 0 = x, \quad x + y' = (x + y)',$$

then is applicable only inasfar as $x + y' \in S_1$. The same holds (mutatis mutandis) for the properties of sums. One can, however, form formal sums $x + y$, $(x + y) + z$, ..., where $x \in S_1$, $y \in S_1$, $z \in S_1$ ... These sums do not exist as natural numbers in the first sense, but only in a new sense. The *proof* of the properties of sums, e.g. $x + y = y + x$ holds only for numbers in the first sense; for those in the second sense these properties must be considered as *postulates*. The fact that nobody doubts that if we *could* enlarge n_1 sufficiently, the proof *would* apply to these formal sums also, does not mean that these relations *have* been proved. In this way formal sums consisting of a limited number of terms can be constructed. Let n_2 be the largest among those, which on a given moment actually have been formed and S_2 the set they form. If n_1 is very large, it does not necessarily contain all natural numbers in the (fictitious) classical sense up to n_2, but only sums consisting of a sufficiently small number of terms for which sufficiently simple abbreviations have been introduced.

5. Similarly the definition of multiplication, of involution [exponentiation?], etc. introduce each a new concept and a new class S_3, S_4, ... of natural numbers, if their constituent parts are allowed to be chosen arbitrarily among the numbers previously constructed. In this sense $10^{10^{10}}$ belongs to S_4, but not to S_3 and still less is a number in the original sense (S_1). Moreover e.g. the statement that

$$10^{10^{10}} + 10^{10^{20}} = 10^{10^{20}} + 10^{10^{10}}$$

can not be said to have been proved, but is only a formal rule for handling formally the symbols, e.g. of S_4.

6. Poincaré's statement that complete induction is the creative principle of mathematics can not be maintained. Such a principle—inasfar as the term is appropriate—is contained in the successive definitions of arithmetical operations, their formal extension outside the class of numbers hitherto obtained, and the formal maintenance of the arithmetical rules proved for those which belong to the first class S_1 .

7. The difference between finite and transfinite numbers can not be defined operationally:* it is possible that always when a mathematician A uses the term "a transfinite number", another mathematician B interprets it as "a finite number" (of course not always the same one) without ever coming to an inconsistency. This will, indeed, be the case if B disposes of a method of defining far larger numbers than A does. B can then construct a natural number (in his sense) Ω surpassing by far all those which A can reach with *his* methods (all of them, of course, applied a limited number of times). If then A, or, if B is "world champion" in the construction of possibly large numbers, any mathematicians, whosoever, speak of the transfinite number ω, B interprets this as the natural number Ω , $\omega + 1$ as $\Omega + 1$, ω^2 as Ω^2 , etc. Note that A will never meet $\Omega - 1$, and that B *need* not interpret ω^2 as the square of his interpretation of ω but may just as well choose a larger number. Being aware of the possibility that another mathematician may find definitions surpassing those he possesses, i.e. that he may lose his "world championship", B will use the terms "transfinite" or "infinite", or symbols like ω and $\aleph_0$ only in the sense of "numbers surpassing everything I can ever obtain" but not as anything essentially different from the numbers he *can* obtain. This implies that the question put in the title of this paper does not admit a unique and unambiguous answer.

8. Brouwer's "Over de grondslagen der wiskunde" (1907) begins with the words (in translation): " One, two, three . . . ; we know this sequence of sounds (spoken ordinal numbers) by heart as a sequence without end, i.e. continuing itself always according to a known law." If one tries to find out what the dots stand for, one sees that Brouwer's statement can not be maintained. All well-known difficulties of defining the well-ordered transfinite numbers of the second class occur among the spoken ordinal numbers; we do *not* know the "whole" sequence by heart, and it does *not* continue according to a known law. Going on, one arrives at million, . . . , billion, . . . , trillion, . . . , quadrillion, quintillion, sextillion, . . . and—knowledge of Latin getting scanty—, millionnillion, . . . , millionnillionnillion, . . . millionnilli . . . illion (million times repeated), etc., corresponding to:

---

* This statement is due in principle to G. Mannoury, *Woord en Gedachte*, 1931, pp. 55–58. The present paper is an effort to relieve the apparent contradiction mentioned above which originally made the statement ununderstandable and unacceptable to most mathematicians, including, till some years ago, myself. More or less similar ideas, however, have been expressed long ago since by E. Borel (cf. his "nombres inaccessibles") and by M. Fréchet.

ω, ω^2, ω^3, ω^4, ω^5, ω^6, ω^ω, ω^{ω^ω}, $\omega^{\omega^{\cdot^{\cdot^{\cdot}}}}$ (ω times repeated), etc.

But: what stands "etc." for?

9. The difference between finite and infinite numbers is not an essential, but a gradual one. According to the successive definitions of "natural numbers" in the successive senses, the individual identifiability and distinguishability disappear gradually if the numbers become larger and larger and can be retained by new definitions only for a scarcer and scarcer class of numbers. We leave here out of consideration, for simplicity, the fact stressed already in 1909 by G. Mannoury* that the identifiability and distinguishability even of the smallest numbers are not absolute.

10. Also the difference between "formalistic" and "intuitionistic" foundation of mathematics is only a gradual one. No "intuitionist" in the world has ever actually "constructed" the number $10^{10^{10}}$ according to its original definition, so that it "only" has a "formal" meaning for him, just like the numbers of the second transfinite class. An intuitionist, if he were consistent, might *not* call $10^{10^{10}}$ a finite natural number. He also would reject (most of) today's socalled intuitionistic mathematics as being too formal. But—luckily perhaps—he is not always so consistent.

11. With a few exceptions like perhaps E. Borel, M. Fréchet and G. Mannoury, most mathematicians, logicians and philosophers have believed in the possibility of obtaining an absolutely unassailable "foundation" of mathematics. This belief must be characterized as an illusion. In particular intuitionist mathematics can not be said to be absolutely "exact", although it can be said to be "more exact" than classical mathematics.

* G. Mannoury, *Methodologisches und Philosophisches zur Elementarmathematik*, P. Visser, Haarlem, (1909), in particular pp. 6–8.

2. David Isles, from "Remarks on the notion of standard non-isomorphic natural number series"

In Chapter 3 we said, "One of the fundamental assumptions of this course will be that we understand how to count and that each of us can continue the sequence 1, 2, 3, ... We know what it means to add 1 and can continue to do so indefinitely."

But though we may be confident that we can always add 1, we also know that some recursive functions give outputs that are too large to be expressed as decimals. All we have are notations for their outputs, such as $10^{10^{10}}$, $67^{(257^{729})}$, or $\psi(5,8)$ where ψ is the Ackermann-Péter function; these we must somehow place within the "unique" ordering of the natural numbers. Perhaps, David Isles argues, we do not have a unique natural number series but many series depending on what notations (functions) we claim give natural numbers.

"The notion of 'formalization' is now to be enlarged; 'to formalize' the use of a notion means for me 'to expose a method of using its name' ".*

To many people, one of the most puzzling of the claims made by Yessenin-Volpin in the article from which the above quote is taken is that it is possible to work consistently with "natural number series" of various lengths each of which may be closed under some but not all primitive recursive functions. This note represents an attempt to make these claims seem more reasonable and to begin an explanation (independent, to some extent, of Volpin's work) of where such a viewpoint might lead.

Since the time of Skolem's work in the 1920's, mathematicians have been aware of non-standard models for arithmetic [see, e.g., Chapter 17 of Boolos and Jeffrey]. Yet the "existence" of such non-standard integers has never really been taken seriously; the faith (dogma) that we possess a clear picture of "the intuitive" natural numbers as well as of the relation of equality between them has never really been shaken. These non-standard integers are usually pictured as having a more complex and richer structure than our familiar natural numbers. In contrast, the view to be advanced here is that what will be called standard natural numbers series or natural number notation systems (NNNS's) are, in general, less rich and complex than "the" so-called "intuitive" series.

I. Skeptical comments

Are "the" intuitive natural numbers categorical [unique up to isomorphism]? That is, is the description of natural number as clear and definitive as we usually take it to be? This was no idle question for Frege, who in the *Foundations of Arithmetic*, attempted to achieve an absolute and clear description of the natural numbers. Any denial of categoricity has important consequences. Whenever we define a class of mathematical objects via an inductive definition and then proceed to establish results about objects in that class we make tacit use of properties of the natural numbers over which the inductive definition is carried out (e.g. we might use the presumed fact that the series is closed under certain functions). One example of such a definition that will be considered at the end of the paper is that of Turing machines and Turing machine computations. Unlike other mathematical constructions, however, our conception of the natural numbers also strongly influences the ways in which we argue and reason about "mathematical objects". For it can reasonably be argued that it is primarily on the basis of our presumed intuitive understanding of the natural numbers that we accept mathematical induction as a valid tool of reasoning. If this is correct, it is a case where the perception (invention, reflective abstraction) of a particular mathematical structure has historically resulted in the acceptance of certain forms of argumentation based on that structure (and, subsequently, to the present social situation where the presence of these forms of argumentation is one of the identifying characteristics of mathematical discourse). The

* A.S. Yessenin-Volpin, "The ultraintuitionistic criticism and antitraditional program for the foundations of mathematics" in *Intuitionism and Proof Theory*, Kino, Myhill, and Vesley eds., North Holland, 1970, p. 30.

acceptance of this form of argumentation quickly leads to that impressive structural richness of the "intuitive" natural numbers which was mentioned earlier. For by its aid we are persuaded that a structure which initially is understood as being closed under a particular unary operation ("successor") is, in fact, closed under every primitive recursive function—and more. . . .

The author is of the opinion that the belief in induction as a method of proof stems from the view that the nature of "natural numbers" is clear, that their generation from 0 is a determinate process, that the equality relation between any two is "immediately graspable," in a word, that the natural numbers are unequivocally defined up to isomorphism. But on what is such a belief in the uniqueness of the natural numbers based? Three arguments are common. The first, because it uses induction to establish an isomorphism between any two natural number series is also circular [cf. Exercise 24.3]. The other two depend on definitions or "constructions" of the natural numbers. The more sophisticated of these is the set-theoretical definition which posits the set of natural numbers as the intersection of all sets which contain a zero and which are closed under a successor function. The first difficulty with this justification is that the class of natural numbers so defined depends on the particular axiomatic set theory adopted. To the author's knowledge there is today no principled reason for selecting one set theory as the preferred one, and thus there would seem to be, at present, no unique axiomatic characterization of the natural numbers attainable in this way. But even if there were, such an impredicative definition [one in which an object is defined by quantification over a set of objects among which is the object to be defined] could be taken as descriptive only if one adopts a realist [platonist] position according to which the definition merely singles out a preexisting set. Consequently, this explanation of natural numbers is only available to those who feel inclined to accept not just realist modes of speaking but in particular realist modes of speaking as providing a contentful reduction of the intuitive notion.

In any case, it is probably true that even most mathematicians whose practice is realist would agree that the set-theoretic explanation fails to be a reductive advance because it is far less clear than the intuitive notion. It is what is called the "counting description" which forms the basis for most mathematicians' belief in the uniqueness of the natural numbers. This description is usually presented in the form of *rules* of which a simple version might be:

R1) Write down a stroke 1;

R2) Given a set of strokes (call it X) write down $X1$.

R3) Now apply R1 once and then apply R2 again and again.

An understanding of the "structure of the natural numbers" thus consists in an understanding of these rules. But what has actually been presented here? Rules R1 and R2 are fairly unambiguous, in fact, one could easily *use* them to write down a few numerals. But rule R3 is in a different category. It does not determine a unique method of proceeding because *that* determination is contained in the words "apply R2 again and again." But these words make use of the very conception of natural number and indefinite repetition whose explanation is being attempted: in other words, this description is circular. Clearly the situation remains unaltered through various elaborations of R3, e.g. "carry out an indefinite loop", etc. (The same point was made by van Heijenoort in 1967, p. 356: "The repeated iteration of the successor operation seems—and perhaps is—very clear to our minds, but it is either circular (to obtain any number we take the successor of 0 a certain number of times) or rests upon hidden and rather complex set-theoretic assumptions ("finitely many times"). The intuitive characterization is so clear because, in fact, no definition at all has been given. A few numbers have been exhibited, and intuition is assisted by words like "repeatedly" and "and so on" or by three dots).

A natural rejoinder to this might be: "Nonsense. I understand R3 perfectly well because I understand how to use it." But to respond thus seems to be saying that the meaning of R3 is given by its use. If this is so, to claim that the natural numbers are unique would be to claim that that use is unique—and that seems palpably false. For the use may be manifested in an enormous variety of forms, using various notation systems, computer, etc., and it may be a difficult matter to see that two such apparently different procedures are, in some sense, isomorphic. I have had someone respond to this criticism by saying that one must stop somewhere, that one simply has to accept that competent users of English understand R3 and its employment. This may be so but it would then seem that any clear description of the natural numbers would have to include discussion of what constitutes competency in this use of English. For example, would someone who took R3 quite literally and applied R2 only twice (obtaining 111) be judged a competent user?

Obviously this argument is subsumed under Wittgenstein's argument that if the meaning of rules is determined by their use, then as usage changes so does the rules' meaning. . . .

II. Natural number notation systems

By way of motivating the alternative description of natural numbers consider again the "counting picture". As a description of a subject's counting behavior, R1 and R2 are probably accurate enough as far as they go. An observer will observe a subject carrying out R1 and R2. But he *will not* observe the subject applying R3 because, as the observation is confined to a finite period of time, the observer will see the subject use R1 and R2 a finite number of times. The observer then uses his own understanding of the natural numbers to interpret the subject's actions as following R1 and R2 in accordance with R3. That is, an understanding of R3 is part of *the observer's* interpretation *of the subject's* "natural number behavior". To say that an understanding of R3 is part of the observer's interpretation does not imply that the observer must be "in possession of" an understanding of the natural numbers. Clearly such a position would merely commit us to the first step of an infinite regress in this analysis. Rather what must be kept clearly in mind at this point is that if we describe the observer studying his subject we have introduced a second observer, observer$_2$, and observer$_2$ can see that the structure of natural numbers which observer$_1$ is using to interpret his study of the subject *is of the same type as the subject's.*

What all of this suggests is an abandonment of Frege's attempt to provide an absolute description of "the" natural numbers and to recognize that we are working with a relative concept. That is, we are working with structures of a "natural number" type and, in general, can only provide a description of one of these by making use of another (perhaps less complex) structure of the same type.

We can get some understanding of the sort of data that Volpin claims characterizes a structure of natural number type by returning to our discussion of the observer of some subject's "natural number behavior". For the observer, the description of the structure of natural numbers as he observes it in the subject includes the following sort of data:

1) Certain explicit generation rules (among these will always be "successor", $+1$);
2) Certain auxiliary rules for distinguishing or identifying various symbols as they are generated;
3) Perhaps a certain collection of notations, e.g. $1, 2, 3, 4, 5, \ldots$ which the subject will actually produce during the observation period.
4) These observed examples $1, 2, 3, \ldots$ will constitute the stage of the subject's natural number series which will have "arrived" or "be completed" at the time of observation. In addition, the subject may describe certain functions applicable to the *arrived* elements (addition, multiplication, exponentiation for example) and indicate his belief that, say 2^{10}, *will be* among his natural numbers. That is, in addition to the "completed" events of the series there will be "expected" or "future" events such as $10 + 10$ or 2^{10}, which will also be taken into attention by the subject (although at the time of observation they are *not* among the arrived members of the series and hence cannot be used as arguments of the indicated functions).

(The reader should resist the temptation to equate "arrived" with some absolute notion of "finite". In this picture it is quite possible that a notation which is arrived, i.e. finite, with respect to one series is in the future, i.e. infinite, with respect to another.)

The elements of this picture seem to be captured by the following:
Tentative Definition 1 A natural number series (or natural number notation system, NNNS) *at a particular stage* consists of

1.1) A collection of notations (the arrived numbers) plus (defined) relations of equality and ordering between them;
1.2) A defined successor operation on the notations which respects the ordering and which permits an enumeration of the arrived elements;
1.3) A collection of function symbols with associated rules for equality and ordering.
  (*Example* $\lambda x\, y\, (x+y)$ with rules i) $a = b$ and $c = d$ $\rightarrow$ $a + c = b + d$, ii) $a < b \rightarrow c + a < c + b$, etc.)

1.4) The notations resulting from *one* application of these function symbols to arrived elements together with the inherited equality and ordering (which will, in general, be a *partial* order) constitute the future elements or those that "will be arrived" *at the next stage of the series.*

Comment 1 The relative character of this definition is indicated by the words "at a particular stage". Here "stage" refers to some arrived number of another natural number notation system Nn . The usual situation is that inductive constructions will be given indicating the definition of the arrived and future elements of an Nn, N_1 , in terms of the arrived and future elements of some Nn, N_0 .

"arrived"   "future"

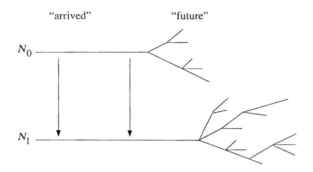

Comment 2 By an "enumeration" of the arrived elements in clause 1.2 is meant the provision of two constructions. First an operation of +1 on the arrived elements of N_1 . Second a procedure E, constructed inductively over the arrived elements of N_0, which can produce a single output at a time of the form 0_{N_1} or $n + 1$ and which outputs all and only the arrived elements of N_1 . Further the inductive construction induces a linear ordering of the output of E such that i) 0_{N_1} is produced first and ii) $n + 1$ is produced only if n is produced at the immediately preceding step.

Comment 3 Because of the partial ordering mentioned in clause 1.4, it follows that +1 induction is only valid when restricted to the *arrived elements of a NNNS.* For this reason, although it is the case by assumption that any NNNS, N_1 say, is closed under $\lambda x (x + 1)$—i.e. that if x has arrived in N_1 $(x \in N_1)$, then $x + 1$ will arrive in N_1 $(x + 1 \in^\Delta N_1)$—it does not follow that it need be closed under $\lambda x (x + 2)$. For if $C(n)$ is the predicate "$n + 2$ will arrive", then, even granted that $(n + 1) + 2 = (n + 2) + 1$, we cannot conclude that $C(n + 1)$ because +1 is only applicable to arrived elements of N_1 and we only have that $n + 2$ *will arrive.*

In this setting the problem of showing that N_1 is closed under a function f takes the following form: x has been constructed in N_1 on the basis of a certain number k_x having arrived in N_0 which itself is closed under a function g (Volpin calls k_x the "genetic support" of x). Then if $g(k_x) \in^\Delta N_0$ one must show that the construction which produces N_1

guarantees that $f(x)$ will have genetic support $g(k_x)$. In a nutshell, to show that an Nn is closed under a function is not a matter which can be established via "internal induction" but only on the basis of its construction from another Nn.

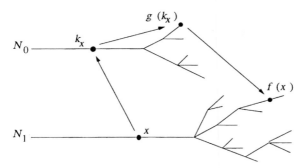

[Isles continues. . .]

IV. Relativization of Inductive Definitions

As a final example, let us consider the changes effected when one uses different NNNS's in place of the intuitive natural numbers in a standard argument from recursion theory. In what follows $\alpha \leftarrow n$ means that Turing machine program α is given input n. Recall the standard

THEOREM (Unsolvability of the Halting Problem)

Let T be the class of Turing machine programs and $|\alpha|$ be the Gödel number of $\alpha \in T$. There is no "test" Turing machine $\beta \in T$ such that

$$\beta \leftarrow |\alpha| \text{ halts in state } \begin{cases} Y & \text{if } \alpha \leftarrow |\alpha| \text{ halts} \\ N & \text{if } \alpha \leftarrow |\alpha| \text{ doesn't halt} \end{cases}$$

Proof: If there were, define the contradictory machine $\beta^* = \beta \cup \{ \,<YSRY> \mid S$ any symbol of $\beta \}$. ∎

In this argument the intuitive natural numbers are used in at least three distinct constructions:
1) in the inductive definition of the class of Turing machine programs. Here a given inductive definition will have a length and we may speak of $l(\alpha)$, the shortest length of the Turing machine program α;
2) the class of inputs to the Turing machines; and
3) to measure the length of Turing machine computations (this is implicit in the words "halts" and "doesn't halt").

Now whatever may be our preconceptions, there is nothing *in this argument* that requires the "same" natural numbers in all three constructions. Indeed all that is required is that if $\alpha \in T$, then $|\alpha|$ should be defined, that is should be available as an input. Hence it is consistent with the structure of the argument to suppose that we have three different NNNS's, N_1, N_2, and N_3 and that for a particular stage k we consider the class of Turing machine programs $T(N_1{}^k)$ (where $\alpha \in T(N_1{}^k)$ means $l(\alpha) \in N_1{}^k$), the class of inputs $N_2{}^k$ and relativize the notion of "halting" to "halting as measured in $N_3{}^k$". If further N_1 and N_2 are so related that if $n \in N_1$ then $2^{2^{2^n}} \in N_2$, it then follows that $|\alpha| \in N_2{}^k$ when $\alpha \in T(N_1{}^k)$. With these changes the theorem now becomes less impressive:

THEOREM Any Turing machine $\beta \in T(N_1{}^k)$ which has the property that for all $\alpha \in T(N_1{}^k)$ (and $|\alpha| \in N_2{}^k$)

$$\beta \leftarrow |\alpha| \text{ halts in state} \begin{cases} Y & \text{if } \alpha \leftarrow |\alpha| \text{ halts } (N_3{}^k) \\ N & \text{if } \alpha \leftarrow |\alpha| \text{ doesn't halt } (N_3{}^k) \end{cases}$$

fails to have this property at stage $k + 1$.

The point of this example is to suggest that the peculiarly "absolute" character of a result such as the unsolvability of the halting problem may be chimerical and have its origin in certain unrecognized assumptions (the uniqueness of the natural numbers). Obviously the same sort of criticism can be brought against the current readings of the first Gödel incompleteness theorem.

Isles, 1981, pp. 111–118 and pp. 131–133

Exercises

1. How does Brouwer's "neo-intuitionism" differ from Kant's intuitionism?

2. a. Does Brouwer accept completed infinities?
   b. Why does Brouwer reject that all the real numbers can be collected into a set? Can that argument be used against the rationals?

3. a. For Brouwer, what is the role of symbolic logic and even ordinary language in mathematics?
   b. Would a proof of the consistency of classical mathematics justify infinitistic reasoning for intuitionists?

4. The theory of recursive functions deals with denumerable sets (cf. our remarks in Chapter 11.B). Would any part of the mathematical development we've given be unacceptable to intuitionists? Do intuitionists accept Church's Thesis?

5. Say whether the following proofs are acceptable to a constructivist such as Brouwer. If not, indicate what classical principle (*PC* or first-order) is being used which is unacceptable, and say what conclusions a constructivist could draw.
   a. The proof in Chapter 8.D that the following function is computable:

$$g(x) = \begin{cases} 1 & \text{if no string of consecutive run of at least } x \text{ 5's in a row} \\ & \text{occurs in the decimal expansion of } \pi \\ 0 & \text{otherwise} \end{cases}$$

   b. The following proof that there are irrational numbers a, b such that a^b is rational: Either $\sqrt{2}^{\sqrt{2}}$ is rational or not. If it is, then we are done. If not, take $a = \sqrt{2}^{\sqrt{2}}$ and $b = \sqrt{2}$. Then $a^b = (\sqrt{2}^{\sqrt{2}})^{\sqrt{2}} = 2$.

6. How does Bishop's view of the role of negation in mathematics differ from Brouwer's?

7. What classical laws of reasoning from Chapters 19 and 21 give rise to Bishop's principle of omniscience? Why doesn't Bishop discuss those instead? Would our first-order logic be acceptable to Bishop if we deleted those?

8. What do you think Bishop means by "reality" when he says that certain formal systems are "detached from reality" ?

9. In Bishop's system prove that $x = (x_n)$ is positive iff there are q and m such that for all $n > m$, $x_n > 1/q$.

10. a. Why are the following theorems of classical mathematics not constructively valid?
    i. If f is a uniformly continuous function from $[0,1]$ to the real numbers and $f(0) < 0$ and $f(1) > 0$, then for some x with $0 < x < 1$, $f(x) = 0$.
    ii. Every continuous function on $[0,1]$ has a maximum.
    iii. For any infinite collection of points in $(0,1)$ there is a point $a \in (0,1)$ such that every interval about a in $(0,1)$ contains a point of the collection.
    iv. Every subset of a finite set is finite.
    b. Present proofs of (i) and (ii) from any calculus textbook and point out where nonconstructive reasoning is used.

11. What does van Dantzig mean when he says that the difference between finite and infinite numbers cannot be defined operationally? Could we just point to some very large finite number and say *this* is what we will take as infinite?

12. Why does van Dantzig say that intuitionist mathematics is not absolutely exact? Do his arguments apply to Bishop's brand of constructivism? Is he correct in saying that at least intuitionism is "more exact" than classical mathematics?

13. a. How does Goodman criticize Brouwer's notion of negation?
    b. Why does Goodman say that Bishop's assumptions should lead him to accept classical logic?

14. When one of the authors of this book does 100 sit-ups in the morning he often counts, "85, 86, 87, 88, 89, 100." Is this what Isles means by a natural number series? If not, why not?

Further Reading

Bridges and Richman in *Varieties of Constructive Mathematics* develop a fair amount of constructive mathematics along the lines set out by Bishop and contrast it with both intuitionistic mathematics and recursive analysis. For a more advanced treatment see Beeson's *Foundations of Constructive Mathematics* which also has an appendix on the history of constructivism, or Troelstra and van Dalen's *Constructivism in Mathematics*.

Heyting's *Intuitionism: an introduction* is a well-known text on intuitionism. "The intended interpretation of intuitionistic logic" by Weinstein has a good discussion of intuitionism with a comparison to Hilbert's views in "On the infinite".

Cantor's struggles to get his work accepted and the resistance put up by Kronecker make a fascinating story as told by Dauben in his book *Georg Cantor*. For Kronecker's influence on Hilbert see Reid's biography of Hilbert.

Philosophy of Mathematics by Benacerraf and Putnam is an excellent collection of essays by Brouwer, Heyting, Gödel, Frege, Russell, and others on the philosophical questions raised in this chapter. *From Frege to Gödel: A Source Book in Mathematical Logic* (ed. van Heijenoort) contains a good selection of the earliest papers in modern logic.

Two papers provide excellent summaries of this course. In "The foundations of mathematics" Goodstein explains his views of the foundations of mathematics and why he prefers to work with only primitive recursive rather than general recursive functions; Gentzen's "The present state of research into the foundations of mathematics", 1938, besides surveying Gödel's theorems, contains a discussion of consistency proofs.

27 Mathematics as Modeling

In this chapter we will look at an article that presents a very different view of mathematics as a process of modeling.

Richard L. Epstein, "On mathematics"

I'd like to tell you a story, a story of how I understand mathematics.

There are so many stories already: the platonist's, the constructivist's, the formalist's, the structuralist's, the humanist's, ... But all of these fail to answer satisfactorily at least one of the following questions:

- How do we create and know mathematics?

- How does mathematics compare to our other intellectual activities, particularly science?

- What is mathematical intuition?

- What is a proof in mathematics?

- Is a good proof in mathematics also a good explanation?

- What is mathematical truth, and are mathematical truths necessary?

- How is it that mathematics is useful in our daily lives and in science?

Any story of mathematics should answer all of these. But a good story should also:

- Be consistent with how we actually do mathematics.

- Be useful to mathematicians, leading to new and interesting work in mathematics.

I hope my story is a good one.

Reasoning by analogy

One of the fundamental ways of reasoning in all our thinking about what passes in our lives is reasoning by analogy.[1]

---

[1] This is a brief summary of the analysis in my *Five Ways of Saying "Therefore"*. In *Critical Thinking* I provide many analyzed examples from ordinary life.

> A comparison becomes *reasoning by analogy* when a claim is
> being argued for: On one side of the comparison we draw a
> conclusion, so on the other side we should draw a similar one.

This situation or thing is just like that; since we can draw this conclusion about the
first, we are justified in drawing a similar conclusion about the second.

Such reasoning is not good until we can say, in each particular case, what we
mean by "is just like" and "similar". No two things, no two situations, no two
experiences are exactly the same. We pay attention to some similarities and ignore the
differences. If the differences don't matter, or rather, if they don't matter too much,
then we are justified in drawing similar conclusions. The point of an analogy is to
force us to be explicit about that justification, setting out, if we can, some general
claim under which the two sides fall and from which the conclusions follow.

Such reasoning is pervasive in our daily lives:

> —"Why shouldn't I hit you? You hit me," says the first-grader,
> invoking the principle that whatever someone does to me that's bad,
> I'm justified in doing back to her. Since it was O.K. there, it should
> be O.K. here, too.

> —We should legalize marijuana. After all, if we don't, what's the
> rationale for making alcohol and tobacco legal?
> Alcohol is legal. Tobacco is legal. Therefore, marijuana
> should be legal. They are sufficiently similar.

We use the similarities to draw conclusions, so long as the differences don't matter.

Analogies and abstractions in science

Reasoning by analogy is used in science, too.

> —Agent Orange has been shown to cause cancer in rats. Therefore,
> there is a good chance that Agent Orange will cause cancer in
> humans.
> Rats are like humans. So if rats get cancer from Agent
> Orange, so will humans. Rats are enough like humans.

Science sometimes proceeds by analogy. But scientists always proceed by
abstracting: choosing some aspect(s) of experience to pay attention to and claiming,
perhaps implicitly, that all other aspects of experience in these kinds of situations
don't matter. What we pay attention to gives us the constraints for saying whether
a claim is true or false; what we pay attention to matters.

A scientific theory is true in the context of what we pay attention to, but it is
false in that it does not take into account all. The hypotheses of scientific theories act
as conditions for where the theory can be applied. When we "falsify" a scientific
theory we do not show that it is false. What we show is that it is not applicable to the
experience described in the falsifying experiment. Then we try to describe carefully
what kind of experience that may be, adding more conditions to our theory in the form
of further premises. Thus, we can continue to use Newton's theory of kinetics even
though, it is said, it is false; we use it so long as the objects we are investigating are

not too large nor too small and are not going too fast; we use it where it is applicable.[2]

> *The abstractions that comprise science are not false, nor are they true.*
> *They are schematic claims until we say what we are paying attention to.*

Mathematics as the art of abstraction

Mathematics abstracts from experience, too, only much more so than any science. That is, mathematical theories ignore a very great deal more than any scientific theory, as we can see in the following examples.

The natural numbers
We first have numbers as adjectives: one dog, two cats, eighteen drops of water, fourteen sonatas, forty-seven ideas about mathematics. Numbers are labels.

With practice, repeating and learning the rules for naming and writing new numerals, the counting numbers become a measuring stick we carry in our head. We count off objects to find how many, as we use a tape to measure off lengths of objects. But there is a definite length we cannot go beyond in measuring with our tape; there is no definite limit we perceive in how far we can go with counting.

Then we take numbers as nouns. We reify our abstracting: the end of the process of abstraction—paying attention to only some of our experience—begins to be treated as a thing, an abstract thing. We abstract from counting to get the natural numbers: 1, 2, 3, ...

Addition of natural numbers
We can combine countings to get what we call addition.

Consider how we learn $3 + 5 = 8$. We take something like pebbles or dots on a piece of paper and count:

$$
\begin{array}{ccccccccc}
\bullet & \bullet & \bullet & & \bullet & \bullet & \bullet & \bullet & \bullet \\
1 & 2 & 3 & & 1 & 2 & 3 & 4 & 5 \\
1 & 2 & 3 & & 4 & 5 & 6 & 7 & 8
\end{array}
$$

We show that when we put the two sequences of counting together into one sequence we get the result that the last item in the new sequence is assigned "8". Then, it seems to us, the same results of counting would apply to any other objects like these; the two ways of counting aren't idiosyncratic to just these dots or pebbles. So we abstract and get a theory of addition.

We've been doing this so long, we learn it so early, it is so much a part of our culture that we don't see this as a model. Surely "$1 + 2 = 3$" is true.

But 1 drop of water + 2 drops of water ≠ 3 drops of water. "$1 + 2 = 3$" is not a truth about the world; it is one of the claims that is needed to apply in a situation in order for arithmetic to be applicable there. We can't use arithmetic for drops of water when we put those together.

---

[2] This section is a brief summary of the analysis of scientific reasoning I first set out in *Five Ways of Saying "Therefore"* and then modified with many examples in "On models and theories, with applications to economics". It is supplemented with a more general theory of truth in my *Essays on Logic as the Art of Reasoning Well.*

Our abstraction from counting is applicable or not. Our "truths" of arithmetic say what follows from our abstraction and hence when our model is applicable, as in any reasoning by analogy or abstraction. Arithmetic is an application of measuring, and we must measure correctly.

Applying this theory is not obvious. We have to learn it. We check with pebbles or dots and verify that $2 + 2 = 4$:

But someone else claims $2 + 2 + 2 = 4$:

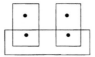

We have to learn (as children) how to apply the model.[3]

It seems to us, once we get the hang of it, that counting-addition is univocal.

---

[3] The example is from Ludwig Wittgenstein, *Remarks on the Foundations of Mathematics*, p. 52.

> George Lakoff and Rafael E. Núñez in *Where Mathematics Comes From* dispute this:
>
> The very idea that babies have mathematical capacities is startling. Mathematics is usually thought of as something inherently difficult that has to be taught with homework and exercises. Yet we come into life prepared to do at least some rudimentary form of arithmetic. Recent research has shown that babies have the following numerical abilities:
>
> 1. At three to four days, a baby can discriminate between collections of two and three items [reference supplied]. Under certain conditions, infants can even distinguish three items from four [reference supplied].
> 2. By four and a half months a baby "can tell" that one plus one is two and that two minus one is one [reference supplied].
> 3. A little later, infants "can tell" that two plus one is three and that three minus one is two [reference supplied].
> [The experiments are described in which babies stare at slides showing various numbers of objects, the length of the stare indicating to the researcher that the baby is discriminating between different configurations.]
> The ability to do the simplest arithmetic was established using similar habituation techniques.   pp. 15–16

Lakoff and Núñez describe it correctly in (1): Babies can discriminate between collections of objects. When they say that babies can discriminate between numbers and do simple arithmetic, consider that most any mammal and many birds can distinguish between collections of two versus three objects. That does not indicate any ability to do mathematics. That does not show that such creatures or that babies have any concept of number. Abstracting the notion of a number as distinct from a number of objects is the first crucial step in mathematics, and there is no evidence that other mammals, birds, or babies can make that step. The familiarity of working with number notations apart from things that are numbered has become so familiar to the authors that they cannot even see that they have begged the question in their description of what the experiments showed.

But that only says we have made it very clear how to apply it. There is no reason to think ahead of time that we will never encounter another situation in which it seems by all we have done so far that our counting-addition is wrong. If we do, we will surely restrict counting-addition not to apply to such a case, saying that what we thought were objects are not things, for only things can be counted.[4] That is, "1 + 2 = 3" cannot be falsified not because it is a necessary truth, but because we preserve it to be true by applying it only in cases in which it is true.[5]

To say that "1 + 2 = 3" is true is to say that numbers—not as adjectives or as arising from counting, or as abstracted from those uses—are actual things about which we reason. That is not incompatible with what I have said, but it gives us no insight into how we create and do mathematics and why mathematics is applicable.

The integers

We have addition of the natural numbers. We can then abstract from the process of taking away objects ("Hey, when you took six sheep from that pen it left me with only five") to get a theory of subtraction. It may sound odd to call addition and subtraction "theories" when they are just part of what we do every day. But they are theories or models just as much as Newton's in that they abstract from our experience.

When we do subtraction along with addition, we find that our calculations—that is, the working out of claims without reference to the things to which they might apply—go a lot more clearly and smoothly if we have some "things" called zero and negative numbers. That's how and why those were first introduced. They flesh out our abstractions. They make the calculations easier. They don't seem to apply to anything.

And hence we feel uneasy about them. If they aren't abstractions from our experience, how can we trust that the calculations in which we use them give results that are applicable? When negative numbers were first introduced in the sixteenth century they were suspect. How can we understand the equality of ratios? How can $1 : -5 = -4 : 20$ when in the first ratio 1 is "larger" than -5, while in the second, -4 is "smaller" than 20? Objections and questions about the legitimacy of their use continued until the 19th century, when they were given a visual/physical interpretation.[6]

As in any scientific theory, if we introduce new "entities" that do not arise by abstraction, there will be, and should be, objections about their use in the theory. When we can see a path of abstraction to them, we begin to have confidence in our theory.

Irrationals

Irrationals were not introduced like negative integers. Irrationals were always part of

---

[4] We recognize this in our language with the distinction between count and non-count (mass) terms. See my *Predicate Logic* for a discussion of the concept of "thing" in our reasoning, and "Concluding Philosophical Remarks" in my *Propositional Logics* for a discussion of other possible arithmetics.

[5] I give a similar analysis of the law of non-contradiction in *Five Ways of Saying "Therefore"*.

[6] See *Philosophy of Mathematics and Mathematical Practice in the Seventeenth Century* by Paolo Mancosu for a discussion of the acceptance of negative integers.

the abstraction of space we call geometry. Or rather points on a line were always part of that abstraction, and some of those, it was discovered, couldn't be measured by ratios of integers. The measuring of those points then became reified into things: irrational numbers.

Complex numbers

The square root of –1 was introduced into algebraic calculations because it facilitated calculations and gave new results that could be checked by older methods. Such calculations were challenged by many mathematicians as being simple fantasy, as having no physical counterpart, as not reliable. Yet mathematicians continued to make those calculations in their work because no contradictions arose when they were used correctly, that is, according to the rules that were eventually determined for them.

Eventually complex numbers became accepted because they were given a visual/physical representation as points on a plane. That clear path of abstraction made us feel confident that their use was legitimate.

Plane geometry

Euclidean plane geometry speaks of points and lines: a point is location without dimension, a line is extension without breadth. No such objects exist in our experience. But Euclidean geometry is remarkably useful in measuring and calculating distances and positions in our daily lives.

Points are abstractions of very small dots made by a pencil or other implement. Lines are abstractions of physical lines, either drawn or sighted. So long as the differences don't matter, that is, so long as the size of the points and the lines are very small relative to what is being measured or plotted, whatever conclusions drawn will be true. Defining a line as extension without breadth is an instruction to use the theory only when we can ignore the breadth of the line.

No one asks (anymore) whether the axioms of Euclidean geometry are true. Rather, when the differences don't matter, we can calculate and predict using Euclidean geometry. When the differences do matter, as in calculating paths of airplanes circling the globe, Euclidean plane geometry does not apply, and another model, geometry for spherical surfaces, is invoked.

Euclidean geometry is a mathematical theory, which, taken as mathematics, would appear to have no application since the objects of which it speaks do not exist in our experience. But taken as a model it has applications in the usual sense, arguing by analogy where the differences don't matter.

Group theory

Consider a square. If we rotate it any multiple of 90° in either direction it lands exactly on the place where it was before. If we flip it over its horizontal or vertical or diagonal axis, it ends up where it was before. Any one of these operations followed by another leaves the square in the same place, and so is the same result as one of the original operations.

To better visualize what we're doing, imagine the square to have labels at the vertices and track which vertex goes to which vertex in these operations. There is one operation that leaves the square exactly as it is: Do nothing. For each of these operations there is one that undoes it. For example: the diagonal flip that takes vertex c to vertex a is undone by doing that same flip again; rotating the square 90° takes vertex a to vertex b, and then rotating 270° returns the square to the original configuration.

Already we have a substantial abstraction. No rotation or flip leaves the square drawn above in the same place, because we can't draw a square with such exact precision (nor can a machine), and were we (or a machine) to cut it out and move it, it wouldn't be exactly where it was before. We are imagining that the square is so perfectly drawn, abstracting from what we have in hand, choosing to ignore the imperfections in the drawing and the movement. If we like, we can then talk about abstract, "perfect" objects of which the picture above is only a suggestion, but that seems to be only a reification of our process of abstracting.

Now consider several objects lined up in a row. We exchange, that is, permute places of some of them, again leaving them lined up in a row. Any permutation followed by another is a permutation, a way to get the objects lined up in a row again. There is one permutation that leaves everything unchanged: do nothing. For any permutation we can undo it by reversing the replacements.

Consider, too, the integers and addition. There is exactly one integer which, when we add it to anything, leaves that thing unchanged, namely, 0. And given any integer there is another which when added to it yields back 0, namely the negative of that integer.

Mathematicians noted similarities among these and many more examples. Abstracting from them, in the sense of paying attention to only some aspects of the examples and ignoring all others, they arrived at the definition of "group". A typical definition is[7]:

A *group* is a non-empty set **G** with a binary operation ∘ such that:

i. For all $a, b \in G$, $a \circ b \in G$.

ii. There is an $e \in G$, called the *identity*, such that for every $a \in G$,
$a \circ e = e \circ a = a$.

iii. For every $a \in G$, there is an $a^{-1} \in G$, called the *inverse* of a,
such that $a \circ a^{-1} = a^{-1} \circ a = e$.

iv. The operation is *associative*, that is, for every $a, b, c \in G$,
$(a \circ b) \circ c = a \circ (b \circ c)$.

Now we can prove claims about groups, such as that there can be only one identity in a group. Such a proof does not show that "There is only one identity in a group" is true—not, that is, unless we reify our abstraction into abstract things called groups. But such a proof does give us a true claim:

"There is only one identity" follows from the assumptions of group theory.

If the assumptions of group theory apply to some thing/situation/part of our experience/process, then the claim about the identity of the group should apply, too. If that claim should fail to apply, then we know that group theory is not applicable in that case, not that group theory is false.

---

[7] From my *Classical Mathematical Logic*.

Ring theory

We have the integers with addition and multiplication. We have the real numbers with addition and multiplication. We have the rationals with addition and multiplication. These and other examples led to the notion of a ring in mathematics. We can abstract from our abstractions.

But don't we mean by "abstract" to ignore and pay attention to certain aspects of our experience?

The practice of doing mathematics is part of our experience, too. Mathematical abstractions are part of our intersubjective mental life, like laws and sonatas. Our intersubjective mental activities are also suitable subjects for abstraction. Mathematics is a human activity.

We can abstract from our abstractions, going further and further in ignoring aspects of the ordinary experience of our daily lives. Some of us have great pleasure in considering and reasoning about such highly abstract subjects. But the pleasure is merely aesthetic until applications of the abstractions are found, relating abstract subjects such as algebra and plane geometry as René Descartes did.

Perhaps you have heard the phrase "the unreasonable applicability of mathematics" used as shorthand for saying that it is very odd that highly abstract theories of mathematics developed solely within the context of other abstractions can so often be applicable.[8] Such a view is mistaken. As mathematicians know, it is rare for a mathematical theory of abstractions of abstractions of abstractions to be applicable to our daily experience, as opposed to just the experience of doing mathematics of a few mathematicians. And when such applications can be found, why should it be so shocking? Yes, the theory was derived from abstracting from abstracting, perhaps many levels. An application is the result of going back along that path of abstraction. And that can only be if what the theory pays attention to and what it ignores is aptly chosen. We can make a theory that ignores almost everything and pays attention to bizarre or abstruse or little-used parts of our experience. Sometimes, by chance, those aspects that are considered turn out to be important in other parts of our experience. In those few cases, the "pure" mathematics is applicable to more of our ordinary lives. We have hit upon a good analogy.[9]

The great mathematicians, those who have some insight into what claims follow in some mathematical theory or who create a theory joining parts of mathematics never before considered similar, who make abstractions of abstractions well, are said to have great intuition, mathematical intuition. That intuition is no different from the intuition that leads a wise person to draw an analogy between dogs and humans in arguing for the humane treatment of animals, or the intuition of the wise person who first "sees" that light can be understood as waves. In my own experience I find that the intuition I had in seeing the general outlines of this paper before I began writing it, and the intuition I had in proving a new theorem about degrees of unsolvability, and the intuition I had in writing one of my plays are the same mental activity, differen-

---

[8] Eugene Wigner, "The unreasonable effectiveness of mathematics in the natural sciences".

[9] Compare: "Why is it that this hammer works so well at pounding nails? " and "How can arithmetic be successful for counting?"

Reuben Hersh in "Inner vision, outer truth" discusses this issue from a similar perspective.

tiated only by the subject matter. I see a general picture, I see a few of the details, I begin with that "vision" or "insight" ahead of me, and I fill in the details, often arriving at something quite different from what I first imagined.

We do not understand how such intuition works. But mathematical intuition is not something different in kind; it is only different in its subject matter.

Transfinite ordinals

We have our theory of counting with the natural numbers. We can count forever and never reach the end.

But we can imagine that there is an end, even if we are not "able" to reach it, just as there is an end to every counting we do in our lives. By analogy we postulate an end to the sequence of natural numbers and call it ω (omega). Then we can continue our counting: $\omega + 1, \omega + 2, \omega + 3, \ldots$. And since we've done it once, we'll do it again: assume an end to that counting, $\omega + \omega$. And then we can continue such counting forever.

This sounds like pure fantasy. What's to tell us that this will "work" in the sense of never leading to contradictions? And why bother? What's the use?

We can describe such ways of counting in a constructive manner, as arrays of natural numbers: one sequence is ω; two sequences, with every number in the first coming before every number in the second is $\omega + \omega$; . . . . When we see a picture of this, we can see a path of abstraction. Moreover, such ways of counting can be seen to correspond to more and more complicated forms of proof by induction.[10]

But when someone considers extending the counting beyond what we can constructively describe, indeed into an infinite that is beyond what could be "counted" in any sense, we have more serious doubts. Why is this acceptable? Why is this theory based on analogy and postulating new kinds of doings consistent?

Mathematical proof

As I mentioned briefly in the discussion of group theory and will make clearer now:

> *A mathematical theorem does not show that a claim is true.*
> *It shows that the claim follows from the assumptions of the theory.*

When we give a proof in Euclidean plane geometry of the claim that there cannot be two distinct lines through a point parallel to another line not through that point we are not showing that claim is true. We've seen that it doesn't even make sense to say it is true, but only true of something, in the sense of applicable. Our proof is good if it is a good argument to establish that the claim about parallel lines follows from the axioms of Euclidean plane geometry.[11]

---

[10] See Chapter 14 above.

[11] Compare the presentation of Euclidean geometry in my *Classical Mathematical Logic*. Albert Einstein in *Relativity* says:

Geometry sets out from certain conceptions such as "plane," "point," and "straight line," with which we are able to associate more or less definite ideas, and from certain simple propositions (axioms) which, in virtue of these ideas, we are inclined to accept as "true." Then, on the basis of a logical process, the justification of which we feel ourselves compelled to admit, all remaining propositions are shown to follow from those axioms, i.e., they are proven. A proposition is then correct

Mathematicians do not say, "There cannot be two distinct lines through a point parallel to another line not through that point." They say, "In Euclidean geometry, there cannot be two distinct lines through a point parallel to another line not through that point." Mathematicians do not say "There is a unique identity." They say, "In any group the identity is unique," invoking thereby that the claim is "true" in the theory of groups.

The truths of mathematics are truths about inferences.
Mathematics is about what follows from what in our abstractions.

Yet many mathematicians and philosophers say not only that mathematical claims are true, but they are necessarily true: there is no possible way such a claim could be false.

There is indeed a necessity in our mathematics, but it is not the necessity of a claim such as "$2 + 2 = 4$". The necessity is that the claim must follow from the assumptions of the theory. There is no way that the axioms of Euclidean geometry could be true and the claim about parallel lines false. There is no way that the assumptions of group theory could be true and the claim that the identity of a group is unique be false. We demand of a mathematical proof that it establish that the inference from the assumptions of the theory to the claim to be proved is a valid inference.

An inference is *valid* means that it is impossible for the premises to be true and the conclusion false (at the same time and in the same way).

This requirement on mathematical proofs goes back to before Euclid. It does not rely on an analysis of the forms of claims. It invokes only the notions of possibility and truth. What the notion of validity requires, and what it guarantees when it holds of an inference, is that if the mathematical claims that are the premises are true in any application, then the conclusion will be true in that application, too.[12]

("true") when it has been derived in the recognised manner from the axioms. The question of the "truth" of the individual geometrical propositions is thus reduced to one of the "truth" of the axioms. Now it has long been known that the last question is not only unanswerable by the methods of geometry, but that it is in itself entirely without meaning. We cannot ask whether it is true that only one straight line goes through two points. We can only say that Euclidean geometry deals with things called "straight lines," to each of which is ascribed the property of being uniquely determined by two points situated on it. The concept "true" does not tally with the assertions of pure geometry, because by the word "true" we are eventually in the habit of designating always the correspondence with a "real" object; geometry, however, is not concerned with the relation of ideas involved in it to objects of experience, but only with the logical connection of these ideas among themselves.

It is not difficult to understand why, in spite of this, we feel constrained to call the propositions of geometry "true." Geometrical ideas correspond to more or less exact objects in nature, and these last are undoubtedly the exclusive cause of the genesis of these ideas. pp. 1–2

[12] This is what is formalized in modern formal logic: a claim in a mathematical theory is true in a particular model; it follows from other claims if it is true in every model in which those claims are true. See my "On valid inferences" and the papers in *Essays on Logic as the Art of*

The following picture summarizes schematically this discussion of mathematical proofs:

<u>A Mathematical Proof</u>

Assumptions about how to reason and communicate.

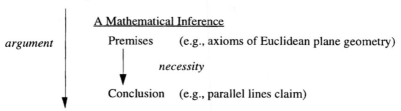

argument

<u>A Mathematical Inference</u>

Premises   (e.g., axioms of Euclidean plane geometry)

necessity

Conclusion   (e.g., parallel lines claim)

The mathematical inference is valid.

The conclusion of the mathematical proof is that the mathematical inference is valid, though mathematicians are rarely so careful as to say that explicitly. Rather, they just show that the inference is valid. The mathematical proof as a whole, then, must be a good argument for that conclusion.

Mathematical proofs are arguments

Reuben Hersh, a mathematician reflecting on his and others' work, characterizes mathematical proofs as follows:

> Mathematical discovery rests on a validation called "proof," the analogue of experiment in physical science. A proof is a conclusive argument that a proposed result follows from accepted theory. "Follows" means the argument convinces qualified, skeptical mathematicians. Here I am giving an overtly social definition of "proof."[13]

Hersh, as many others, has conflated the two arrows in the picture above. The mathematical proof must be a good argument, and that can be very loosely described as one that convinces qualified, skeptical mathematicians. But the "follows from" that must be established is that the mathematical inference is valid.

We can say a great deal about what constitutes a good argument.[14] To begin:

> An *argument* is an attempt to convince someone, possibly yourself, that a particular claim, called "the conclusion" is true. The rest of the argument is a collection of claims called "premises", which are given as the reason for believing the conclusion is true.

The following are necessary conditions for an argument to be good:

> The premises are plausible.
> The premises are more plausible than the conclusion.
> The conclusion follows from the premises.

Reasoning Well for the background on truth, inferences, and possibilities discussed here.

[13] *What is Mathematics, Really?*, p. 6. Though I criticize Hersh here, there is much good in his book, and reading it stimulated me to write this paper.

[14] This and the ensuing discussion of necessary conditions for an argument to be good come from my *Five Ways of Saying "Therefore"* and are applied to many examples in my *Critical Thinking*.

Plausibility is a measure of how much reason we have to believe that a claim is true. In mathematical proofs it is not the assumptions of the mathematical inference that must be plausible. After all, how can we say that the axioms of Euclidean plane geometry are plausible when they aren't even true? Some of them aren't even simpler than the claim about parallel lines.[15]

The plausibility conditions apply to the mathematical proof as a whole: The assumptions about the nature of reasoning and abstractions and how we communicate must be plausible and more plausible than the conclusion that the mathematical claim follows from the assumptions of the theory.

Many times in the history of mathematics the assumptions of mathematical proofs—the part labeled "argument"—have been questioned. At one time it seemed (or still seems) very dubious that we could reason using negative numbers, or the square root of –1, or that we could invoke infinities, or use the method of proof by contradiction, or use the law of double negation, or invoke the axiom of choice. In these cases, though no one doubted that the mathematical inference was indeed valid if given those assumptions, they doubted that the proof, the argument using those assumptions, really established that the inference was valid. They found the premises of the argument—not of the inference—dubious.

The premises of mathematical proofs, the claims about reasoning and communication, the assumptions about mathematics that we use in our everyday work in mathematics, are usually left unsaid; they are treated briefly in undergraduate texts and even more briefly in undergraduate mathematics courses. They become of interest to mathematicians only when paradoxes or wrong proofs arise in some area of mathematics. Uncovering and making those assumptions explicit so that we can debate them is what logicians and philosophers of mathematics do. Logicians and philosophers then join in the debates with working mathematicians, for it is the working mathematicians who will finally decide if some method or assumption is acceptable in the canon of mathematics.

Two examples about what assumptions are currently acceptable in mathematical proofs illustrate these points.

Some mathematicians have given proofs that certain very large numbers are prime by setting out probabilistic analyses. They show that within a very small possibility of being wrong, a particular number is prime. But they then want to say more: This suffices to show that the number is prime. After all, they say, if it is certainty we want from our proofs, many mathematical proofs that are very long and have many steps left to the reader are less certain than the small probability of error of their straightforward probabilistic proofs.[16]

But it is not certainty that is at issue. The issue is whether the inference in a mathematical proof must be shown to be valid, not whether the argument, the mathematical proof, is convincing. If we accept probabilistic inferences in mathematics, then mathematics is no different from any science, for all sciences use strong arguments and not just valid ones to show that a claim follows from the assumptions of their theories.[17] At present there is little division in the mathematical community

---

[15] See the axiomatization due to Lesław Szczerba of Euclidean plane geometry in my *Classical Mathematical Logic*.

[16] See Chapter 5.G.1 for a description of this method.

about this issue: almost everyone agrees that a mathematical proof must show that the inference is valid.

The other new method of proof utilises computers to evaluate many complicated cases that could not be done by hand, eliminating each as a possible counterexample, and concluding a claim such as the four-color theorem. Mathematicians are hesitant to accept such proofs because they rely on our trusting that the computer software is right and that the computer itself is functioning correctly. How can that be part of mathematics? In response, it is again said that very long proofs that leave many steps to the reader and are accepted on the word of one or two referees are much more dubitable than such computer proofs.

Here the issue is not whether a mathematical proof should establish that an inference is valid, for the computer proof is claimed to do just that. The issue is about what counts as a good argument in mathematics.

Mathematical arguments, just like arguments in our daily lives, leave much unsaid. And of what is said, much is only hints or sketches, with lots explicitly left to the reader. We accept such arguments because they are a form of communication. We can see how to understand and evaluate an argument made by a person, filling in the gaps when needed. When we cannot fill in the gaps, when questions cannot be answered, we reject the argument as a mathematical proof. In contrast, a proof by computer can only be followed step by step in the hopes that we can see how each step is used in the proof, and that is impossible when there are so many steps that the prover had to have recourse to a computer. We cannot imagine the intention of the computer as a guide for how to repair a proof, for computers do not have intentions.

We can, however, try to verify that the program run by the computer does what it is intended to do. We can perform tests on a few inputs where the outputs are already known, we can examine how the program is written, and declare that the program is correct. It is possible that the program might not be, but the chances of that, it is believed, are small enough considering the cost of more extensive checking. But that, still, leaves us only with accepting or rejecting a proof by a computer, not understanding it as one would a human communication. The debate on the acceptability of computer programs in mathematical proofs continues to divide the mathematical community.[18]

A comparison of two proofs of a simple claim in arithmetic

Theorem $1 + 2 + \cdots + n = 1/2\, n \cdot (n + 1)$

Proof 1 $\quad 1 = 1/2\, 1 \cdot (1 + 1)$. This is called the *basis of the induction*.

Suppose that $1 + 2 + \cdots + n = 1/2\, n \cdot (n + 1)$. This is called the *induction hypothesis*. Then:

$$1 + 2 + \cdots + n + (n + 1) = \left[1/2\, n \cdot (n + 1) \right] + (n + 1).$$

So $\quad 1 + 2 + \cdots + n + (n + 1) = 1/2\, (n^2 + n) + 1/2\, (2n + 2),$

---

[17] An inference is strong if it is not impossible but it is highly unlikely that the premises could be true and conclusion false. See my *Five Ways of Saying "Therefore"* or "Plausibility, and the strength of inferences" for a discussion of the nature of strong inferences.

[18] See Brian Davies' "Whither mathematics?" for fuller discussions of these and more examples of methods of proof that are currently in debate in the mathematical community.

so $\quad 1 + 2 + \cdots + n + (n + 1) = \frac{1}{2}\,(n^2 + 3\,n + 2)$,

so $\quad 1 + 2 + \cdots + n + (n + 1) = \frac{1}{2}\,(n + 1) \cdot (n + 2)$.

That is, $1 + 2 + \cdots + n + (n + 1) = \frac{1}{2}\,(n + 1) \cdot ((n + 1) + 1)$, which was to be proved.

Some proof like this is what all mathematicians have encountered in learning mathematics. We are told that anything less rigorous doesn't count as a proof. We are told that it establishes that "$1 + 2 + \cdots + n = \frac{1}{2}\,n \cdot (n + 1)$" is true.

But the proof does not establish that "$1 + 2 + \cdots + n = \frac{1}{2}\,n \cdot (n + 1)$" is true. What it shows is that the claim follows if we accept the method of proof by induction. But isn't that true? True of what—our abstraction? No. The method of proof by induction is a condition of our theory: it is part of what establishes the subject of arithmetic. That method is not obvious to students until they have been drilled on it; it is very hard to use. Any claim that follows from the method of proof by induction on the natural numbers must be applicable to any instantiation of the theory. If such a claim fails, then that's not what we are talking about. We can't falsify the method of proof by induction.

This is what the ultra-constructivists who deny induction miss.[19] Their program is not about what the natural numbers are, nor the right way to reason about them, nor disproofs of induction. They are going back to counting and deciding not to make the abstraction from our abilities to say that we can "count" forever.[20] They propose a more "realistic" theory of counting and arithmetic, where a theory can be said to be more realistic than another if it abstracts less from the same experience(s).

Proof 2

$$1 + 2 + \cdots + n = \frac{1}{2}\,n \cdot (n + 1)$$

Is this a proof? When I saw it I felt for the first time that I understood and could believe "$1 + 2 + \cdots + n = \frac{1}{2}\,n \cdot (n + 1)$". Before, I only knew from memory that there is a proof using a symbolic manipulation of symbols that I could reconstruct

---

[19] See, for example, D. van Dantzig "Is $10^{10^{10}}$ a finite number?" in Chapter 26.E.

[20] Or in the case of David Isles, "Remarks on the notion of standard non–isomorphic natural number series" (Chapter 26.E above) and "Questioning articles of faith", that arithmetical functions such as multiplication or exponentiation do not obviously lead to places on the list of numbers we can count to by 1's.

fairly easily. Each time I did that algebraic proof I saw indeed that the claim followed by induction. But here I could see that it was true, really true.[21]

True? The picture convinced me, without any recourse to induction, that the equality will apply to any things to which our basic model of counting and arithmetic apply. I see from the picture that any things I can count up to some number, call it *n*, can be put, along with other such things, in an array that justifies the equality by counting again. And a 6×7 array was enough to convince me.[22]

But how do we know that we'll always get the same result, regardless of how large we expand the diagram? We know in the same way that we know "3 + 5 = 8" will be true for any selection of objects to which we wish to apply our methods of counting and addition. We can give a proof of "3 + 5 = 8" from axioms for arithmetic, but that is less convincing and assumes a great deal more as a theory of addition.

The first proof shows that the equality follows from the assumptions of our more general theory of counting and arithmetic in which we also accept proof by induction; the assumptions about reasoning, though not about the nature of our abstractions, are relatively explicit. This second proof by picture leads us to believe that the equality applies to anything to which our more basic theory of counting and arithmetic apply, without invoking induction; the assumptions about the modeling are relatively clear, but the methods of proof are not. In neither case are we showing that a claim is true; we are showing that it follows, that it is part of our theory. The truths of mathematics are truths about inferences and applications of theories.

A proof of Pythagoras' theorem, and progress in mathematics

Consider another proof by diagram, this time of Pythagoras' theorem.

First, we can represent the product of two numbers **a** and **b** by the rectangle with sides of length **a** and **b**. Then we have the identity: $(a + b)^2 = a^2 + b^2 + 2ab$.

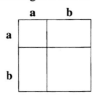

Now we can note that the area of a right triangle with sides **a** and **b** is $1/2$ **ab**.

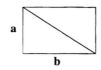

Then we can get Pythagoras' theorem: the square of the length of the hypotenuse

---

[21] The picture comes from *Proofs without Words* by Roger B. Nelsen, p. 69. Martin Gardner in "Mathematical Games", from which Nelsen takes this example, says:

> The first *n* consecutive positive integers can be depicted by dots in triangular formation. Two such triangles fit together to form a rectangular array containing *n*(*n*+1) dots. Because each triangle is half of the rectangle, we see at once that the formula for the number of dots in each triangle is *n*(*n* + 1)/2. This simple proof goes back to the ancient Greeks. p. 114

[22] One mathematician said to me that the picture is convincing for *n* = 6, but not for any larger numbers. But then why not say it is good only for circles colored and laid out in this manner?

in a right triangle is the sum of the squares of the lengths of the legs. All we need do is calculate the area of the large square below by the two different methods:

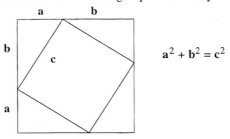

$$a^2 + b^2 = c^2$$

It is clear to me once I have understood the diagrams that it is impossible to come up with a triangle that would not satisfy the identity.[23]

Yet some say that these diagrams do not constitute a proof. Rather, it is only in Euclid that we find a real proof of Pythagoras' theorem. That view, I believe, is a reflection of the desire to make our implicit assumptions explicit. Geometric assumptions necessary for proving Pythagoras' theorem are explicitly set out in Euclid. Or at least they seemed to be for over two millennia. But now the same criticism of Euclid can be given as is given of the diagrams: too much was left as unstated assumptions. The first time we had a real proof of Pythagoras' theorem, it is said, was when Hilbert formalized Euclidean geometry.

What is a proof depends on what we tolerate as implicit assumptions. If progress in mathematics means (at least in part) making more explicit the assumptions by which we prove results, then it is not clear that Hilbert's system is progress over the diagrams, because the assumptions he has made explicit seem to go far beyond what was in our use of the diagrams. What progress there is in mathematics depends more on the assumptions that are being made explicit having greater generality and applicability. That is, progress in mathematics in this sense depends on greater generality through abstraction.[24]

This is not, however, the exclusive kind of progress of mathematics. Solving a difficult and long-known problem counts as progress, too. But universally it is acknowledged that such a solution is not of much interest, that is, is not really considered progress, if it does not employ new methods that allow for the solution of other problems or a generalization of the original problem.[25]

---

[23] We can easily see that the method does not depend on these specific triangles and rectangles. Indeed, in ancient times often just one example, such as using "5" and "8" here (which is the ratio as close as the printer or computer monitor can manage), was given and the reader was expected to see that the method of proof or calculation was quite general. See I. G. Bashmakova and G. S. Smirnova, "Geometry: The first universal language of mathematics".

[24] See Herbert Breger "Tacit knowledge and mathematical progress" for a discussion of this point and much more on the nature of abstraction in mathematics.

[25] This is what is frustrating about evaluating the use of computers in the proof of the four-color theorem: nothing new is used in the proof, no new generalities or ideas, just the brute force of the computer working through many, many cases.

Formal proofs

Some mathematicians think that an objective standard can be given for what counts as a proof in mathematics. They say an argument counts as a mathematical proof only if it could be formalized, by which they mean within a system of formal logic.[26]

In my *Classical Mathematical Logic* I provide a derivation in first-order logic of the claim "The identity is unique" from the axioms of group theory. But that is not a mathematical proof. I give the mathematical proof first, and then argue that the formalization is apt.

A proof that a claim follows from some axioms in a fully formal system of logic is not a proof in mathematics. It is evidence that can be used in a mathematical proof: Why should I believe that this claim follows from these others? Because—and we point to the formal proof. It is the pointing that is crucial. It relies on many assumptions, most particularly that the formal system chosen for the formalization is an apt model of reasoning and a good one for this mathematical proof, and that the steps that have been added—for there are always steps that have to be filled in—are appropriate.[27]

A Mathematical Proof

    Assumptions about how to reason and communicate.

        A Mathematical Inference

This has not been formalized.   Premises   (e.g., axioms of group theory)

        This is formalized.

        Conclusion   (e.g., the identity is unique)

    The mathematical inference is valid.

We cannot formalize the argument that constitutes the mathematical proof without leading to an infinite regress.

We cannot replace proofs in mathematics with formal proofs, though we can use formal proofs as evidence in mathematical proofs. Formalizing mathematical proofs can lead to uncovering or clarifying assumptions behind such informal proofs and seeing how or whether such assumptions are needed.

Mathematics as pure intuition

At the other extreme there are those who say that mathematics is entirely subjective. The noted mathematician R. L. Wilder says,

> What is the role of proof? It seems to be only a testing process that we apply to these suggestions of our intuition.

---

[26] See Don Fallis' review of papers by Reuben Hersh and others.

[27] As I show with examples in my *Classical Mathematical Logic*, first-order logic is rarely appropriate because much mathematics requires second-order assumptions. And second-order logic and set-theory give no unique standard because there are many systems of those that differ too much. Further, as I point out in "The metaphysical basis of logic", all those systems are based on a metaphysics that denies a mathematics of processes as distinct from things.

> Obviously, we don't possess, and probably will never possess, any standard of proof that is independent of time, the thing to be proved, or the person or school of thought using it.[28]

The celebrated mathematician G. H. Hardy seems to concur:

> There is strictly speaking no such thing as a mathematical proof; we can, in the last analysis, do nothing but point; . . . proofs are what Littlewood and I call *gas,* rhetorical flourishes designed to affect psychology, pictures on the board in the lecture, devices to stimulate the imagination of pupils.[29]

And the mathematician and philosopher L. E. J. Brouwer says,

> In the construction of [all mathematical sets of units which are entitled to that name] neither the ordinary language nor any symbolic language can have any other rôle than that of serving as a nonmathematical auxiliary, to assist the mathematical memory or to enable different individuals to build up the same set.[30]

To do mathematics we must use our intuition. But that does not mean that mathematics is subjective. It is intersubjective, like law, like drama, like etiquette. To appeal to each individual mathematician's intuition leaves us no criteria at all for judging whether we have a proof, just as appealing to only a judge's intuition gives no standard for what is legal. And we know at least one clear standard for mathematical proofs: they must establish that an inference is valid.

The view of mathematics as subjective introspection leaves it difficult for us to explain how we learn mathematics, how we judge whether what we have is a proof, what mathematical truth is, ... All the questions that I began this paper with remain unanswered. Though intuitionists who followed Brouwer have created a full and rich mathematical theory, they did so by denying his quote above: they use and work with proofs just as all other mathematicians do.

Set theory and the existence of infinities

Given any few small objects we can collect them together. Given any things that we can describe, we can make up a description of all of them at once. So proceeding by analogy, mathematicians in the 19th century assumed that we can "collect" any things whatsoever into a new entity called the "set" of those things.

We can describe a beautiful theory that way that has many applications. But it leads to contradictions, such as the set of all sets that are not elements of themselves. Not every beautiful way of postulating new things in analogy with old ones is good.

But the utility of such a theory is so desirable that mathematicians worked to rescue it, modifying the analogy to say that only certain ways of collecting are acceptable. And thus we have modern set theory. And we hope that it is consistent, for it is a very useful high-level abstraction in which we can codify and which we can apply to many areas of mathematics.

The assumptions of our new set theory countenance our "collecting" all natural numbers into one set. And we can also "collect" all points on the line into one set—as

---

[28] "The nature of mathematical proof."

[29] "Mathematical proof," p. 18.

[30] "Intuitionism and formalism", p. 81 (see Chapter 26.A above).

if a line that is finite but forever extendible were actually extended and completed as an infinite thing.

That makes a lot of mathematicians uneasy, from the ancients, through the 17th century, and continuing today. Intuitionists and constructivists deny that such abstractions are legitimate in mathematics. They can see no path of abstraction that leads to such a fantastical analogy. Those who use set theory say it should be accepted because it is fruitful and (so far, it seems) consistent.

The utility and consistency of a mathematical theory, some argue, are sufficient for us to investigate it. Indeed, not even utility but just a sense that the theory is beautiful has been enough in the past century for mathematicians to publish papers on new theories.

But, as we've learned from Kurt Gödel, it is rare that we can prove a theory to be consistent. Instead, we try to relate it to other theories we know arise from a path of abstraction and for which we have inductive evidence of consistency: No contradiction has arisen in the many years that many mathematicians have worked in the theory.

Some go farther. They say that consistency of a mathematical theory is all that is needed for us to conclude that the things of which it speaks exist.[31] The infinities beyond infinities of set theory exist, they say. And they often say so without the qualifier "if our set theory is consistent". They have gotten used to working in the theory and thinking of these objects, so how could they not exist? If the theory is contradictory, they feel that they can modify it once again to retain their world of abstract objects; they do not consider that if the theory is modified, the objects of which it speaks might not be the ones they had been thinking of earlier.

But we do not need that mathematical objects postulated by our theories exist in order for our theories to be used and to be useful. We only need that the theory is consistent, for then we can act *as if* they exist: It is not logically impossible for them to exist. And possibilities are all we need in order to reason about mathematical inferences in our mathematical proofs: an inference is valid if there is no *possible way* for the premises to be true and conclusion false.

Some find it remarkable that our theories are consistent and cite that as evidence

---

[31] In particular, in introducing new numbers, mathematics is only obliged to give definitions of them, by which such a definiteness and, circumstances permitting, such a relation to the older numbers are conferred upon them that in given cases they can definitely be distinguished from one another. As soon as a number satisfies all these conditions, it can and must be regarded as existent and real in mathematics.

> Georg Cantor, *Grundlagen einer allgemeinen Mannigfaltigkeitslehre*, p. 182.

A mathematical entity exists, provided its definition implies no contradiction.

> Henri Poincaré, *The Foundations of Science*, p. 61.

If the arbitrarily given axioms do not contradict one another with all their consequences, then they are true and the things defined by the axioms exist. This is for me the criterion of truth and existence.

> David Hilbert, letter to Frege of December 29, 1899, in *Gottlob Frege: The Philosophical and Mathematical Correspondence*, pp. 39–40.

Frege strongly disputes Hilbert's view in that volume (pp. 43–47).

that mathematical claims are indeed true or false.[32] But the process of abstraction, ignoring some of our experience and focusing on just part of it, is not likely to lead to an inconsistency. It is only when we postulate something in addition to our experience that we risk inconsistency. The great difficulty in analyzing processes of abstraction is distinguishing between those cases where we are only ignoring certain aspects of our experience, as with addition and multiplication of counting numbers, and those cases where we postulate some additional ability or capacity of ourselves or some extension of our experience, as in set theories that allow for infinite collections.[33]

Mathematical proofs as explanations

It is often said that a good mathematical proof does more than just show that a mathematical claim is true; it provides a good explanation of why the claim is true. We want to know not just that "$1 + 2 + \cdots + n = 1/2 \, n \cdot (n + 1)$" is true, but why it is true.[34]

An explanation in this sense is characterized as follows:

An *inferential* explanation is a collection of claims that can be understood as "E because of A, B, C, . . .". The claims A, B, C, . . . are called the *explanation* and E is the claim being explained. The explanation is meant to answer the question "Why is E true?"

For an inferential explanation to be good, the inference from A, B, C, . . . to E must be valid or strong. And the claim being explained must be highly plausible: we do not explain anything we do not already believe.[35]

To view a proof of a mathematical claim as an explanation raises the same problems that viewing a mathematical proof as an argument for that claim: we must accept that mathematical claims are true or false, not just true or false in application, and we have to come up with a way to understand how one mathematical claim is more or less plausible than another. It is this latter that has particularly stymied attempts at analyzing mathematical proofs as explanations.

Further, for an explanation to be good, at least one of the claims doing the explaining must be no more plausible than the claim being explained. Otherwise, we would have an argument for the conclusion. Thus we would have to have that some inferences to "$1 + 2 + \cdots + n = 1/2 \, n \cdot (n + 1)$" are to be judged as arguments for that claim, establishing the truth of it, and some are to be judged as explanations, telling us why the claim is true. There would not be a uniform standard by which to judge mathematical proofs..

None of these problems arise if we go back to the schematic diagram of mathematical proofs presented above (p. 283). The mathematical inference to the mathematical claim is neither an argument nor is it an explanation; it is a pure inference, to be judged solely as to whether it is valid or not. The mathematical proof

---

[32] See Hilary Putnam, "What is mathematical truth?", p. 73.

[33] See my "On models and theories, with applications to economics" for a fuller discussion of this.

[34] See, for example, Paolo Mancosu's *Philosophy of Mathematics and Mathematical Practice in the Seventeenth Century* and "On mathematical explanation".

[35] See my "Arguments and explanations" for a discussion of inferential explanations.

is an argument that the inference is valid. Part of the feeling that some proofs are better at showing "why a claim is true" has to do with what criteria we have for such an argument to be good.

We have seen necessary conditions for an argument to be good (p. 283). We can also say that one argument is better than another if its premises are more plausible and it is more clearly valid or strong. For some kinds of arguments, such as generalizations and analogies, more can be said about what constitutes a good or better argument.[36] But consideration of what conditions are needed for one argument to be better than another in mathematics has been obscured by seeing it in terms of explanations.

For example, Mark Steiner says,

> Now I see no reason, except dogmatism, not to accept this story at face value: The embedding of the reals in the complex plane yields explanatory proofs of otherwise unexplained facts about the real numbers. The explanatory power of such proofs depends on our investing the complex numbers with properties they were never perceived as having before: length and direction.[37]

There is more than dogmatism as a motive to reject Steiner's story, for it depends on our assuming that mathematical claims have truth-values. Yet if mathematical claims have truth-values, we cannot "invest" the complex numbers with properties. Either they have those properties or they don't. In any case, mathematicians don't claim that the complex numbers have those properties: we represent them using those quantities.

To see better what Steiner might be getting at, consider the following quotation from Philip Kitcher:

> And as in other sciences, explanation can be extended by absorbing one theory within another. It is customary to praise scientific theories for their explanatory power when they forge connections between phenomena which were previously regarded as unrelated. Within mathematics the same is true and it has become usual to defend the "abstract" approach to mathematics by appealing to the connections which are revealed by studying familiar disciplines as instantiations of general algebraic structures.[38]

It is not clear what Kitcher means by "explanation" and "explanatory power", for we are not explaining anything in his examples. Rather, we are setting out further analogies, connecting our abstractions to show that they have instantiations we didn't previously see, and showing that some of our abstractions can be abstracted further to relate them by analogy to other abstractions. As always, good analogies help us "see" the relationships in the sides of the analogy. It's not different from marijuana compared to alcohol, or humans compared to dogs: We don't explain anything with such an analogy, but we do see common aspects and reason to similar claims based on those aspects when the differences don't matter.

We need such further analogies or instantiations of our abstractions because our abstractions have become too abstract to reason about well, or because they are so abstract that we are not sure they are related to anything in experience beyond what

---

[36] See my *Five Ways of Saying "Therefore"*.

[37] "Penrose and Platonism", p. 137.

[38] "Bolzano's ideal of algebraic analysis", pp. 259–260.

they have been abstracted from, or because we are postulating new entities that need to be shown to have an instantiation in something less abstract, or because we have run out of ways to conceive of further progress in the area and have need of some other way to visualize the subject, as with the algebraization of geometry. All these help us understand our analogies better. But they are not explanations.

Consider what Ernest Nagel says about a particular mathematical claim that needs explaining:

> Why is the sum of any number of consecutive odd integers beginning with 1 always a perfect square (for example, $1 + 3 + 5 + 7 = 16 = 4^2$)? Here the "fact" to be explained (called the *explicandum*) will be assumed to be a claimant for the familiar though not transparently clear label of "necessary truth," in the sense that its denial is self-contradictory. A relevant answer to the question is therefore a demonstration which establishes not only the universal truth but also the necessity of the explicandum. The explanation will accomplish this if the steps of the demonstration conform to the formal requirements of logical proof and if, furthermore, the premises of the demonstration are themselves in some sense necessary. The premises will presumably be the postulates of arithmetic; and their necessary character will be assured if, for example, they can be construed as true in virtue of the meanings associated with the expressions occurring in their formulation.[39]

The fact to be explained cannot be "the sum of any number of consecutive odd integers beginning with 1 is always a perfect square" because that is not obviously true. It has not to be explained but demonstrated. The following proof of that claim does all one could hope for in making clear why the claim is "true".

$$1 + 3 + 5 + \ldots + ((2n) + 1) = (n + 1)^2$$

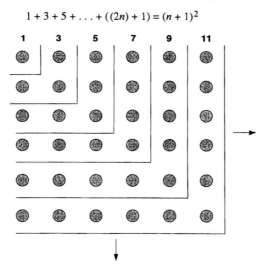

As Martin Gardner says,

> Think of the pattern as extending any desired distance to the right and down. Each reversed *L*-shaped strip contains the odd number of circles indicated at the top. It

---

[39] *The Structure of Science*, p. 16.

is obvious that each additional strip, that is, each new odd number in the series $1 + 3 + 5 + \ldots$, enlarges the square by one unit on a side, and that the total number of dots in each square bounded by the nth odd number is n^2. [40]

The picture convinces us that the claim is true in any application of the assumptions of our theory of addition and multiplication. To say that it is thus true by virtue of the meaning of the expressions requires us to pack a very great deal into what is understood by "meaning". When a mathematical theory can be shown not to apply to some situation/experience/thing/process, such as addition to drops of water, we do not say the theory is false, but that this is a bad application. Has the meaning of the expressions changed? Or are we only clarifying what we implicitly assumed were the meanings? The appropriateness of an application seems hard to assimilate to implicit meanings of expressions in a theory.

Finally, consider what Michael Resnik and D. Kushner say about proofs as explanations:

> We can account for what is probably the most basic intuition behind the idea that there must be explanatory proofs as such, namely that all proofs convince us that the theorem proved is true but only some leave us wondering why it is true. We have this intuition, we submit, because we have observed that many proofs are perfectly satisfactory as proofs but present so little information concerning the underlying structure treated by the theorem that they leave many of our why questions unanswered. In reflecting on this, we tend to conflate these unanswered why questions under the one form of words "why is this true?" and thus derive the mistaken idea that there is an objective distinction between explanatory and non-explanatory proofs. [41]

The distinction between explanatory proofs and non-explanatory ones is no different for arguments in daily life, such as in newspaper editorials. Sometimes we can follow the steps of the argument very well but remain unconvinced. We are not being irrational: we just don't "see" the connections that make the argument valid or strong. Similarly, we sometimes need to "see" the connections in mathematics, too. It requires us not only to see the deductive connection but the relation to what we already know or to something familiar, perhaps through a higher-level abstraction that creates an analogy. Asking "Why is this true?" leads to the mistaken idea of looking for the grounds of the truth of a mathematical claim. Often what is explanatory replaces symbolic manipulation (e.g., an induction proof) with something more concrete in our experience (a picture).

The utility of this story

A good story of mathematics should lead to new and interesting work in mathematics.

By considering the path of abstracting in a mathematical subject rather than focusing only on the final abstraction, we can see where we have chosen to ignore certain aspects of our experience. Then, when the development of the subject becomes stuck, when we cannot adapt our abstractions to accommodate problems that resist solution, we can look to what we have ignored and see if it is possible to take more

---

[40] "Mathematical Games", p. 114, from which this diagram is taken.
[41] "Explanation, independence, and realism in mathematics", p. 154.

into account. This is the usual method of scientists, and it works equally well for mathematicians.[42]

By considering how we as people develop and use mathematics, we are no longer left grasping at abstractions hoping that they will somehow reveal to our intellects new solutions, abstractions, or analogies through their ineffable nature. By remembering that mathematics is a human activity, we can solve more.

We can encourage this view with our teaching. By introducing the process of abstracting rather than focusing only on the final abstraction, we can help students grasp new concepts and make use of them, not only as a subject to be learned, but as a tool for modeling further.[43]

Grounding our stories of mathematics

The view of mathematics that I present is grounded in mathematics as a human activity. Any story of mathematics has to account for that. The assumptions about the world that I invoke, such as that there are people who do mathematics who communicate and who have intuitions, are not controversial. In that sense, I have offered a story that has minimal metaphysics.

I model what I see mathematicians doing. I do not try to account for all. Abstract things arise—or come to our attention if you are a platonist—via our abstracting. All mathematics is applied mathematics; we've just forgotten that it's applied because of its familiarity. Mathematics is abstraction, proofs, and applications, repeated over and over, limited only by our experience and imagination.

But the question can still arise: Why these abstractions?

We can ask for ultimate explanations. Perhaps we have intersubjective work in mathematics because there are platonic objects that ground our insights; or because mental constructions and subjective thoughts are somehow shared by all thinking creatures; or because . . . . These ultimate explanations are the basis of other views of mathematics. They are beyond testing, except for whether they answer the questions that a good story of mathematics should answer.

We can view various philosophies of mathematics with their accompanying

---

[42] This is how I developed the work in my *Propositional Logics*. Rather than viewing propositional logics as being about abstract things called "propositions", I saw them as ways to reason using ordinary language and abstractions of that. Rather than looking for the "right" logic that captures exactly the properties of such abstract things, I saw that what we pay attention to in our abstracting, what aspects of ordinary language claims we deem important, determines the appropriateness of the logic we choose. As we vary the aspect we deem important, we vary the logic. And that variation, I saw, can be described quite generally by devising an abstraction of our propositional logics—an abstraction of our abstractions. The general structures that arise are then worth investigating not only for their own interest but for the relations among logics they illuminate and for the assumptions about how to reason well that they uncover. I did not deny the abstract, for how could I show that there are no such things as abstract propositions? Rather, I focused on the process of abstracting, and that gave rise to new mathematics that is grounded in experience.

[43] In this book Carnielli and I have given the historical development of the theory of computable functions in order to recapitulate the path of abstraction to the theory. This is the "genetic" method of teaching, long favored by George Pólya, as discussed in his *Mathematical Methods in Science*.

metaphysics as providing us with some ultimate account behind why we do mathematics as I have described. Or we can see such stories of abstract objects and mental constructions as psychological props that allow us to reason better about our abstractions.

As for me, I am content to do mathematics and to reflect on how we communicate when we do that. Come, let us reason together.

Appendix: Other views of mathematics

One of the virtues of the story I have told is its seeming familiarity. But it is quite new. In this appendix I'll compare it to several well-known views of mathematics other than platonism and various brands of constructivism, which I discussed above.[44]

Mathematical claims are true

Any view of mathematics that takes mathematical claims such as "2 + 2 = 4" or "In every group there is an identity" to be true is different from what I suggest. Often the reason given for why mathematical claims must be understood as true or false is that they are indispensable for science. Our scientific theories are true, so the mathematics that is essential for them must also be true. Or, it is said, sciences produce true claims, so mathematics, to be a science, has to produce true claims. But sciences produce no true claims in the usual sense, only claims that are true upon proper application. So for mathematics to be a science, it needs to do only that, too. Indeed, to be a science, mathematics must do only that and nothing more, for if mathematical claims are indeed true prior to any application, then that alone distinguishes mathematics from all sciences.

Mathematics is no different than science, proceeding by the same method of abstraction and analogy to yield theories that are more or less applicable.

Logicism

This is the view that mathematics is part of logic. If we develop logic sufficiently, we will be able to see mathematical theories as consequences of the logical axioms. This, then, will explain their certainty and objectivity.

There are two main problems with this view.

First, in the working out of the program of logicism it always turns out that in order to get mathematical theories certain axioms have to be added that are not "logical". At least one axiom about the existence of an infinite collection seems always to be needed. And some axiom or axioms that reduce one part of the formalism to another are always needed, which seems ad hoc. This makes the claim that mathematics is part of logic look very suspect.

---

[44] This is not meant to be a comprehensive or even fair discussion of others' views. See J. Ferreirós and J. J. Gray's introduction to *The Architecture of Modern Mathematics* for an excellent survey.

After writing this paper I discovered the mathematician W. W. Sawyer's *Mathematician's Delight* in which he shows more clearly and thoroughly than I the development of mathematical theories by a process of abstraction. That part of my views, at least, has been commonplace among mathematicians for a long time.

Second, if the reduction is to show that mathematics is certain and objective, that must come from logic being certain and objective. But logic as developed in this vein has no clear claim to that.

The problem is that the formal theories of logic that are used are either descriptive or prescriptive. Normally, logical systems are understood as prescriptive theories. It does not count against the theory that some or even almost all people do not reason in accord with the theory. We say they are reasoning incorrectly. This is the view of logic as prescribing how to reason well. If logic is taken as presciptive, then mathematics is prescriptive, too, which seems wrong.

If the logicist claims that his or her logical theories are descriptive, we have to ask: Descriptive of what? Certainly they are not descriptive of how people actually reason. So they must be descriptive of some other kind of experience. Boole claimed that his logical theories were laws of thought. But neither he nor any of the logicists ever invoked psychological experiments to demonstrate this. And in any case, it surely must be how people should think, not how they actually think, for we know that people do not follow the "laws" of logic in their thinking (very often). The only other possibility I can see is that the logicist takes his or her logical theories to describe some abstract reality. Then logicism is a particular program of platonism.

Structuralism

This is the view that mathematics is not about objects but about relationships or, it is said, structures. Michael Resnik says:

> In mathematics, I claim, we do not have objects with an "internal" composition arranged in structures, we have only structures. The objects of mathematics, that is, the entities which our mathematical constants and quantifiers denote, are structureless points or positions in structures. As positions in structures, they have no identity or features outside a strucure.[45]

Structuralists reify structures rather than mathematical objects, and in doing so they seem to be platonists. They, too, have a problem of how mathematics can be applied. Charles Chihara says:

> I shall not attempt to explain here in any detail how geometry is applied, but a few generalities may be helpful. Let us begin with the fact that the axioms of this geometry characterize a type of structure. Since the rules of inference by which theorems are derived yield sentences that must hold in any structure that is characterized by the axioms, and since physical space itself can be represented as having a mathematical structure of the sort that is characterized by the axioms—Hilbert proves that his geometry is identical to the ordinary "Cartesian geometry"

---

[45] "Mathematics as a science of patterns: ontology and reference", p. 530. Compare:

> "All right," said the Cat; and this time it vanished quite slowly, beginning with the end of the tail, and ending with the grin, which remained some time after the rest of it had gone.
>   "Well! I've often seen a cat without a grin," thought Alice; "but a grin without a cat! It's the most curious thing I ever saw in all my life!"
>   Lewis Carroll, *Alice in Wonderland*

Carroll got it right: he started with the cat, rather than starting with the grin and wondering if it was the grin of something.

[reference supplied]—it follows that the theorems proved in the geometry (when given an appropriate interpretation) must hold of the represented structure. Thus, one is justified in drawing conclusions about the lines and points constructed in physical space from the theorems of Hilbert's geometry.[46]

That physical space can be represented as having the structure of the axioms of some extant geometry is really the issue: what is the nature of that representation? I understand it as a process of abstraction. If we ignore enough, and that which we ignore doesn't matter to the issues involved, then we can view our experience as interpreting Euclidean geometry. Hilbert was a little closer:

> [T]he application of a theory to the world of appearances always requires a certain measure of good will and tactfulness; e.g., that we substitute the smallest possible bodies for points and the longest possible ones, e.g., light rays, for lines. We also must not be too exact in testing the propositions, for these are only theoretical propositions.[47]

But it isn't good will and tact. That's the issue: the nature of abstraction and application.

Hilbert's views
Hilbert wanted to show that the truths of mathematics could be grounded in the finite, allowing non-finitary claims only as ideal statements whose value came from their utility in deriving further finitary claims. Hilbert took mathematics as formal systems, speaking then of their interpretations.[48] But he did not see, or at least did not stress the emergence of mathematics from experience and its uniformity with all other sciences.

Deductivism
Deductivism, or variants of it called "if-thenism", is an outgrowth of the demise of logicism. Mathematics isn't a part of logic. Rather, it is said, any substantive claim of mathematics is really an inference with the assumptions of the theory as premises, and that claim as conclusion.[49]

---

[46] *A Structural Account of Mathematics*, p. 45.

[47] Letter to Frege, December 29, 1899, in *Gottlob Frege: The Philosophical and Mathematical Correspondence*, pp. 40.

[48]   It is surely obvious that every theory is only a scaffolding or schema of concepts together with their necessary relations to one another, and that the basic elements can be thought of in any way one likes. If in speaking of my points I think of some system of things, e.g. the system: love, law, chimney-sweep . . . and then assume all my axioms as relations between these things, then my propositions, e.g. Pythagoras' theorem, are also valid for these things. In other words: any theory can always be applied to infinitely many systems of basic elements.
>   David Hilbert, letter to Frege, December 29, 1899, in *Gottlob Frege: The Philosophical and Mathematical Correspondence*, pp. 40–41.

[49] Or a conditional with the assumptions conjoined as antecedent and the substantive claim the consequent. Gottfried Wilhelm Leibniz says:

> As for "eternal truths:", it must be understood fundamentally they are all conditional; they say, in effect: given so and so, such and such is the case. For

This sounds very much like what I said above:

The truths of mathematics are truths about inferences.
Mathematics is about what follows from what in our abstractions.

But in all views of deductivism I have seen, the deductions or conditionals are understood to be part of a system of formal logic. Because of that, deductivists divide mathematics into pure and applied and find it a significant issue how pure mathematics can be applied. They do not see mathematics as arising by a process of abstraction that is no different from what is used in the sciences, where the inferences need not be formalized to count as good.[50]

---

instance, when I say: *Any figure which has three sides will also have three angles*, I am saying nothing more than that given that there is a figure with three sides, that same figure will have three angles.
    New Essays on Human Understanding, Bk IV, Ch. xi, §14Bk IV, Ch. xi, §14.

P. H. Nidditch in *Elementary Logic of Science and Mathematics* says:

[I]t has become more and more widely accepted during the past hundred years, with the result that it is now the orthodox doctrine, that to say of a mathematical proposition *p* that it is true is merely to say that *p* is true in some mathematical system S, and that this in turn is merely to say that *p* is a theorem in S. . . . This view of the nature of mathematical truth . . . was first put forward with full explicitness and clarity by the Scottish philosopher Dugald Stuart. "Whereas in all other sciences," he says, "the propositions which we attempt to establish express fact, real or supposed—in mathematics, the propositions which we demonstrate only assert a connection between certain suppositions and certain consequences. Our reasonings, therefore, in mathematics, are directed to an object essentially different from what we have in view, in any other employment of our intellectual faculties—not to ascertain *truths* with respect to actual existence, but to trace the logical filiation of consequences from our assumed hypotheses." p. 287

[50] Alan Musgrave in "Logicism revisted" says:

Russell sought a way to bring geometry into the sphere of logic. And he found it in what I shall call the *If-thenist manoeuvre*: the *axioms* of the various geometries do not follow from the logical axioms (how *could* they, for they are mutually inconsistent?), nor do geometrical *theorems*; but the *conditional statements linking axioms to theorems* do follow from logical axioms. Hence, geometry, *viewed as a body of conditional statements*, is derivable from logic after all. . . . Russell argued that the discovery of non-Euclidean geometries forced us to distinguish *pure geometry*, a branch of pure mathematics whose assertions are all conditional, from *applied geometry*, a branch of empirical science. pp. 109–110

If-thenism has nothing to say about un-axiomatised or pre-axiomatised mathematics, in which creative mathematicians work. Therefore, *even if* its account of axiomatised mathematics is acceptable, as an account of mathematics as a whole it is seriously defective. p. 119

Bertrand Russell, in "Mathematics and the metaphysicians," says:

Pure mathematics consists entirely of assertions to the effect that, if such and such a proposition is true of anything, then such and such another proposition is

Hilary Putnam presents a version of deductivism more similar to what I suggest:

> There is another way of doing mathematics, however, or at any rate, of viewing it.
> This way, which is probably much older than the modern way, has suffered from
> never being explicitly described and defended. It is to take the standpoint that
> mathematics has *no* objects of its own at all. You can prove theorems about
> anything you want—rainy days, or marks on paper, or graphs, or lines or spheres—
> but the mathematician, on this view, makes no existence assertions at all. What he
> asserts is that certain things are *possible* and certain things are *impossible*—in a
> strong and uniquely mathematical sense of "possible" and "impossible". In short,
> mathematics is essentially *modal* rather than existential, on this view, which I have
> elsewhere termed "mathematics as modal logic".
>   Let me say a few things about this standpoint here.
>   (1) This standpoint is not intended to satisfy the nominalist. The nominalist,
> good man that he is, cannot accept modal notions any more than he can accept the
> existence of sets. We leave the nominalist to satisfy himself.[51]

But the differences are essential. First, Putnam understands the modal nature of
mathematics as justified, legitimated, or somehow essentially explicated by formal
modal logic.[52] Second, there is nothing in our experience that can count as a possible
way the world could be in which Euclidean plane geometry is true. What counts as a
possibility, and what justifies our use of inferences with claims that are neither true
nor false, is an application of the theory where what we count as true or false is
constrained by our agreements, explicit or not, as to what we will pay attention to in
our reasoning.[53]

A version of deductivism described by Alan Musgrave seems closer to what I
have presented:

> true of that thing. It is essential not to discuss whether the first proposition is
> really true, and not to mention what the anything is, of which it is supposed to be
> true. Both these points would belong to applied mathematics. p. 75

[51] "What is mathematical truth?", p. 70.

[52] He says that his view of mathematics as modal logic is equivalent to taking mathematics
as based on set theory; see his "Mathematics without foundations". In "What is mathematical
truth?" he says:

> The main question we must speak to is simply, *what is the point*? Given that
> one can either take modal notions as primitive and regard talk of mathematical
> existence as derived, or the other way around, what is the advantage to taking
> the modal notions as the basic ones? p. 72

Putnam's comments about the nominalist show that his conception is based on some
more ample metaphysics than mine, for there is nothing in my view that prevents a nominalist
from accepting it—at least in those cases where the abstraction is from sufficently clear
experience—so long as the nominalist agrees to reason with us. Reasoning commits us to
some notion of possibility, but that notion need not be unacceptable to a nominalist; see my
"On valid inferences".

[53] These comments are based on the work I have done in my other works taking the notion of
a proposition or claim to be: a declarative sentence that we agree to view as true or false. I
stress that the agreements need not be explicit or conventional. A platonist could agrue, for
example, that our agreements are, perhaps unconsciously, forced on us by our perception of a
platonic reality. See my article "Truth".

I think, for example, that the sophisticated *evolutionary* Platonism of Popper need not trouble an If-thenist. Popper tries to combine a Platonistic view of the *objectivity* of human knowledge with the Darwinian view that human knowledge is an *evolutionary product*. Thus he insists that the natural numbers are a human creation (part and parcel of the creation of descriptive languages with devices for counting things), but that once created they become *autonomous* so that objective discoveries can be made about them and their properties are not at the mercy of human whim [Karl Popper, *Objective Knowledge*, pp. 158–161]. An If-thenist could agree with much of this. We create, first of all, languages in which to express certain *empirical* claims: "Two apples placed in the same bowl as two other apples give you four apples"; "Two drops of water placed together give you one bigger drop of water"; *etc*. Then we come to treat numbers and their addition in a more abstract way (so that the second statement just given does not count as an empirical refutation of "1 + 1 = 2"). This is, at bottom, to create a more or less explicit collection of 'axioms' for the natural number sequence. And then we find that, once these are granted, we must also grant other statements about numbers like "There are infinitely many prime numbers". We *discover*, in other words, that our axioms have certain unintended logical consequences. The objectivity of mathematics is guaranteed by the fact that *what follows from what* is an objective question, and we need not postulate a realm of 'abstract mathematical entities' to ensure it.[54]

The objectivity of math comes from the objectivity of the inference relation *relative to our assumed metaphysics*. In any case, neither Popper nor Musgrave developed these ideas into an analysis of mathematical proof and applications of mathematics.

John Stuart Mill

John Stuart Mill is supposed by many to claim that mathematics is an induction from experience. But when he uses the words "induction" and "generalization" in discussions of mathematics he seems to mean what I mean by "abstraction". Read that way his views are very similar to mine:

> We can reason about a line as if it had no breadth; because we have a power, which is the foundation of all the control we can exericse over the operations of our mind; the power, when a perception is present to the senses, or a conception to our intellects, of *attending* to a part only of that perception or conception, instead of the whole.

> Since, then, neither in nature, nor in the human mind, do there exist any objects exactly corresponding to the definitions of geometry, while yet that science can not be supposed to be conversant about nonentities; nothing remains but to consider geometry as conversant with such lines, angles, and figures, as really exist; and the definitions, as they are called, must be regarded as some of our first and most obvious generalizations concerning these natural objects. The correctness of these generalizations, *as* generalizations, is without a flaw: the equality of all the radii of a circle is true of all circles, so far as it is true of any one; but it is not exactly true of any circle; it is only nearly true; so nearly that no error of any importance in practice will be incurred by feigning it to be exactly true. When we have occasion to extend these inductions, or their consequences, to cases in which the error would be appreciable—to lines of perceptible breadth or thickness, parallels which

---

[54] "Logicism revisited", footnote, p. 123

deviate sensibly from equidistance, and the link—we correct our conclusions, by combining them with a fresh set of propositions relating to the aberration; just as we also take in propositions relating to the physical or chemical properties of the material, if those properties happend to introduce any modification into the result.

When, therefore, it is affirmed that the conclusions of geometry are necessary truths, the necessity consists only in this, that they correctly follow from the suppositions from which they are deduced.[55]

Whether Mill's views really are similar to or even compatible with mine awaits further study.[56]

---

[55] *System of Logic*, Book II, Chapter V, §1.

[56] John Skorupski in "Later empiricism and logical positivism" discusses Mill's views and suggest a program for empiricists that is similar to what I have done here.

Computability and Undecidability— A Timeline

The story of the development of the theory of computable functions and the undecidability of arithmetic to 1970

The person who is credited with a discovery is the first one who explains it in such a way that it doesn't need to be discovered again.

Apocryphal

Computability and Undecidability

The story of the development of
the theory of computable functions
and the undecidability of arithmetic

Charles Babbage's
design for his
Analytical Engine

Augusta Ada Lovelace's
description of Babbage's
work; Babbage's programs

Hermann Grassmann:
recursive definitions
of + and ·

$x + 0 = x$
$x + (y + 1) = (x + y) + 1$
$x \cdot 0 = 0$
$x \cdot (y + 1) = (x \cdot y) + x$

Cantor's first paper on
transfinite set theory

1834   ⋯   1843   ⋯   1861   ⋯   1872   ⋯   1879

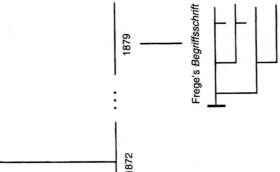

Frege's *Begriffsschrift*

$$[\vdash (b \to a) \to (\neg a \to \neg b)]$$

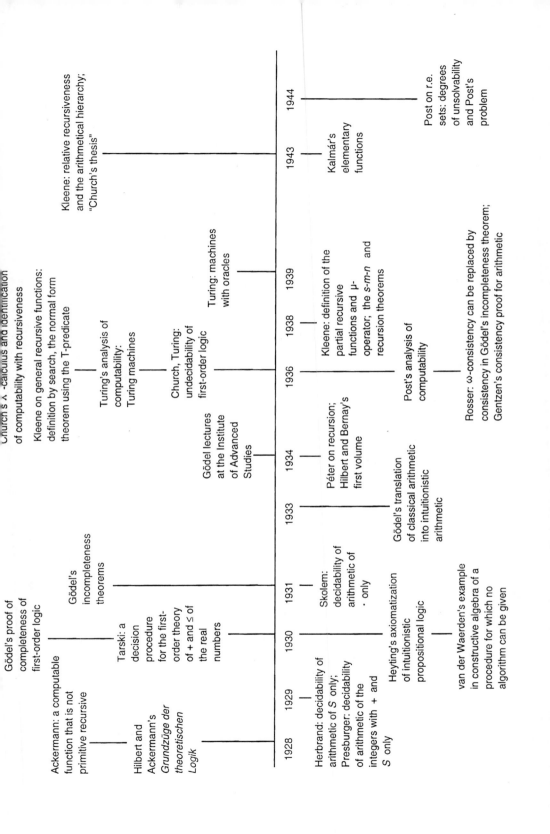

Church's λ-calculus and identification of computability with recursiveness

Gödel's proof of completeness of first-order logic

Kleene on general recursive functions: definition by search, the normal form theorem using the T-predicate

Kleene: relative recursiveness and the arithmetical hierarchy; "Church's thesis"

Ackermann: a computable function that is not primitive recursive

Gödel's incompleteness theorems

Turing's analysis of computability: Turing machines

Church, Turing: undecidability of first-order logic

Turing: machines with oracles

Hilbert and Ackermann's *Grundzüge der theoretischen Logik*

Tarski: a decision procedure for the first-order theory of + and ≤ of the real numbers

Gödel lectures at the Institute of Advanced Studies

| 1928 | 1929 | 1930 | 1931 | 1933 | 1934 | 1936 | 1938 | 1939 | 1943 | 1944 |

Kalmár's elementary functions

Post on r.e. sets: degrees of unsolvability and Post's problem

Herbrand: decidability of arithmetic of *S* only; Presburger: decidability of arithmetic of the integers with + and *S* only

Skolem: decidability of arithmetic of · only

Péter on recursion; Hilbert and Bernay's first volume

Gödel's translation of classical arithmetic into intuitionistic arithmetic

Kleene: definition of the partial recursive functions and μ-operator; the *s-m-n* and recursion theorems

Post's analysis of computability

Heyting's axiomatization of intuitionistic propositional logic

van der Waerden's example in constructive algebra of a procedure for which no algorithm can be given

Rosser: ω-consistency can be replaced by consistency in Gödel's incompleteness theorem; Gentzen's consistency proof for arithmetic

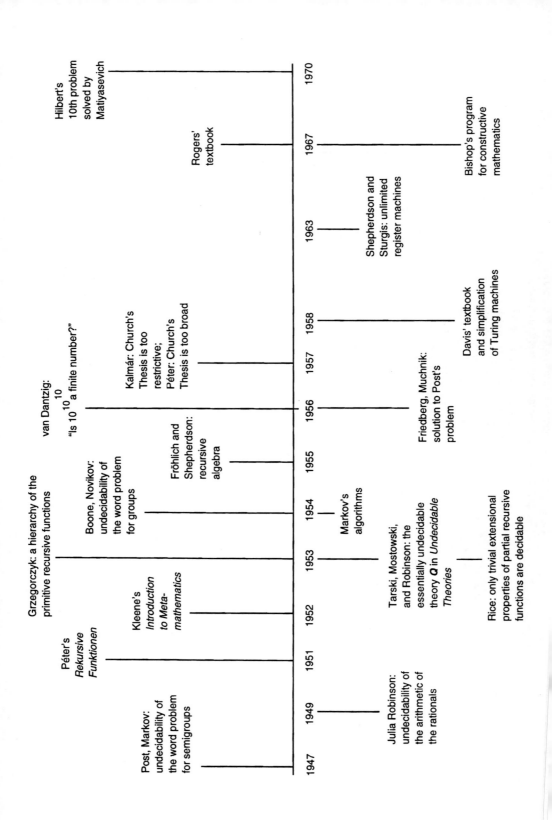

1834

The English mathematician **Charles Babbage**, 1791–1871, designs his Analytical Engine, the first general purpose digital computer. It is to consist of a store, an arithmetic unit, punched card input and output, and a mechanism to control the sequence of cards that allows for iteration and conditional branching.

> It appears that the whole of the conditions which enable a *finite* machine to make calculations of *unlimited* extent are fulfilled in the Analytical Engine . . . I have converted the infinity of space, which was required by the conditions of the problem, into the infinity of time.

Babbage's ideas are far in advance of the available technology, and he and others build no more than parts of his machine, now on display in the Science Museum in London. Babbage's influence is minimal until World War II when his work is rediscovered.

1843

Byron's daughter, **Augusta Ada**, 1815–1852, translates into English an article on Babbage's work by Luigi Ménabréa, later Prime Minister of Italy. She includes her own comments which are several times longer than the original paper and, with Babbage's help, appends sample programs, including one for computing the sequence of Bernoulli numbers by means of a loop within a loop. Babbage, anxious to have someone of the aristocracy promote his work, is outraged that after all his tutoring she resists including under her own name his exaggerated claims and requests for funding for his engine (see *Stein, 1985*).

1861

Hermann Grassmann, 1809–1877, publishes his *Lehrbuch der Arithmetik für höhere Lehranstalten* in which he gives recursive definitions of addition and multiplication:

$$x + 0 = x \qquad\qquad x \cdot 0 = 0$$
$$x + (y + 1) = (x + y) + 1 \qquad\qquad x \cdot (y + 1) = (x \cdot y) + x$$

Though also a founder of the study of vector algebra, Grassmann's application to be a university professor is rejected because in E. E. Kummer's judgement his best essay contains "commendably good material expressed in a deficient form"—his writing is too formal and abstract. He continues throughout his life as a *gymnasium* teacher and becomes a reknowned linguist and specialist in Sanskrit literature.

1872

In an extraordinary departure from previously accepted mathematics, **Georg Cantor**, 1845–1918, advocates using completed infinite totalities in mathematics and creates a mathematical theory of the infinite. His publications beginning this year culminate with his *Beiträge zur Begrundung der transfiniten Mengenlehre* of 1895–1897. But his use of classical logic and nonconstructive existence proofs, as well as various levels of infinities, is not universally accepted, and later the "paradoxes" of set-theory lead to serious reaction. Opposition is greatest from Leopold Kronecker, who blocks him from advancing beyond a position at the University of Halle. From 1884 onwards he suffers from a series of nervous breakdowns and dies in a mental institution.

1879

Gottlob Frege, 1848–1925, publishes his *Begriffsschrift, a formula language, modeled upon that of arithmetic, for pure thought* in which he establishes much of what we now know as mathematical logic: the use of variables, quantifiers, relations, propositional logic, functions, an axiomatization of logic. He proves the principle of induction from his logical principles.

Frege's syntactical approach and his new notation are at odds with the current algebraic approach to logic so that, working alone at Jena, his ideas receive little attention until the turn of the century.

$[\![\,Law\ of\ Transposition\ \vdash (b \rightarrow a) \rightarrow (\neg a \rightarrow \neg b)\,]\!]$

1887

Leopold Kronecker, 1823–1891, publishes his "Über den Zahlbegriff", the first manifesto for constructivist mathematics: all mathematics must be based on a finite number of operators involving only the integers, such that each operation can always be evaluated in a finite number of steps. A member of the Berlin Academy and later Professor of Mathematics at the University of Berlin, he is vehemently opposed to Cantor's introduction of actual infinities into mathematics, having earlier prevented Cantor from publishing one of his papers because of its "meaningless" proofs, and later blocking Cantor's advancement to a better university position. He is reputed to have said:

> God created the positive integers; all the rest is the work of man.

1888

Richard Dedekind, 1831–1916, publishes his *Was sind und was sollen die Zahlen?* in which he states the principle of the definition of functions by induction, what is now called *primitive recursion*:

> If there is given an arbitrary transformation θ of a system Γ in itself, and besides a determinate element α in Γ, then there exists one and only one transformation ψ of the number series N, which satisfies the conditions
>
> I. $\psi(N) \subseteq \Gamma$
>
> II. $\psi(1) = \alpha$
>
> III. $\psi(n') = \theta\psi(n)$ where n represents every number.

The existence of such a function he justifies by recourse to the principle of (what we now call) first-order induction, which he states and proves as a corollary to his theorem justifying (full) second-order induction. He uses the latter to prove that the natural numbers are unique.

When only 22 Dedekind received his *doctorate* in mathematics at Göttingen. Though a good friend and collaborator of Cantor, he nonetheless turns down a position at Halle for family reasons, remaining in Braunschweig where he teaches at the Technische Hochschule.

Relative levels of advanced academic degrees in continental, especially German-speaking countries (italics), vs. the English and U.S. system:

> Master (Thesis)
> *Doktorate* (Dissertation)
> Ph. D. (Thesis, Dissertation)
> *Habilitation* (*Habilitationschrift*)

1889

Giuseppe Peano, 1858–1932, publishes in Latin his axiomatization of arithmetic. His clear notation is popular and eventually becomes standard, but his excessive formalism makes his work less and less accessible to mathematicians outside his circle at the University of Turin, and eventually he is asked to stop teaching. Just as he intends his notation to be used to unify the presentation of all mathematics, he sets out to create a new popular universal language, "Latino sine flexione", to facilitate communication on all subjects.

> Because the usual spoken or written languages do not in the least satisfy the requirements of consistency demanded of this symbolic logic, formalists try to avoid the use of ordinary language in mathematics. How far this may be carried is shown by the modern Italian school of formalists, whose leader, Peano, published one of his most important discoveries concerning the existence of integrals of real differential equations in the *Mathematische Annalen* in the language of symbolic logic; the result was that it could only be read by a few of the initiated and that it did not become generally available until one of these had translated the article into German.
>
> *Brouwer, 1912*, p. 79

Peano's axioms

1. $1 \, \varepsilon \, N.$
2. $a \, \varepsilon \, N \, . \supset . \, a = a .$
3. $a, b \, \varepsilon \, N . \supset : a = b \, . = . \, b = a .$
4. $a, b, c \, \varepsilon \, N . \supset \, \therefore \, a = b . \, b = c \, : \supset . \, a = c .$
5. $a = b . \, b \, \varepsilon \, N : \supset . \, a \, \varepsilon \, N.$
6. $a \, \varepsilon \, N \, . \supset . \, a + 1 \, \varepsilon \, N.$
7. $a, b \, \varepsilon \, N \, . \supset : a = b . = . \, a + 1 = b + 1.$
8. $a \, \varepsilon \, N \, . \supset . \, a + 1 - \, = 1.$
9. $k \, \varepsilon \, K \, \therefore \, 1 \, \varepsilon \, k \, \therefore \, x \, \varepsilon \, N. \, x \, \varepsilon \, k \, : \supset_x . \, x + 1 \, \varepsilon \, k \, : : \supset . \, N \supset k .$

1891

Cantor introduces his diagonal method to show that the reals are not countable: If the reals were countable then the reals between 0 and 1 would be, too. So we could list them, using only representations that do not terminate in a tail of 9's. And then the real number b defined below must be between 0 and 1, yet could not be on the list.

$$
\begin{aligned}
a_0 &= .a_{00} \ a_{01} \ a_{02} \ \cdots \\
a_1 &= .a_{10} \ a_{11} \ a_{12} \ \cdots \\
a_2 &= .a_{20} \ a_{21} \ a_{22} \ \cdots \\
&\vdots \qquad \vdots \\
a_n &= .a_{n0} \ a_{n1} \ a_{n2} \ \cdots \ a_{nn} \ \cdots
\end{aligned}
$$

$$
b = .b_0 b_1 \ldots b_n \ldots \quad \text{where} \quad b_n = \begin{cases} a_{nn} + 1 & \text{if } a_{nn} < 8 \\ a_{nn} - 1 & \text{if } a_{nn} \geq 8 \end{cases}
$$

In 1874 Cantor had already proved that the reals are not countable using an argument based on nested intervals and properties of the irrationals. The point of his new method is its generality: now he shows that the cardinality of *any* set must be less than the cardinality of its power set and hence that there must be infinitely many levels of infinities.

1897

Methods for checking the validity of deductions in aristotelian logic have become well-known. This year the English mathematician and author of *Alice in Wonderland*, **Lewis Carroll** (Charles Lutwidge Dodgson), 1832–1898, presents three of them in his popular text on logic *Symbolic Logic*: Euler-Venn diagrams, Carroll's diagrams and counters, and a method of subscripts (algebraic).

1899

The program to axiomatize geometry begun by Euclid more than two millenia earlier comes to fruition with Hilbert's axiomatic development of Euclidean geometry in his book *The Foundations of Geometry*. The formal axiomatic development and the clear distinction between the formal language and models for it influence Hilbert towards a formalistic conception of mathematics.

> One must be able to say at all times—instead of points, straight lines, and planes— tables, chairs, and beer mugs.
>
> Hilbert in *Reid, 1970*, p. 57.

David Hilbert, 1862–1943, having received his doctorate at Königsberg in 1885, is now permanently at Göttingen.

1900

The Second International Congress of Mathematicians is held in Paris, and **Hilbert**, one of the preëminent mathematicians of the day, states 23 problems as signposts for the development of mathematics.

The first problem asks whether every subset of the real numbers can be put into one-one correspondence with either the entire set of real numbers or the set of natural numbers and, further, whether it is possible to well-order the set of all real numbers. That is, is it possible to order the real numbers so that every subset has a least element?

> [The second problem is] to prove that "the arithmetical axioms" are not contradictory, that is, that a finite number of logical steps based upon them can never lead to contradictory results.

> The tenth problem is to determine of an arbitrary polynomial equation with integer coefficients in any number of unknowns whether there is a solution for it in the integers.

> The conviction of the solvability of every mathematical problem is a powerful incentive to the worker. We hear within us the perpetual call: There is the problem. Seek its solution. You can find it by pure reason, for in mathematics there is no *ignorabimus*.

1902
Bertrand Russell, 1872–1970, writes to Frege of his discovery of a contradiction inherent in the first volume of Frege's *The Basic Laws of Arithmetic*, the work in which Frege is attempting to develop mathematics as part of logic.

> Let *w* be the predicate of being a predicate which cannot be predicated of itself. Can *w* be predicated of itself? From either answer follows its contradictory. We must therefore conclude that *w* is not a predicate. Likewise, there is no class (as a whole) of those classes which, as wholes, are not members of themselves. From this I conclude that under certain circumstances a definable set does not form a whole.
>
> Russell in *Frege, 1980*, pp. 130–131

The second volume is nearly ready for press and Frege can only add an appendix explaining the problem. He writes to Russell:

> Your discovery of the contradiction has surprised me beyond words and, I should almost like to say, left me thunderstruck, because it has rocked the ground on which I meant to build arithmetic. . . . Your discovery is at any rate a very remarkable one, and it may perhaps lead to a great advance in logic, undesirable as it may seem at first sight.
>
> *Frege, 1980*, p.132

This and paradoxes in the theory of sets create concern about the foundations of mathematics among mathematicians and logicians. Hilbert fears that the finitists such as Kronecker may be right and that mathematics based on actual infinities is indeed incoherent.

1904
Ernst Zermelo, 1871–1953, working with Hilbert's group in Göttingen, "solves" part of Hilbert's first problem by showing that *every* set can be well-ordered. He does so by explicitly formulating and using a principle that he claimed was implicit in the work of many mathematicians, what comes to be known as the *axiom of choice*. The mathematical community becomes divided as to whether it is a legitimate principle. (See *Moore, 1982* for the history.)

Hilbert, in an address to the Third International Congress of Mathematicians at Heidelberg, presents a sketch for proving the consistency of axiomatic arithmetic, the method of which becomes widely adopted: exhibit a combinatorial property that the axioms possess and which theorems inherit by the rules of proof, but which no contradiction has.

1910–1913
Alfred North Whitehead, 1861–1947, and **Russell**, with a grant from the Royal Society and £100 of their own money, publish their *Principia mathematica*. Deeply influenced by the work of Peano and Frege, they attempt to show that all of mathematics can be developed as a part of logic. Their axiomatization and formal development serve as a basic reference and establish notation for most logical research for more than twenty years.

The infinite, however, cannot be deduced from solely logical principles:

> This assumption [[the axiom of infinity]], like the multiplicative axiom, will be adduced as a hypothesis whenever it is relevant. It seems plain that there is nothing in logic to necessitate its truth or falsehood, and that it can only be legitimately believed or disbelieved on empirical grounds.
>
> *Principia mathematica*, vol. 2, p. 183

⊢ : $p \supset q$. $\supset$. $\sim q \supset \sim p$

⟦*Law of Transposition*⟧

After many years of work on logic and the philosophy of mathematics, Russell turns to the study of ethics, epistemology, and metaphysics after World War I because, "All the high-flown thoughts that I had had about the abstract world of ideas seemed to me thin and rather trivial in view of the vast suffering that surrounded me." Whitehead turns mainly to metaphysics, and in 1924 accepts a chair in philosophy at Harvard, where he will continue for the rest of his life.

1912

Luitzen Egbertus Jan Brouwer, 1881–1966, is appointed Professor in set theory, function theory, and axiomatics at the University of Amsterdam. His inaugural address is "Intuitionism and formalism," the first public manifesto for his vision of constructive mathematics. He argues against Hilbert's formalistic understanding of mathematics and rejects the use of classical logic outside the realm of the finite.

> The intuitionist recognizes only the existence of denumerable sets, i.e., sets whose elements may be brought into one-to-one correspondence either with the elements of a finite ordinal number or with those of the infinite ordinal number ω. And in the construction of these sets neither the ordinary language nor any symbolic language can have any other role than that of serving as a non-mathematical auxiliary, to assist the mathematical memory or to enable different individuals to build up the same set.

1915

Leopold Löwenheim, 1878–1957, working in the algebraic tradition of Boole and Schröder in which truth and validity, not methods of proof, are central, publishes a paper on the "calculus of relatives", that is, algebraic identities involving quantifiers over individuals and relations on a domain, where a quantifier is taken to be a possibly infinite sum or product. He shows that validity for formulas involving quantifiers over individuals and not relations depends only on finite and countably infinite domains, and he gives a procedure to decide the validity of any such formula if it contains only unary predicates. He is not able to give such a procedure for the entire calculus of formulas involving quantification only over individuals, but reduces that problem to deciding the validity of formulas involving only binary relations.

From 1903 to 1933 Löwenheim teaches at a secondary school in Berlin, but is dismissed that year because he is "quarter-Jewish". Surviving the war in Berlin, he returns to his previous post in 1946 until his retirement in 1949.

1918

Paul Bernays, 1888–1977, invited the previous year from Zurich where he was a Privatdozent to be an assistant to Hilbert in the foundations of mathematics at Göttingen, establishes in his Habilitationsschrift that for the propositional part of Russell and Whitehead's calculus "every valid formula is a provable formula, and conversely". The proof is not disseminated outside Hilbert's circle until its publication in 1926.

1919

The Norwegian mathematican **Thoralf Skolem**, 1887–1963, recasts Löwenheim's decision procedure to apply to the first-order monadic predicate calculus, explaining the test for validity in terms of successive elimination of quantifiers in formulas.

1921

Emil Post, 1897–1954, publishes his Ph. D. thesis from Columbia University in which he formalizes the truth-table method for establishing validity for propositional logic and proves that the propositional part of Russell and Whitehead's axiomatic logic is complete and hence consistent.

Troubled by ill-health throughout his life, Post is able to take up a permanent faculty position only in 1935 at the City University of New York. When Alfred Tarski, meeting him many years later, congratulates him on being the only non-Pole to contribute significant work on propositional logic, Post responds that he *is* Polish, having come to the United States from Augustów in 1904.

1923

Skolem publishes "The foundations of elementary arithmetic established by means of the recursive mode of thought, without the use of apparent variables ranging over infinite domains." As part of a program of constructive mathematics, he sets forth primitive recursive definitions of many of the basic functions of number theory.

> The justification for introducing apparent [[bound]] variables ranging over infinite domains therefore seems very problematic; that is, one can doubt the justification of the actual infinite or the transfinite. . . .
>
>   [[A further paper simplifying this one, too,]] is built upon Kronecker's principle that a mathematical definition [[Bestimmung]] is a genuine definition if and only if it leads to the goal by means of a *finite* number of trials.
>
> *Skolem, 1923,* pp. 332–333

In 1947 he says:

> [[The paper *1923*]] is so far as I know the first investigation in recursive number theory. The utterance . . . that recursive arithmetic can be traced back to Dedekind and Peano seems to me rather strange, because the researchs of these men had another purpose than to avoid the use of quantifiers.

1924

The legitimacy of infinitistic reasoning and particularly the axiom of choice becomes more pressing in the face of a new paradox. The Polish mathematicians **Stefan Banach**, 1892–1945, and **Alfred Tarski**, 1901–1983, use the axiom of choice to give a nonconstructive proof that a sphere can be cut into finitely many pieces and the parts rearranged so as to form two spheres each the same size as the first.

Tarski, who uses his family name Tajtelbaum, receives his Ph.D. this year. He continues working with the logic seminar at Warsaw University, first as Docent then as Adjunct Professor, but these positions are not sufficient to support him and his family. At the start of the war in 1939 he is separated from his family while lecturing in the United States. After several temporary jobs he takes up a position as lecturer at the University of

California at Berkeley in 1942, where in 1946 he is made full professor. His family, surviving the war in Poland, joins him.

Wilhelm Ackermann, 1896–1962, one of Hilbert's students and collaborators at Göttingen, publishes a constructive consistency proof for what he claims to be all of analysis (arithmetic with quantification over functions). But just before publication he finds an error and adds that significant restrictions on the system must be made for the proof to work. In 1927 **John von Neumann**, 1903–1957, revises the ideas to apply correctly to first-order arithmetic in which the induction schema is restricted to quantifier-free formulas.

1925

Hilbert delivers a lecture entitled "On the infinite" at a congress in honor of Karl Weierstrass. It is the fullest expression of his program for the foundations of mathematics: The infinite has a legitimate place in mathematics as an ideal element, corresponding to nothing in the world, but fleshing out the system to simplify and clarify finitistic mathematics, in the same sense that $i = \sqrt{-1}$ can be added to the real numbers and leads to no contradiction with what can be proved without it. But a proof is needed that indeed the infinite and infinitistic reasoning lead to no contradiction with finitistic mathematics; Hilbert claims to be able to do just that. He gives the broad outlines of a proof of the consistency of arithmetic which uses only finitistic methods.

> No one shall drive us out of the paradise that Cantor has created for us.

1927

The American **Cooper Harold Langford**, 1895–1964, building on the work of Skolem's 1919 paper and improvements given in 1922 in Göttingen by **Heinrich Behmann**, 1891–1970, uses the method of successively eliminating quantifiers to show that various theories of orderings ("sets of properties") determine the truth-value of every propositional function in the first-order language of $\leq$. If the truth-value of each sentence of the language is determined, then the theory must be consistent.

1928

Ackermann revises Hilbert's 1917 lectures on logic which he and **Hilbert** publish as *Grundzüge der theoretischen Logik*. This small and succinct presentation of mathematical logic has great influence: much more accessible than Whitehead and Russell's *Principia mathematica*, it poses the problem of whether every valid formula of first-order logic can be proved from the axioms of *Prinicipia mathematica*.

Ackermann also publishes an example of a function that is not of Hilbert's type 1; that is, its definition uses, in Hilbert's system, functions of functions. He shows it can be expressed by a definition using "simultaneous recursion":

$$\varphi\,(\,a, b, 0\,) \,=\, a + b$$

$$\varphi(\,a, 0, n + 1) \,=\, \begin{cases} 0 & \text{if } n = 0 \\ 1 & \text{if } n = 1 \\ a & \text{if } n \neq 0 \text{ and } n \neq 1 \end{cases}$$

$$\varphi(\,a, b + 1, n + 1) \,=\, \varphi(a,\, \varphi(a, b, n + 1),\, n)$$

(*Calude and Marcus* show that the Romanian G. Sudan independently defined in 1927 a function that is computable and not primitive recursive.)

Later when Ackermann marries, Hilbert is opposed, feeling it will keep Ackermann from his scientific obligations. He refuses to do anything to advance Ackermann's career, who is thus unable to find a university position and teaches secondary school to support himself. On hearing that the Ackermanns are expecting a child, Hilbert says, "That is wonderful news for me. Because if this man is so crazy that he gets married and then even has a child, it completely relieves me from having to do anything for such a crazy man." (*Reid, 1970*, p. 173.)

1929

Jacques Herbrand, 1908–1931, having already published several papers, completes his thesis for the doctorate in mathematics at the Sorbonne. In Chapter 4 he uses the method of elimination of quantifiers to show how to determine for each statement of the theory of arithmetic with only successor whether it is true or false of the natural numbers. That is, the theory is decidable, or as he says, "resoluble" (see *van Heijenoort*, pp. 580–581).

In 1931 he extends these ideas to prove the consistency of a much larger fragment of arithmetic. The day that paper is received by the journal editors he dies in an accident climbing in the Alps.

In Warsaw Tarski has been presenting his work on metamathematics at seminars, emphasizing decision problems. This year his master's degree student **Mojzesz Presburger**, 1904–1943?, presents in a talk at the First Congress of Slavic Mathematicians an axiomatization of the sentences true of the integers in which the only operation is addition (in modern notation):

$$x = y \rightarrow (x + z) = (y + z)$$
$$(x + z) = (y + z) \rightarrow x = y$$
$$x + (y + z) = (x + y) + z$$
$$x + 0 = x$$
$$x + y = y + x$$
$$\exists y \, (x + y = z)$$

and for each natural number $n > 1$, taking "nx" to mean $\underbrace{x + \cdots + x}_{n \text{ times}}$

$$nx = ny \rightarrow x = y$$
$$\exists y \, (ny = x \lor ny + 1 = x \lor \cdots \lor ny + (n - 1) = x)$$
$$nx + 1 \neq 0$$

Using elimination of quantifiers, he establishes that the truth-value of every formula in this language can be decided.

After receiving his master's degree Presburger works at an insurance company until 1939. He is last seen in 1943, almost certainly perishing in the destruction of the Warsaw Ghetto. Ironically, it is not his axiomatization but Tarski's earlier first-order system for addition and successor on the natural numbers that becomes known as "Presburger Arithmetic", following Hilbert and Bernays demonstration in 1934 that Presburger's methods apply to that, too.

1930

Tarski announces that the theory of the real numbers in the language of $+$, $\leq$, and 1 is decidable. He publishes a sketch of the proof in 1931, using elimination of quantifiers, where provability and truth seem to be identified:

> In the system of arithmetic described in §1, every sentence of order 1 ⟦i.e., first-order sentence⟧ can be proved or disproved. Moreover, by analysing the proof of this result, we see that there is a mechanical method which enables us to decide in each particular case whether a given sentence (of order 1) is provable or disprovable. (p.134)
>
> In order to carry out the construction of the formal system of arithmetic which I shall sketch here, it would be necessary to formulate explicitly those sentences which are to be regarded as *axioms* (both the general logical ones and those which are specifically arithmetical), and then to formulate the *rules of inference* (*rules of proof*) with the help of which it is possible to derive from the axioms other sentences called *theorems* of the system. The solution of these problems occasions no great difficulty. If I omit their analysis here, it is because they are not of much importance for what is to follow. (p. 116)
>
> *Tarski, 1931*

(For the story of decision problems by Tarski and others at this time see *Doner and Hodges, 1988.*)

In 1927 a prize had been offered by the Dutch Mathematical Association to anyone who could formalize Brouwer's intuitionistic mathematics. **Arend Heyting**, 1898–1980, having already received his doctorate and now teaching at a secondary school, collects the prize in 1928 and this year publishes his axiomatization of the propositional logic of intuitionism.

In his textbook *Modern Algebra* the Dutch mathematician **Bartel Lennart van der Waerden**, 1903–1996, freely uses the axiom of choice to develop field theory on the basis of the well-ordering theorem. In a separate paper influenced by the intuitionists, he considers constructive algebra and shows that apparently there can be no general splitting algorithm applicable to all "explicitly given" fields, for if there were then it would lead to a decision procedure for any question of the form "Is there an n such that $E(n)$?" where E is *any* property of integers. In the second edition of his book (1937) he abandons the axiom of choice:

> We say "a field is given explicitly" if its elements are uniquely represented by distinguishable symbols with which addition, subtraction, multiplication, and division can be performed in a finite number of steps.

Kurt Gödel, 1906–1978, publishes the proof from his doctoral dissertation at the University of Vienna that the first-order predicate calculus is complete.

> Whitehead and Russell, as is well known, constructed logic and mathematics by initially taking certain evident propositions as axioms and deriving the theorems of logic and mathematics from these by means of some precisely formulated principles of inference in a purely formal way (that is, without making further use of the meaning of the symbols). Of course, when such a procedure is followed the question at once arises whether the initially postulated system of axioms and principles of inference is complete, that is, whether it actually suffices for the derivation of every true logico-mathematical proposition, or whether, perhaps, it is conceivable that there are true propositions (which may even be provable by means of other principles) that cannot be

derived in the system under consideration. For the formulas of the propositional calculus the question has been settled affirmatively; that is, it has been shown that every correct formula of the propositional calculus does indeed follow from the axioms given in *Principia mathematica.* The same will be done here for a wider realm of formulas, namely, those of the "restricted functional calculus" "first-order logic"; that is, we shall prove

Theorem 1: *Every valid formula of the restricted functional calculus is provable.*

This and his work in 1931 establish him as the premier mathematical logician of his time, not least because of the exceptional clarity with which he expounds his results. Though he has been attending the seminars of the Vienna Circle of philosophers, scientists, and mathematicians, his platonist conception of mathematics is so at odds with their views that he gradually loses contact with them. As only a Privatdozent he is paid by his students and must supplement his income with visiting positions in the United States. During the academic year 1933–1934 he visits the Institute for Advanced Study at Princeton; following his return to Europe he has a nervous breakdown, and in 1935 his return visit is cut short because of mental illness. In 1938-1939 he again lectures in the United States and then returns to Austria. Plans to return to the U. S. in the fall of 1939 are disrupted by the onset of war. By that time most of his colleagues have fled or been forced from the country, the position of Privatdozent has been abolished and Gödel, being thoroughly apolitical, applies for a new paid position, *Dozent neuer Ordnung.* Questions are raised about his association with Jewish professors and his lack of support for the Nazi regime, and when he is declared fit for military service and realizes he may be called up to serve, he flees in 1940 to the United States via the Trans-Siberian railway and Yokohama. He remains bitter throughout his life about his treatment in 1939–1940, blaming it more on Austrian sloppiness than the Nazi regime.

In the United States he takes up a permanent position at the Institute for Advanced Study. Later in his life Gödel's mental condition deteriorates further and, despite an active interest in logic and mathematics, he ceases to publish after 1958.

1931

Skolem uses the method of elimination of quantifiers to establish that arithmetic restricted to multiplication without successor and addition is decidable.

This sentence is not true.	The Liar Paradox
This sentence is not provable.	Gödel's undecidable sentence

Gödel turns a paradox into a powerful tool. In "On formally undecidable propositions of *Principia mathematica* and related systems I" he writes:

These two systems [[that of *Principia mathematica* and Zermelo-Fraenkel set theory]] are so comprehensive that in them all methods of proof today used in mathematics are formalized, that is, reduced to a few axioms and rules of inference. One might therefore conjecture that these axioms and rules of inference are sufficient to decide *any* mathematical question that can at all be formally expressed in these systems. It will be shown below that this is not the case, that on the contrary there are in the two systems mentioned relatively simple problems in the theory of integers that cannot be decided on the basis of the axioms. This situation is not in any way due to the special nature of the systems that have been set up, but holds for a wide class of formal systems.

His method is first to number the formulas of the formal language of *Principia mathematica*, a procedure that comes to be called "Gödel numbering". Gödel then views various predicates that apply to formulas, such as "is a well-formed-formula", "is an axiom", "is a consequence of via *modus ponens*" as relations on the numbers associated with the formulas. And those relations he shows are representable in the formal theory, that is, for each there is a formula with the appropriate number of free variables that is provable for those numerals corresponding to the wffs of which the predicate is true, and whose negation is provable for those numerals corresponding to wffs of which the predicate is false.

By a diagonal argument he is able to construct a formula of the formal language which, when interpreted in this way, asserts that it itself is not provable. But that does not entail a contradiction: it proves instead that *if* the formal theory is consistent, then the formula cannot be provable. And hence the formula is true but not provable, that is, first-order arithmetic is incomplete. Actually, something stronger than consistency is used in the argument to get that the sentence is not provable: the formal theory is assumed to be ω-consistent, that is, there is no formula with one free variable, $A(x)$, such that both $A(n)$ is provable for every numeral n and $\exists x \neg A(x)$ is also provable.

But more, a predicate can be constructed in the formal theory which allows one in effect to assert that the formal theory is itself consistent. Gödel shows that it, too, cannot be proved within the theory so long as the theory is ω-consistent. Gödel points out that this does not depend on the idiosyncracies of the system of *Principia mathematica*: all that is needed is enough of formal arithmetic to represent the appropriate metamathematical formulas.

Gödel is careful not to claim that the consistency of the system cannot be proved by *any* finitistic methods, because to do so there must be a clear idea of what "finitistic" means. Gödel investigates the class of arithmetical functions necessary for the representation, dubbing them the "recursive functions", later called the "primitive recursive functions". He also briefly discusses an idea of Herbrand concerning a generalization of the schema of primitive recursion.

(See *Dawson* for Gödel's first verbal announcement of his theorems in 1930.)

1933

Gödel shows that if intuitionistic arithmetic is consistent so is classical number theory by exhibiting a translation of classical number theory into intuitionistic arithmetic:

Classical logic	*Intuitionistic logic*
p	p
$\sim p$	$\neg p$
$p \cdot q$	$p \wedge q$
$p \rightarrow q$	$\neg(p \wedge \neg q)$
$p \vee q$	$\neg(\neg p \wedge \neg q)$
$\forall x$	$\forall x$
$\exists x$	$\neg \forall x \neg$

The system of intuitionistic arithmetic and number theory is only apparently narrower than the classical one, and in truth contains it, albeit with a somewhat deviant interpretation. . . . The above considerations, of course, provide an intuitionistic consistency proof for classical arithmetic and number theory. This proof, however, is not "finitary" in the sense in which Herbrand, following Hilbert, used the term.

(*Epstein, 1990* has a history of translation of classical into intuitionistic logic.)

1934

Rózsa Péter, 1905–1977, studies the class of functions called "recursive" by Gödel, which she dubs the *primitive recursive functions*, presenting normal forms and demonstrating that definitions by nested recursion and course-of-values recursion do not lead out of the class. The following year she publishes a simplification of Ackermann's function, showing that it is not simultaneous recursion but *nested* recursion on two variables that leads out of the class of primitive recursive functions:

$$\theta(0, n) = 2n+1$$
$$\theta(m+1, 0) = \theta(m,1)$$
$$\theta(m+1, n+1) = \theta(m, \theta(m+1, n))$$

Unable to find work teaching, Péter has been supporting herself by tutoring since her graduation in 1927, turning more to writing and translating poetry. Her interest in mathematics is reawakened by her classmate **László Kalmár**, 1905–1976, who suggests problems from Gödel's work. From this year onwards she publishes solely under the name "Péter" in place of her original German-Jewish name "Politzer". In 1945 she finally obtains a teaching position at the Budapest Teachers Training College where she stays until its closure in 1955, taking up a position as full professor at Loránd Eötvös (Budapest) University.

Hilbert and **Bernays** also examine the nature of recursion in the first volume of their *Grundlagen der Mathematik*. In the introduction to that work Hilbert defends his program in the light of Gödel's work: the incompleteness results only indicate that better methods must be found to give more widely applicable finitary consistency proofs.

The previous year Bernays had been fired from his position at the Mathematical Institute because he is a "non-Aryan", i.e., Jewish. He returns to Zurich, where he has citizenship and received his doctorate, eventually obtaining a position at the Eidgenössische Technische Hochschule.

Gödel delivers a series of lectures on his incompleteness theorems at the Institute for Advanced Study, simplifying the presentation of 1931. He finds an active and receptive audience, von Neumann having already lectured there on the 1931 paper. The notes for the lectures are written up by two students of **Alonzo Church**, 1903–1995: **Stephen C. Kleene**, 1909–1994, who this year receives his Ph. D., and **John Barkley Rosser**, 1907–1989. In mimeographed form the notes are widely distributed and serve as the basis for study for a whole generation of logicians, although remaining unpublished until 1965 (see *Davis, 1965*).

In the lectures Gödel advances an idea of Herbrand about the most general form of recursion:

> One may attempt to define this notion as follows: If φ denotes an unknown function, and $\psi_1, \ldots, \psi_k$ are known functions, and if the ψ's and the φ are substituted in one another in the most general fashions and certain pairs of the resulting expressions are

equated, then if the resulting set of functional equations has one and only one solution for φ, φ is a recursive function.

Gödel, 1934, p. 368

1936

Church develops an analysis of computability with his system of the *λ-calculus,* a method for defining and deriving values of functions in an equational calculus (see, e.g., *Rosser, 1984*).

> We introduce the following infinite list of abbreviations [where → means "stands for"],
>
> $1 \rightarrow \lambda ab \cdot a(b),$
> $2 \rightarrow \lambda ab \cdot a(a(b)),$
> $3 \rightarrow \lambda ab \cdot a(a(a(b)))$
>
> and so on, each positive integer in Arabic notation standing for a formula of the form $\lambda a\, b \cdot a(a(\cdots a(b) \cdots))$.

He observes that every function recursive in Gödel's 1934 sense is also λ-definable.

At the same time, **Kleene**, working in collaboration with Church, modifies the Herbrand-Gödel notion of recursiveness by defining the notion of a *general recursive function*:

> Thus the extension of general over primitive recursive functions consists only in that to substitutions and primitive recursions is added the operation of seeking indefinitely through the series of natural numbers for one satisfying a primitive recursive relation.

He introduces the notation "$\in x[A(x)]$" for the least natural number satisfying $A(x)$ if there is one and 0 otherwise.

This method of defining a function, however, may require an infinite search, and the question arises: when is it legitimate to claim that a function has actually been defined? Kleene shows by a diagonal argument that this question cannot be decided by a recursive procedure: a numbering can be given of all possible definitions of functions, say φ_1, φ_2, $\ldots$, φ_n, $\ldots$, and Kleene presents a normal form for the functions in terms of what comes to be called the *Kleene T-predicate.* But the numbering and normal form calculations are recursive, too, so that the function $\psi(x) = \varphi_x(x) + 1$ is recursive and hence is φ_m for some m. Yet the apparent contradiction, $\psi(m) = \varphi_m(m) + 1 = \psi(m) + 1$, is no contradiction at all: It shows rather that $\varphi_m(m)$ is not be defined, nor can it be recursively determined for which x is $\varphi_x(x)$ defined. This is the first example of a class of computable functions for which the computable diagonalization procedure does not lead outside the class.

Kleene observes that a function is general recursive if and only if it is λ-definable. This, and the fact that the diagonal function does not lead out of the class, and the wide applicability of these notions convince Church to propose in his paper a formal counterpart to the intuitive notion of a computable function:

> We now define the notion, already discussed, of an *effectively calculable* function of positive integers by identifying it with the notion of a recursive function of positive integers (or of a λ-definable function of positive integers). This definition is thought to be justified by the considerations which follow, so far as positive justification can ever be obtained for the selection of a formal definition to correspond to an intuitive notion.

Church, too, exhibits an arithmetical predicate which diagonalizes the class of
λ-definable functions. By virtue of his new definition he then claims that it is undecidable,
not in the sense of not derivable in his or some other system but in the sense of not being
computable by any effective means.

In a short paper later this year Church extends his conclusions to show there is no
effectively calculable procedure to determine which formulas are theorems of a theory of
arithmetic based on Hilbert and Ackermann's system of first-order predicate logic, nor of the
system without arithmetical terms and axioms. That is, the *Entscheidungsproblem* is
undecidable. Church maintains reservations about concluding that there is no effective
procedure for determining validity, for that would require invoking Gödel's nonconstructive
proof of completeness.

Working independently in Cambridge, England, **Alan Turing**, 1912–1954, develops his
own analysis of computability. He breaks the notion of computation into its smallest
components:

> Let us imagine the operations performed by the computer to be split up into "simple
> operations" which are so elementary that it is not easy to imagine them further divided.
> Every such operation consists of some change of the physical system if we know the
> sequence of symbols on the tape, which of these are observed by the computer (possibly
> with a special order), and the state of mind of the computer. We may suppose that in a
> simple operation not more than one symbol is altered. Any other changes can be split up
> into simple changes of this kind. The situation in regard to the squares whose symbols
> may be altered in this way is the same as in regard to the observed squares. We may
> therefore, without loss of generality, assume that the squares whose symbols are changed
> are always "observed" squares.

The intuitive arguments lead to his definition of *Turing machines.*

> The "computable" numbers may be described briefly as the real numbers whose
> expressions as a decimal are calculable by finite means. . . . According to my
> definition, a number is computable if its decimal can be written down by a machine.

Having just received a copy of Church's paper, Turing adds as an appendix a proof
that a function is Turing machine computable if and only if it is λ-definable. This proves
decisive: agreement in the mathematical community quickly follows that these definitions
adequately formalize the notion of computability.

The problem that Kleene addresses of when a least search operation actually yields a
definition of a function appears in Turing's work as the *halting problem for Turing
machines*: On what inputs does a Turing machine halt? Turing, by a diagonal argument,
shows that the question cannot be decided by a Turing machine. And, like Church, he
applies his methods to logic:

> I shall show that there is no general method which tells whether a given formula A is
> provable in **K** [[the first-order predicate logic of Hilbert and Ackermann]], or, what
> comes to the same, whether the system consisting of **K** with −A adjoined as an extra
> axiom is consistent.

During World War II Turing works at Bletchley Field in England on cracking the
German Enigma codes, and is active then and later designing computers. In 1950 he
addresses the question of whether a machine can "think" and proposes a test:

Can a person, unaware of who is giving the responses, determine whether a person or machine is answering him?

> I believe that at the end of the century the use of words and general educated opinion will have altered so much that one will be able to speak of machines thinking without expecting to be contradicted.

In 1952 Turing reports a theft to the police and upon questioning admits that he suspects his homosexual lover. He is arrested, convicted of the crime of homosexuality, and forced to undergo hormone treatments. In 1954 he dies, apparently a suicide.

Post, influenced by the work in Princeton but working independently of Turing, presents an analysis of computability that is strikingly similar to Turing's. His view of the identification of computability with recursiveness, however, is very different from that of Church and Kleene:

> The writer expects the present formulation to turn out to be logically equivalent to recursiveness in the sense of the Gödel-Church development. Its purpose, however, is not only to present a system of a certain logical potency but also, in its restricted field, of psychological fidelity. In the latter sense wider and wider formulations are contemplated. On the other hand, our aim will be to show that all such are logically reducible to formulation 1. We offer this conclusion at the present moment as a *working hypothesis.* And to our mind such is Church's identification of effective calculability with recursiveness.
>
> 〚*Footnote*〛 Actually the work already done by Church and others carries this identification considerably beyond the working hypothesis stage. But to mask this identification under a definition hides the fact that a fundamental discovery in the limitations of the mathematicizing power of Homo Sapiens has been made and blinds us to the need of its continual verification.

Rosser extends Gödel's incompleteness theorem by showing it is enough to assume that a formal system of arithmetic is consistent, rather than ω-consistent, in order to establish that it is not possible to prove its own consistency within it.

Yet **Gerhard Gentzen**, 1909–1945, working as an assistant to Hilbert at Göttingen, proves that first-order Peano arithmetic is consistent. His method, however, goes beyond what was envisioned as finitistic in Hilbert's program. Péter had already exhibited orderings of not only ω^2, ω^3, ..., ω^n, ..., but also of ω^ω, ω^{ω^ω}, ..., for each of which there is a principle of induction that can be reduced to ordinary induction. Gentzen's proof uses an ordering ε_0 of the natural numbers that incorporates *all* these orderings. The method is adopted by others to prove that various subsystems of second-order arithmetic and of set theory are consistent.

At the outbreak of World War II, Gentzen is conscripted into the armed forces and works in telecommunications, but due to ill health is released. On May 5th, 1945, along with other professors from the University of Prague, he is interned by the Soviets, and on August 4th dies in his cell of malnutrition (at least according to one account given in *Kreisel, 1971*, pp. 255–256).

1938

Kleene accepts the inherent partial nature of definitions using search procedures and defines the class of *partial recursive functions*, based on a new operator:

> If $R(\mathsf{x}, y)$ is a relation, then $\mu y\, R(\mathsf{x}, y)$ denotes the function of x $[\![x_1, \ldots, x_n]\!]$
> which, for each fixed x, takes as value the least y such that $R(\mathsf{x}, y)$ is true,
> provided such a y exists and $R(\mathsf{x}, y)$ is defined for the preceding values of y,
> and is undefined otherwise.

Two partial functions are equal, for which he introduces the notation $\varphi(x) \simeq \psi(x)$, if they are defined for the same values of x and when defined yield the same value.

Kleene points out that his normal form theorem generalizes to partial recursive functions and, in just a few short sentences, establishes his *s-m-n* theorem and recursion theorem, justifying the most general form of definition of a function in terms of itself:

> There is a primitive recursive function $S_n^m(z, y_1, \ldots, y_m)$ such that, if e defines
> recursively $\varphi(y_1, \ldots, y_m, \mathsf{x})$ as a function of $m + n$ variables, and $k_1, \ldots, k_m$
> are fixed numbers, then $S_n^m(e, k_1, \ldots, k_m)$ defines recursively $\varphi(k_1, \ldots, k_m, \mathsf{x})$
> as a function of the n remaining variables. (The construction of S_n^m is based on the
> relation $\varphi(y_1, \ldots, y_m, \mathsf{x}) \simeq \Phi_{m+n}(e, k_1, \ldots, k_m, \mathsf{x})$.) If $\psi(z, \mathsf{x})$ is any
> partial recursive function, there is a number f which defines $\psi(f, \mathsf{x})$ recursively.
> (Let e define $\psi(S_n^1(y, y), \mathsf{x})$ recursively, and set $f = S_n^1(e, e)$.)

The point of his definitions is to investigate systems of notations for ordinals, and Kleene shows there is a least nonconstructive one, the ordinal ω_1, as formulated in 1936 by Church and him. Kleene is now at the University of Wisconsin where, except for service in the Navy during the war, he remains throughout his career.

1939

Turing introduces the idea of an "oracle" for a nonrecursive set being made available to a Turing machine in a study of the possibility of avoiding Gödel's incompleteness theorem by replacing one logic by a hierarchy of logics.

> Let us suppose that we are supplied with some unspecified means of solving number-
> theoretic problems; a kind of oracle as it were. We shall not go any further into the
> nature of this oracle apart from saying that it cannot be a machine. With the help of the
> oracle we could form a new kind of machine (call them *o*-machines), having as one of
> its fundamental processes that of solving a given number-theoretic problem. More
> definitely these machines are to behave in this way. The moves of the machine are
> determined as usual by a table except in the case of moves from a certain internal
> configuration **o**. If the machine is in the internal configuration **o** and if the sequence of
> symbols marked with *l* is then the well-formed formula **A**, then the machine goes into
> the internal configuration **p** or **t** according as it is or is not true that **A** is dual. The
> decision as to which is the case is referred to the oracle.

1943

Kalmár investigates the class of functions that can be obtained by using the four elementary operations of arithmetic: addition, subtraction, multiplication, and division restricted to the natural numbers, using composition and *bounded recursion*, i.e., a function can be defined by primitive recursion only if its values are less than one already obtained. This class, the *elementary functions*, contains most of the usual functions of number theory and in particular coding and uncoding functions for numbering the partial recursive functions.

Kleene, influenced by Turing's 1939 paper, sets out a hierarchy of classes of arithmetic predicates based on the number of alternations of quantifiers, starting with the recursive predicates. In doing so, he defines the notion of *relative recursiveness*, whereby a function is recursive in $g_1, \ldots, g_n$ if it can be defined from them and successor, the constant zero function, and the projections by substitution, primitive recursion, and the least-search operator. Much of the paper is devoted to the motivation and intuitive basis of his ideas. He argues at length for Church's identification of recursiveness with computability, which he dubs *Church's thesis*.

The same hierarchy was developed by **Andrzej Mostowski**, 1913–1975, in occupied Poland, but, as he says:

> I had a nice, very big, wonderful notebook with all these discoveries—and then in 1944 there was an uprising in Warsaw and I remember the soldiers came to our house and ordered us to leave. So I was with my mother in this house and I hesitated whether I had to take the notebook with me or some bread. I decided to take some bread, so all my notes were burnt. So then I reconstructed the papers some time in 1945.
>
> *Crossley*, p. 32

1944

Post gives the first detailed analysis of *recursively enumerable sets*, sets which can be enumerated as the output of a recursive function. As an example he shows that the problem of finding integral solutions to integral diophantine equations, Hilbert's 10th problem of 1900, can be viewed as the decision problem for a particular recursively enumerable set.

To investigate these sets he defines the notion of a degree of unsolvability, as influenced by Turing, and states what comes to be known as *Post's problem*: Is there a recursively enumerable set whose degree of unsolvability is intermediate between that of the recursive sets and the halting problem?

1947

In 1914 the Norwegian **Axel Thue**, 1863–1915, proposed the problem of determining for an arbitrary semigroup whether two words on its alphabet are equal, the *word problem for semigroups*. This year **Post** and the Russian mathematician, **Andrei Andreevich Markov**, 1903–1979, independently show that the problem is undecidable by exhibiting one semigroup whose word problem is not recursive, the first example of an open mathematical problem whose solution consists of showing that á particular set is not decidable. The word problem for groups remains open.

1949

Of the major number systems it has been shown that the first-order arithmetic of the natural numbers and hence of the integers is undecidable, while according to Tarski the first-order theory of the reals and hence complex numbers is decidable. This year **Julia Robinson**, 1919–1985, publishes the proof from her doctoral dissertation under Tarski that the arithmetic of the rationals is undecidable. The set of integers can be defined within the rationals via:

A rational number q is an integer iff

$$\forall a\, b\, (\varphi(a, b, 0) \wedge \forall m(\varphi(a, b, m) \rightarrow \varphi(a, b, m+1))) \rightarrow \varphi(a, b, q))$$

where $\varphi(a, b, k)$ is the formula: $\exists x\, y\, z\, (2 + abk^2 + bz^2 = x^2 + ay^2)$

Robinson's husband, **Raphael Robinson**, 1911–1995, is a Professor in the Department of Mathematics at the University of California, Berkeley, and nepotism rules prevent her from teaching there too. She nonetheless pursues her research and becomes known for her work on Hilbert's 10th problem (*Davis, Putnam, and Robinson*). In 1975 she is hired as a Lecturer in the department and in 1976, the same year she is elected to the National Academy of Sciences, she is promoted to full professor (see *Reid, 1986* for a biography).

> I have been designated the official token woman mathematician and have to shrink from too much distinction. Anyway, *I* know my considerable limitations and would like to be left alone in obscurity.
>
> J. Robinson in *Smorynski,* p. 77

1951

Péter publishes *Rekursive Funktionen*, the first book on the recursive functions. A thorough analysis of virtually all that is known about the subject, it contains her hierarchy of computable functions based on nested induction on n variables and induction on ordinal types. A superb text and compendium, the English translation is delayed until 1967 due to difficulties in obtaining the rights.

1952

Kleene publishes his textbook *Introduction to Metamathematics* on classical and intuitionistic mathematical logic with considerable emphasis on recursive function theory. It becomes the main reference for mathematical logic.

1953

The Polish logician **Andrzej Grzegorczyk**, 1922– , establishes a hierarchy of primitive recursive functions. Each class after the third, which is Kalmár's elementary functions, arises by first using one unlimited recursion on the previous class to obtain a function which dominates all the functions of that class, and closing under composition and bounded recursion. The diagonalization functions are in essence the levels of Ackermann's function of 1934, $\lambda\, n\varphi(n, x, y)$, which dominates all the primitive recursive functions.

The American **Henry Gordon Rice**, 1920–?, shows that the unsolvability of the halting problem is not an isolated phenomenon, for only the most trivial of extensional properties of the partial recursive functions are decidable: those true of either all functions or none.

Tarski, Mostowski, and **Raphael Robinson** publish a collection of three papers which codify and extend the basic procedures for establishing the undecidability of mathematical theories. Central to their analysis is Robinson's finitely axiomatized subtheory Q, every consistent extension of which is undecidable:

$$Sx = Sy \rightarrow x = y$$
$$0 \neq Sy$$
$$x \neq 0 \rightarrow \exists y(x = Sy)$$
$$x + 0 = x$$
$$x + Sy = S(x + y)$$
$$x \cdot 0 = 0$$
$$x \cdot Sy = (x \cdot y) + x$$

1954

In the first book in Russian on computable functions, **Markov** proposes a new analysis of computability, what are later dubbed *Markov algorithms*. They, too, are equivalent to the recursive functions. Unlike his predecessors, however, Markov gives a truly finitistic analysis of algorithms: He begins with alphabets understood as concrete inscriptions on paper, for which he gives a mathematical theory of strings, and only then considers identifications of alphabets or strings, that is, alphabets and words as types.

The American **William Werner Boone**, 1920–1983, and the Russian **Petr Sergeevich Novikov**, 1901–1975, independently show that the word problem for groups is undecidable. (See *Davis, 1958.*)

1955

Albrecht Fröhlich, 1916–2001, and **John Cedric Shepherdson**, 1926–, working in London, apply the ideas of recursive function theory to algebra, formalizing and extending van der Waerden's work on explicitly given fields.

1956

As early as 1935 Bernays criticized the intuitionists' analysis of mathematics:

> Intuitionism makes no allowance for the possibility that, for very large numbers, the operations required by the recursive method of constructing numbers can cease to have a concrete meaning. From two integers k, l one passes immediately to k^l; this process leads in a few steps to numbers which are far larger than any occurring in experience, e.g., $67^{(257^{729})}$.
>
> Intuitionism, like ordinary mathematics, claims that this number can be represented by an Arabic numeral. Could not one press further the criticism which intuitionism makes of existential assertions and raise the question: What does it mean to claim the existence of an arabic numeral for the foregoing number, since in practice we are not in a position to obtain it?

This year the Dutch mathematician **David van Dantzig**, 1900–1959, presses this criticism further and asks where the line between the finite and the infinite can be drawn:

> The difference between finite and transfinite numbers can not be defined operationally: it is possible that always when a mathematician A uses the term "a transfinite number", another mathematician B interprets it as "a finite number" (of course not always the same one) without ever coming to an inconsistency.

Subsequently, the Russian mathematician **Alexander S. Yessenin-Volpin**, 1924– , son of the famous poet Sergei Yessenin, develops this kind of analysis into a critique of current mathematical practice. Unable to find a university position in the USSR due to his previous imprisonment on political charges, Yessenin-Volpin translates logic texts from English to support himself until he emigrates to the U.S. in 1972. His ideas are taken up by several Americans, notably **David Isles**, 1935– , who turn Yessenin-Volpin's general remarks into a mathematical theory, strict finitism.

The American undergraduate student **Richard Friedberg**, 1935– , and the Russian **Albert Abramovich Muchnik**, 1931– , independently solve Post's problem by exhibiting a nonrecursive set that is recursively enumerable and yet has a lower degree of unsolvability than that of the halting problem (see *Odifreddi, Volume 1*).

1957

Church's thesis comes under attack for being too restrictive. **Kalmár** objects to the limitations that the thesis imposes and gives several mathematical and quasi-mathematical arguments against it. He concludes:

> There are pre-mathematical concepts which must remain pre-mathematical ones, for they cannot permit any restriction imposed by an exact mathematical definition. Among these belong, I am convinced, such concepts as that of effective calculability, or of solvability, or of provability by arbitrary correct means, the extension of which cannot cease to change during the development of Mathematics.

At the same conference **Péter** argues that Church's Thesis is too broad. For functions of one variable the thesis states:

> A function f is computable if and only if $\exists$ an index e such that $\forall x, \exists$ a computation that results in $\varphi_e(x)$ being defined and equal to $f(x)$ [[where $\varphi_1, \varphi_2, \ldots, \varphi_e, \ldots$ is a computable listing of the partial recursive functions]].

But how are we to take the existence operator? It cannot be constructive, for then the definition would be circular. But the alternative is equally unacceptable according to Heyting, who recasts Péter's argument from an intuitionist perspective in 1960:

> The notion of a recursive function, which had been invented in order to make that of a calculable function more precise, is interpreted by many mathematicians in such a way, that it loses every connection with calculability, because they interpret non-constructively the existential quantifier which occurs in the definition.
>
> Of course every finite set is primitive recursive. But is every subset of a finite set recursive? Who can calculate the Gödel number of the characteristic function of the set of all non-Fermat exponents less than 10^{10} or of the set
>
> $$P_n = \{ x \mid x < n \ \& \ (Ey)T_1(x,x,y) \}$$
>
> [[i.e., $\{x: \varphi_x(x)$ is defined and $x < n\}$]], where n is a given natural number? The answer depends upon the logic which is adopted. If recursiveness is interpreted non-constructively, then P_n constitutes a counter-example to the converse of Church's thesis.

1958

The American **Martin Davis**, 1928–, publishes *Computability and Undecidability,* the first undergraduate textbook on recursive function theory and the undecidability results in which he simplifies the presentation of Turing machines.

1963

The theory of computable functions was developed before there were computers. Now **Shepherdson** and the American **H. E. Sturgis**, 1936–, recast the notion of computability in terms of an ideal digital computer with unlimited memory and unlimited time to compute.

1967

Ever since Kronecker's time ordinary mathematicians unconcerned with the foundations of mathematics have been attempting to replace nonconstructive existence proofs with constructive ones. The American **Abraham Seidenberg**, 1916–1988, for example, with no background in logic or recursive function theory showed that much of algebraic geometry can be developed constructively. Invariably the point is to exhibit constructions; proofs that a problem cannot be solved are of no concern.

This year the American **Errett Bishop**, 1928–1983, in his book *Foundations of Constructive Analysis* organizes that positive, constructive viewpoint into a major program by developing a significant portion of real analysis without any infinitistic, nonconstructive assumptions. Even negation is defined in a positive way: two real numbers are said to be unequal only if another can be constructed between them.

> Van Dantzig and others have gone as far as to propose that negation could be entirely avoided in constructive mathematics. Experience bears this out. In many cases where we seem to be using negation—for instance, in the assertion that either a given integer is even or it is not—we are really asserting that one of two finitely distinguishable alternatives actually obtains. Without intending to establish a dogma, we may continue to employ the language of negation but reserve it for situations of this sort, at least until experience changes our minds, and for counterexamples and purposes of motivation.

Recursive function theory and Church's thesis are rejected by Bishop as requiring a formalization of that which cannot be formalized: constructivity.

The American **Hartley Rogers**, 1926–, publishes his textbook *Theory of Recursive Functions and Effective Computability*, bringing into one place the accumulated research on relative recursiveness and orderings of nonrecursive sets. For 20 years it serves as the standard reference. It is noteworthy, though, for invoking Church's Thesis at almost all places where a function is claimed to be computable.

1970

The 22-year-old Russian **Yuri Matiyasevich**, 1947–, fits in the last piece of the puzzle to solve Hilbert's Tenth Problem of 1900: There is no general algorithm which can determine of any polynomial equation in any number of unknowns with integer coefficients whether it has a solution in the integers (see *Davis, 1958,* 2nd edition).

Bibliography

In quote from books and papers our editorial remarks are in square brackets [ ] or double brackets ⟦ ⟧. When only a portion of an article is quoted the author's footnotes are usually not included.

Page numbers for papers reprinted in Davis, *The Undecidable*, are from that book.
All other quotations in the text are from the latest English edition cited.

General references for the historical development of computability and undecidability
Overview of the early development of first-order logic: *Edwards, Goldfarb*, and *Moore, 1988*.
Original articles with historical essays: *Benacerraf and Putnam, Gödel, 1986, Davis, 1965, Mancosu, 1988* and *van Heijenoort, 1967b*.
Oral histories: *Crossley* and *Dawson*.
Biographies: *Edwards, Mittelstrass*, and the *Dictionary of Scientific Biography*, eds. Carl Boyer et al., American Council of Learned Societies, 1970.

ABELSON, Raziel
   1967   Definitions
                In *Edwards*, vol. 2, pp. 314–324 .
ACKERMANN, Hans Richard
   1983   Aus dem Briefwechsel Wilhelm Ackermanns
                History and Philosophy of Logic, vol. 4, pp. 181–202.
ACKERMANN, Wilhelm
   (See *Hans Richard Ackermann* for a biography.)
   1924   Begründung des "tertium non datur" mittels der Hilbertischen Theorie der Widerspruchsfreiheit
                Mathematische Annalen, vol. 93, pp. 1–36.
                (See *van Heijenoort, 1967b*, pp. 485–486 for a summary and commentary.)
   1928   Zum Hilbertschen Aufbau der reellen Zahlen
                Mathematische Annalen, vol. 99, pp. 118–133. Translated as "On Hilbert's construction of the real numbers" in *van Heijenoort, 1967b*, pp. 495–507.
ADA, Augusta
   (See *Stein* for a biography.)
   1843   Sketch of the Analytical Engine invented by Charles Babbage Esq., by L. F. Ménabréa of Turin, Officer of the Military Engineers
                Reprinted in *Morrison and Morrison, 1961*.
BABBAGE, Charles
   (See also *Campbell-Kelly, Morrison and Morrison*, and *Stein*.)
   1864   *Passages from the life of a Philosopher*
                Reprinted by Augustus M. Kelley Publishers, 1969.
                (An autobiography with a description of the Analytic Engine.)

BANACH, Stefan and Alfred TARSKI

    1924   Sur le décomposition des ensembles de points en parties respectivement congruents
            Fundamenta Mathematicae, vol. 6, pp. 244–277.

BARWISE, J., H. J. KEISLER, and K. KUNEN, eds.

    1980   *The Kleene Symposium*
            North-Holland.
            (Contains an unsigned biography of Kleene.)

BASHMAKOVA, I. G. and G. S. SMIRNOVA

    2000   Geometry: The first universal language of mathematics
            In *Grosholz and Breger*, pp. 331–340.

BAUM, Robert J.

    1973   *Philosophy and Mathematics*
            Freeman Cooper and Co.

BEESON, Michael J.

    1980   *Foundations of Constructive Mathematics*
            Springer–Verlag.

BEHMANN, Heinrich

    1922   Beiträge zur Algebra der Logik, insbesondere zum Entscheidungsproblem
            Mathematische Annalen, vol. 86, pp. 163–229.

BENACERRAF, Paul, and Hilary PUTNAM, eds.

    1983   *Philosophy of Mathematics*
            2nd edition, Cambridge University Press.
            (Contains papers by Hilbert, Brouwer, and others about foundations of mathematics.)

BERNAYS, Paul

    1926   Axiomatische Untersuchung des Aussagen-Kalkuls der "Principia Mathematica"
            Mathematische Zeitschrift, vol. 25, pp. 305–320.

    1935   Sur le platonisme dans les mathématiques
            L'Enseignement mathématique, 1st ser., vol. 34, pp. 52–69. Translated as "On platonism
            in mathematics" in *Benacerraf and Putnam*, pp. 258–271.

BREGER, Herbert

    2000   Tacit knowledge and mathematical progress
             In *Grosholz and Breger*, pp. 221–230.

BEZBORUAH, A., and J. C. SHEPHERDSON

    1976   Gödel's second incompleteness theorem for Q
            Journal of Symbolic Logic, vol. 41, pp. 503–512

BISHOP, Errett

    1967   *Foundations of Constructive Analysis*
            McGraw–Hill. A revised edition is *Bishop and Bridges*.

BISHOP, Errett, and Douglas BRIDGES

    (See also *Bridges and Richman*.)

    1985   *Constructive Analysis*
            Springer-Verlag.
            (A revised version of *Bishop, 1967*.)

BOOK, Ronald V.

    1980   Review of Garey and Johnson
            Bulletin of the American Mathematical Society, vol. 3, no. 2, pp. 898–904.

BOOLOS, George S., and Richard C. JEFFREY
   1980   *Computability and Logic*
         2nd edition, Cambridge University Press.
BOONE, W. W.
   1954   Certain simple, unsolvable problems in group theory V, VI
   –1957 *Indagationes mathematicae*, vol. 16, pp. 231–237, 492–497, 1954; vol. 17, pp. 252–256,
         571–577, 1955; vol. 19, pp. 22–27, 227–232, 1957.
         (See also *Davis, 1958.*)
BOYER, Carl A.
   1968   *A History of Mathematics*
         Wiley.
BRIDGES, Douglas, and Fred RICHMAN
   1987   *Varieties of Constructive Mathematics*
         London Mathematical Society, Lecture Notes Series, no. 97, Cambridge University Press.
         (A textbook on constructive algebra and analysis with a comparison of Bishop's
         constructivism, Russian constructivism, and intuitionism.)
BROUWER, L. E. J.
   1913   Intuitionisme en formalisme
         Wiskundig tijdschrift 9, pp. 199–201. Translated as "Intuitionism and formalism" by
         Arnold Dresden, *Bulletin of the American Mathematical Society*, vol. 20, pp. 81–96,
         which is reprinted in *Benacerraf and Putnam*, pp. 77–89.
BUCK, R. C.
   1963   Mathematical induction and recursive definitions
         The American Mathematical Monthly, vol. 70, Feb., pp. 128–135.
CALUDE, Christian and Solomon MARCUS
   1979   The first example of a recursive function which is not primitive recursive
         Historia Mathematica, vol. 6, pp. 380–384.
CAMPBELL-KELLY, Martin
   1989   *Babbage's Calculating Engines*
         Charles Babbage Institute for the History of Computing, Volume 2, Tomash Publishing,
         Los Angeles.
         (Contains reprints of Babbage's writings.)
CANTOR, Georg
   (See *Dauben* and *Purkert and Ilgauds* for biographies.)
   1872   Über die Ausdehung eines Satzes aus der Theorie der trignometrischen Reihen
         Mathematischen Annalen, vol. 5, pp. 123–132.
   1883   *Grundlagen einer allgemeinen Mannigfaltigkeitslehre*
         Teubner, Leipzig. (The translation in our text comes from *Dauben, 1979*, pp. 128–129.)
   1891   Über eine elementare Frage der Mannigfaltigkeitslehre
         Jahresbericht der Deutschen Mathematiker-Vereinigung, vol. 1, pp. 75–78.
   1895   *Contributions to the Founding of the Theory of Transfinite Numbers*
   –1897 Translated from the German with an introduction by Philip E. B. Jourdain, Open Court,
         1915. Reprinted by Dover, 1955.
CARROLL, Lewis
   1897   *Symbolic Logic*
         MacMillan and Co., London. Reprinted by Dover, 1958.

CHANG, Chin-Liang, and Richard C. LEE
    1973   *Symbolic Logic and Mechanical Theorem Proving*
           Academic Press.

CHIHARA, Charles S.
    2004   *A Structural Account of Mathematics*
           Oxford University Press.

CHURCH, Alonzo
    1933   A set of postulates for the foundation of logic (second paper)
           Annals of Mathematics, vol. 34, pp. 839–864.
    1935   Abstract of Church, 1936
           Bulletin of the American Mathematical Society, May, abstract 205, pp. 332–333.
    1936   An unsolvable problem of elementary number theory
           The American Journal of Mathematics, vol. 58, pp. 345–363, reprinted in *Davis,*
           1969, pp. 89–107 .
    1936a A note on the Entscheidungsproblem
           Journal of Symbolic Logic, vol. 1, pp. 40–41; correction ibid., pp. 101–102.
           Reprinted in *Davis, 1965*, pp. 110–115.
    1937   Review of *Turing, 1936*
           Journal of Symbolic Logic, vol. 2, pp. 42–43
    1937a Review of *Post, 1936*
           Journal of Symbolic Logic, vol. 2, p. 43
    1938   The constructive second number class
           Bulletin of the American Mathematical Society, vol. 44, pp. 224–232 .
    1956   *Introduction to Mathematical Logic*
           Princeton University Press.

CHURCH, Alonzo, and S. C. KLEENE
    1936   Formal definitions in the theory of ordinal numbers
           Fundamenta Mathematicae, vol. 28, pp. 11–21.

COHEN, Daniel
    1987   *Computability and Logic*
           Ellis Horwood and Wiley.

CROSSLEY, J. N. ed.
    1975   Reminiscences of logicians
           In *Algebra and Logic*, J. N. Crossley ed., Springer–Verlag, Lecture Notes in
           Mathematics, no. 450, pp. 1–62.
           (A recorded discussion of several of the founders of recursion theory.)

DAUBEN, Joseph Warren
    1979   *Georg Cantor: His mathematics and philosophy of the infinite*
           Harvard University Press. Reprinted 1990, Princeton University Press.

DAVIES, Brian
    2005   Whither mathematics?
           Notices of the American Mathematical Society, vol. 52, pp. 1350–1356.

DAVIS, Martin
    1958   *Computability and Unsolvability*
           McGraw-Hill. 2nd edition (with Matiyasevich's solution to Hilbert's Tenth Problem),
           Dover, 1982.

1965   *The Undecidable* (ed.)

       Raven Press.

       (Contains papers in the history of computability. For corrections to some of the translations see a review by S. Bauer-Mengelberg, *J. Symbolic Logic*, vol. 31, 1966, pp. 484–494.)

DAVIS, Martin, Hilary PUTNAM, and Julia ROBINSON

    1961   The decision problem for exponential diophantine equations

          Annals of Math., vol. 74, pp. 425–436.

DAWSON, John W., Jr.

    1984   Discussion on the foundations of mathematics

          History and Philosophy of Logic, vol. 5, no. 1, pp. 111–129.

          A translation with introduction and commentary of *Erkenntnis*, vol. 2 (1931), pp. 135–151.

DEDEKIND, Richard

    1888   *Was sind und was sollen die Zahlen?*

          Translated in *Essays on the Theory of Numbers*, Open Court, 1901. Reprinted by Dover, 1963.

DONER, John and Wilfrid HODGES

    1988   Alfred Tarski and decidable theories

          The Journal of Symbolic Logic, vol. 52, pp. 20–35.

DUMMETT, Michael, and Roberto MINIO

    1977   *Elements of Intuitionism*

          Clarendon Press, Oxford.

EDWARDS, Paul

    1967   *The Encyclopedia of Philosophy*

          Macmillan and The Free Press.

EINSTEIN, Albert

    1915   *Relativity: The special and general theory*

          Fifteenth edition, 1961, Crown Publishers, Inc.

EPSTEIN, Richard L.

    1979   *Degrees of Unsolvability: Structure and Theory*

          Springer–Verlag, Lecture Notes in Mathematics, no. 759.

    1990   *Propositional Logics (The Semantic Foundations of Logic)*

          Kluwer, 2nd ed. Oxford University Press, 1995, corrected 2nd ed. Wadsworth, 2001.

    1994   *Predicate Logic (The Semantic Foundations of Logic)*

          Oxford University Press. Reprinted Wadsworth, 2001.

    1999   *Critical Thinking*

          1st edition, Wadsworth. 3rd edition with Carolyn Kernberger, 2005.

    1999   The metaphysical basis of logic

          Manuscrito, vol. 22, no. 2, pp. 133–148. Reprinted in *Epstein, 2001A*.

    2001   Arguments and explanations

          Bulletin of Advanced Reasoning and Knowledge, 1, pp. 3–17.

          (Also at www.AdvancedReasoningForum.org.)

    2001   *Five Ways of Saying "Therefore"*

          Wadsworth.

    2004   On models and theories with applications to economics

          Bulletin of Advanced Reasoning and Knowledge, 2, pp. 77–98.

          (Also at www.AdvancedReasoningForum.org.)

2006   *Classical Mathematical Logic*
       Princeton University Press.

2007   Plausibility and the strength of arguments
       In *Epstein, 2008a*.

2007   Truth
       In *Epstein, 2008a*.

2008   On valid inferences
       To appear 2008, *Proceedings of the 2nd Colloquium on Metaphysics*, João Pessoa, Brazil.

2008a   *Essays on Logic as the Art of Reasoning Well*
       Typescript.

FALLIS, Don

    1998   Review of *Hersh, 1993* et al.
           The Journal of Symbolic Logic, vol. 63, pp. 1196–1200.

FEFERMAN, S.

    1960   Arithmetization of metamathematics in a general setting
           Fundamenta Mathematicae, vol. 49, pp. 35–92.

FERREIRÓS, José and Jeremy J. GRAY

    2006   *The Architecture of Modern Mathematics*
           Oxford University Press.

FRAENKEL, Abraham, Yehoshua BAR–HILLEL, and Azriel LEVY

    1973   *Foundations of Set Theory*
           2nd revised edition, North–Holland.

FREGE, Gottlob

    (See *Sluga* for a biography.)

    1879   *Begriffsschrift, eine der aritmetischen nachgebildete Formelsprache des reinen*
           Denkens, Halle.
           Translated as "*Begriffsschrift*, a formula language, modeled upon that of arithmetic, for
           pure thought", in *van Heijenoort, 1967b*, pp. 5–82.

    1893   *Grundgesetze der Arithmetik, begriffsschriftlich abgeleitet, Vol. 1*
           H. Pohle, Jena. Translated in part by M. Furth as *The Basic Laws of Arithmetic*,
           University of California Press, 1967.

    1903   *Grundgesetze der Arithmetik, begriffsschriftlich abgeleitet, Vol. 2*
           H. Pohle, Jena. Translated in part as by M. Black and P. Geach in *Translations from the*
           Philosophical Writings of Gottlob Frege, Basil Blackwell, Oxford, 1970.

    1980   *Gottlob Frege: The Philosophical and Mathematical Correspondence*
           eds. G. Gabriel, H. Hermes, F. Kambartel, C. Thiel, and A. Veraart, U. Chicago Press.

FRIEDBERG, Richard

    1957   Two recursively enumerable sets of incomparable degrees of unsolvability
           Proceedings of the National Academy of Sciences, vol. 43, pp. 236–238.

FRÖHLICH, A. and J. C. SHEPHERDSON

    1955   Effective procedures in field theory
           Philosophical Transactions of the Royal Society of London, ser. A, vol. 248, pp. 407–432.

GARDNER, Martin

    1973   Mathematical games
           Scientific American, vol. 229 (October), p. 114–118.

GAREY, Michael R., and David S. JOHNSON
1979   *Computers and Intractability*
       W. H. Freeman and Co., San Francisco.

GENTZEN, Gerhard
(See *Kreisel* for a biography.)
1936   Die Widerspruchsfreiheit der reinen Zahlentheorie
       Mathematische Annalen, vol. 112, pp. 493–565. English translation in *Gentzen, 1969.*
1938   The present state of research into the foundations of mathematics
       English translation in *Gentzen, 1969,* pp. 234–251.
1969   *The Collected Papers of Gerhard Gentzen*
       ed. M. E. Szabo, North-Holland.

GÖDEL, Kurt
1930   Die Vollständigkeit der Axiome des logischen Funktionenkalküls
       Monatschefte für Mathematik und Physik, vol. 37, pp. 349–360. Translated as "The
       completeness of the axioms of the functional caluculus of logic" in *van Heijenoort,
       1967b,* pp. 583–591, and reprinted in *Gödel, 1986,* pp. 103–123.
1931   Über formal unentscheidbare Sätze der *Principia mathematica* und verwandter Systeme I
       Monatschefte für Mathematik und Physik, vol. 38, pp. 173–198. Translated as "On
       formally undecidable propositions of *Principia mathematica* and related systems I" in
       Gödel, 1986, pp. 144–195.
1931a Discussion on providing a foundation for mathematics
       In *Gödel, 1986,* pp. 201–205.
1933   Zur intuitionistischen Arithmetik und Zahlentheorie
       Ergebnisse eines mathematischen Kolloquiums, vol. 4 (1931–32), pp. 34–38.
       Translated as "On intutionistic arithmetic and number theory" in *Davis, 1965,* pp. 75–81,
       and in *Gödel, 1986,* pp. 287–295.
1934   *On Undecidable Propositions of Formal Mathematical Systems*
       Mimeographed lecture notes. Published with revisions by Gödel in *Davis, 1965,*
       pp. 41–74, and reprinted with further emendations in *Gödel, 1986,* pp. 346–371.
1944   Russell's mathematical logic
       In *The Philosophy of Bertrand Russell,* ed. P. Schlipp, The Library of Living
       Philosophers, pp. 125–153. Reprinted in *Benacerraf and Putnam,* pp. 447–469.
1946   Remarks before the Princeton Bicentennial Conference on Problems in Mathematics
       In *Davis, 1965,* pp. 84–88 .
1986   *Collected Works, Volume 1: Publications 1929–1936.*
       S. Feferman et al. eds., Oxford University Press.
       (Contains a biography and historical articles by the editors.)

GOLDFARB, Warren D.
1979   Logic in the Twenties: the nature of the quantifier
       The Journal of Symbolic Logic, vol. 44, pp. 351–368.

GOODMAN, Nicolas D.
1981   Reflections on Bishop's philosophy of mathematics
       In *Richman, 1981,* pp. 135–145.

GOODSTEIN, R. L.
1951   *Constructive Formalism*
       University College, Leicester.

1951a The foundations of mathematics
Inaugural lecture, University College, Leicester.

1961 *Recursive Analysis*
North–Holland.

GRASSMANN, Hermann

1860 *Lehrbuch der Arithmetik für höhere Lehranstalten*
Berlin.

GRATTAN-GUINNESS, I.

1975 The Royal Society's financial support of the publication of Whitehead and Russell's *Principia mathematica*
Notes and Records of the Royal Society of London, vol. 30 (1975-1976), pp. 89–104.

GROSHOLZ, E., and H. BREGER

2000 *The Growth of Mathematical Knowledge*
Kluwer.

GRZEGORCZYK, Andrzej

1953 *Some classes of recursive functions*
Rozprawy Matematyczne IV, Mathematical Institute of the Polish Academy of Sciences.

1974 *An Outline of Mathematical Logic*
Reidel.

HAND, Michael

1990 Hilbert's iterativistic tendencies
History and Philosophy of Logic, vol. 11, no. 2, pp. 185–192.

HARDY, G. H.

1928 Mathematical proof
Mind, vol. 38, pp. 1–25.

HARKLEROAD, Leon and Edie MORRIS

1990 Rózsa Péter: Recursive function theory's founding mother
The Mathematical Intelligencer, vol. 12, pp. 59–61.

HAUSDORFF, Felix

1962 *Set Theory*
2nd edition, Chelsea.

HERBRAND, Jacques

1929 Non-contradiction des axiomes arithmétiques
Comptes rendus hebdomadaires des séances de l'Academie des sciences (Paris),
vol. 186, pp. 303–304. Translated as "The consistency of the axioms of arithmetic" in
Logical Writings, by Jacques Herbrand, edited by W. D. Goldfarb, Harvard U. Press,
1071, pp. 35–37. (The latter contains a biography.)

1931 Sur la non-contradiction de l'arithmétique
Journal für die reine und angewandte Mathematik, vol. 166, pp.1–8. Translated as
"The consistency of arithmetic" in *van Heijenoort, 1967b*, pp. 620–628.

HERKEN, Rolf, ed.

1988 *The Universal Turing Machine, A Half-Century Survey*
Oxford.

HERMES, Hans

1969 *Enumerability, Decidability, Computability*
2nd ed., Springer–Verlag. Translated from German by G. T. Hermann and O. Plassmann.

HERSH, Reuben

  1993 Proving is convincing and explaining
        Educational Studies in Mathematics, vol. 24, pp. 389–399.

  1997 *What is Mathematics Really?*
        Oxford University Press. Reprinted 1998, Vintage.

  2006 Inner vision, outer truth
        In Hersh, *18 Unconventional Essays on the Nature of Mathematics,* Springer.

HEYTING, A.

  1930 Die formalen Regeln der intuitionistischen Logik
        Sitzungsberichte der Preussischen Akad. der Wissenschaften, Phys.-Math. Kl., pp. 42–56.
        Translated as "The formal rules of intuitionistic logic" in *Mancosu, 1998,* pp. 311–327.

  1956 *Intuitionism: an introduction*
        North–Holland.

  1959 *Constructivity in Mathematics, Proceedings of the Colloquium held at Amsterdam, 1957*
        (Editor) North–Holland.

  1962 After thirty years
        In *Logic, Methodology, and Philosophy of Science, Proceedings of the 1960 International Congress,* eds. E. Nagel, P. Suppes, and A. Tarski, Stanford U. Press, pp. 194–197.

HILBERT, David

  (See also *Frege, 1980,* and, for a biography, *Reid.*)

  1899 *Grundlagen der Geometrie*
        Translated as *Foundations of Geometry,* Open Court, 1902.

  1900 Mathematische Probleme
        Proc. Int. Congr. Math. (1900), pp. 58–114. Translated into English in *The Bulletin of the American Mathematical Society,* vol. 8, 1901–1902, pp. 437–479.

  1904 Über die Grundlagen der Logik und der Arithmetik
        Proc. Int. Congr. Math. (1904), pp. 174–185. Translated as "On the foundations of logic and arithmetic" in *van Heijenoort, 1967b,* pp. 130–138.
        (Reviews the then current philosophical views about the foundations of mathematics.)

  1925 Über das Unendliche
        Mathematische Annalen, vol. 95, pp. 161–190. Translated in part by E. Putnam and G. Massey as "On the Infinite" in *Benacerraf and Putnam,* pp. 183–201; the entire paper translated by S. Bauer-Mengelberg in *van Heijenoort, 1967b,* pp. 369–392.

HILBERT, David, and Wilhelm ACKERMANN

  1928 *Grundzüge der theoretischen Logik*
        Springer, Berlin. Translated as *Principles of Mathematical Logic,* Chelsea Publ., 1950.

HILBERT, David, and Paul BERNAYS

  1934 *Grundlagen der Mathematik, Vol. I*

  1939 *Grundlagen der Mathematik, Vol. II*
        Springer, Berlin.

HODGES, Andrew

  1983 *Alan Turing: The Enigma*
        Simon and Schuster.

HUGHES, Patrick, and George BRECHT

  1975 *Vicious Circles and Infinity*
        Penguin.

ISLES, David
  1981   Remarks on the notion of standard non–isomorphic natural number series
         In *Richman, 1981*, pp. 111–134.
  2004   Questioning articles of faith: A re-creation of the history and theology of arithmetic
         Bulletin of Advanced Reasoning and Knowledge, 2, pp. 51–59.
         (Also at www.AdvancedReasoningForum.org.)

JOWETT, Benjamin
  1892   *The Dialogues of Plato*
         Third edition, MacMillan Co.

KALMÁR, László
  1935   Über die Axiomatisierbarkeit des Aussagenkalküls
         Acta Scientiarum Mathematicarum 7, pp. 222–243.
  1943   Egyszeru példa eldönthetetlen aritmetikai problémára
         (A simple example of an undecidable arithmetical problem). (Hungarian with German
         abstract.) *Matematikai és Fizikai Lapok*, vol. 50, pp. 1–23.
  1957   An argument against the plausibility of Church's thesis
         In *Heyting, 1959*, pp. 72–80.

KITCHER, Philip
  1975   Bolzano's ideal of algebraic analysis
         Studies in the History and Philosophy of Science, vol. 6, pp. 229–269.

KLEENE, Stephen C.
  (See also *Church and Kleene* and *Kleene and Post*. For a biography see *Barwise, et al.*)
  1936   General recursive functions of natural numbers
         Mathematischen Annalen, vol. 112, pp. 727–742. Reprinted with an addendum in *Davis,
         1965*, pp. 237–253.
  1938   On notations for ordinal numbers
         The Journal of Symbolic Logic, vol. 3, pp. 150–155.
  1943   Recursive predicates and quantifiers
         Transactions of the American Mathematical Society, vol. 53, no.1, pp. 41–73. Reprinted
         in *Davis, 1965*, pp. 255–287.
  1952   *Introduction to Metamathematics*
         North–Holland. Sixth reprint with corrections, 1971.
  1981   Origins of recursive function theory
         Annals of the History of Computing, vol. 3, no.1, pp. 52–67.

KLEENE, S. C. and Emil POST
  1954   The upper semi-lattice of degrees of recursive unsolvability
         Annals of Mathematics, ser. 2, vol. 59, pp. 379–407.

KOLATA, Gina Bari
  1976   Mathematical proofs: the genesis of reasonable doubt
         Science, vol. 192, pp. 989–990.

KREISEL, Georg
  1965   Mathematical logic
         In *Lectures on Modern Mathematics*, vol. III, ed. T. L. Saaty, Wiley, pp. 95–195.
  1971   Review of *The Collected Papers of Gerhard Gentzen*
         Journal of Philosopy, vol. 68, pp. 238–265.

1980 Obituary of K. Gödel

Biographical Memoirs of Fellows of the Royal Society (London), vol. 26, pp. 149–224 .

KRONECKER, Leopold

(See also *Dauben.*)

1887 Uber den Zahlbegriff

Journal für die reine und angewandte Mathematik, vol. 101, pp. 337–355.

LAKOFF, George and Rafael E. NÚÑEZ

2000 *Where Mathematics Comes From*

Basic Books.

LANGFORD, C. H.

1927 Some theorems on deducibility

vol. 28, pp. 16–40 and 459–471.

LEIBNIZ, Gottfried Wilhelm

1981 *New Essays on Human Understanding*

Cambridge University Press.

LERMAN, Manuel

1983 *Degrees of Unsolvability*

Springer–Verlag.

LIPSCHUTZ, Seymour

1964 *Set Theory*

Schaum's Outline Series, McGraw-Hill.

LÖWENHEIM, Leopold

(See *Thiel* for a biography.)

1915 Über Möglichkeiten im Relativkalkül

Mathematische Annallen, vol. 76, pp. 447–470. Translated as "On possibilities in the calculus of relatives" in *van Heijenoort, 1967b,* pp. 228–251.

MAL' CEV, A. I.

1970 *Algorithms and Recursive Functions*

Walters–Noordhof. Translated from the Russian by Leo F. Boron.

MANCOSU, Paolo

1996 *Philosophy of Mathematics and Mathematical Practice in the Seventeenth Century*

Oxford University Press.

1998 *From Brouwer to Hilbert: The debate on the foundations of mathematics in the 1920s*

Oxford University Press.

(Contains translations of many papers written in the 1920s.)

2000 On mathematical explanation

In *Grosholz and Breger,* pp. 103–119.

MARKOV, A. A.

1947 On the impossibility of certain algorithms in the theory of associative systems I, II

(Russian) *Doklady Akademii Nauk S. S. S. R.,* n.s., vol. 55, pp. 587–590. English translation in *Comptes rendus de l'academie des sciences de l'U. R. S. S.,* n.s., vol. 55, pp. 583–586. Part II, Ibid., vol. 58, pp. 353–356.

1954 *Teoriya Algorifmov*

*Trudy Math. Inst. Steklov,*vol. 42, Akademia Nauk S.S.S.R, Moscow. Translated as *Theory of Algorithms,* Israel Program for Scientific Translations, Jerusalem, 1971.

MATES, Benson

    1981   *Skeptical Essays*

          University of Chicago Press.

MATIYASEVICH, Yuri

    1970   Enumerable sets are diophantine

          Doklady Akad. Nauk. SSSR, vol. 191, pp. 279–282. Translated into English in

          Soviet Math. Doklady, 1970, pp. 354–357. (See also *Davis, 1958,* 2nd edition.)

MENDELSON, Elliott

    1963   On some recent criticism of Church's Thesis

          Notre Dame J. Formal Logic, vol. IV, no. 3, pp. 201–204.

    1964   *Introduction to Mathematical Logic*

          D. Van Nostrand. 3rd edition, Wadsworth & Brooks/Cole, 1987.

MILL, John Stuart

    1874   *A System of Logic, ratiocinative and inductive, being a connected view of the principles*

          of evidence and the methods of scientific investigation

          Eighth edition, Harper & Brothers, New York.

MITTELSTRASS, J., ed.

    1984   *Enzyklopädie Philosophie und Wissenschaftstheorie*

          Bibliographisches Institut, Mannheim.

MOORE, Gregory

    1982   *Zermelo's Axiom of Choice: Its Origins, Development, and Influence*

          Springer–Verlag.

    1988   The emergence of first-order logic

          In *History and Philosophy of Modern Mathematics,* eds. W. Aspray and P. Kitcher,

          University of Minnesota Press.

MORRISON, Phillip and Emily MORRISON

    1961   *Charles Babbage and his Calculating Engine*

          Dover.

          (Reprints some of Babbage's writings with a short biography and index).

MOSTOWSKI, Andrzej

    1944   On definable sets of positive integers

          Fundamenta Mathematicae, vol. 34, pp. 81–112.

    1966   *Thirty Years of Foundational Studies*

          Barnes and Noble.

          (Lectures on the history and then current status of mathematical logic.)

MUCHNIK, A. A.

    1956   Negative answer to the problem of reducibility in the theory of algorithms

          Doklady Akad. Nauk, vol. 108, pp. 194–197.

MUSGRAVE, Alan

    1977   Logicism revisted

          British Journal of Philosophy of Science, vol. 28, 1977, pp. 90–127.

NAGEL, Ernest

    1961   *The Structure of Science*

          Harcourt, Brace & World. Reprinted by Hackett Publishing Company, 1979.

NELSEN, Roger B.

1993   *Proofs without Words*
     Mathematical Association of America.

NIDDITCH, P. H.

1960   *Elementary Logic of Science and Mathematics*
     The Free Press of Glencoe, Illinois.

NOVIKOV, P.

1955   On the algorithmic unsolvability of the word problem in group theory (Russian)
     Trudy Math. Inst. Steklov,vol. 44, Akademiya Nauk S.S.S.R, Moscow.
     (See *Davis, 1958.*)

ODIFREDDI, Piergiorgio

1989   *Classical Recursion Theory, Volume 1*
     North-Holland.
     (A textbook with historical annotations and extensive bibliography.)

PARRIS, J. and L. HARRINGTON

1977   A mathematical incompleteness in Peano arithmetic
     In *Handbook of Mathematical Logic*, ed. J. Barwise, North-Holland, pp. 1133–1142.

PEANO, Giuseppe

1889   The principles of arithmetic, presented by a new method.
     Turin. Translated from the Latin in *van Heijenoort, 1967b*, 1967, pp. 85–97.

PÉTER, Rózsa

(See *Harkleroad and Morris* for a biography.)

1934   Über den Zusammenhang der verschiedenen Begriffe der rekursiven Funktion
     Mathematische Annalen, vol. 110, pp. 612–632.

1935   Konstruktion nichtrekursiver Funktionen
     Mathematische Annalen, vol. 111, pp. 42–60.

1951   *Rekursive Funktionen*
     Akadémai Kiadó, Budapest. Translated as *Recursive Functions*, Academic Press, 1967.

1957   Rekursivität und Konstrutivität
     In *Heyting, 1959*, pp. 226–233.
     Translated as "Recursivity and Constructivity" by L. Harkleroad, typescript.

POINCARÉ, Henri

1921   *The Foundations of Science*
     Translated by G. B. Halstead, The Science Press, New York.

POLYA, George

1963   *Mathematical Methods in Science*
     Vol. IX of *Studies in Mathematics*, School Mathematics Study Group, National Science
     Foundation. Revised edition edited by Leon Bowden, The Mathematical Association of
     America, 1977.

POPPER, Karl

1972   *Objective Knowledge*
     Oxford University Press.

POST, Emil

1921   Introduction to a general theory of elementary propositions
     American Journal of Mathematics, vol. 43, pp. 163–185.

1936   Finite combinatory processes—Formulation I
       Journal of Symbolic Logic, vol. 1, pp. 103–105, reprinted in *Davis, 1965*, pp. 288–291.

1944   Recursively enumerable sets of positive integers and their decision problems
       Bulletin of the American Mathematical Society, vol. 50, pp. 284–316. Reprinted in
       Davis, 1965, pp. 304–337.

1947   Recursive unsolvability of a problem of Thue
       The Journal of Symbolic Logic, vol. 12, pp. 1–11.

PRESBURGER, Mojzesz

(See *Zygmunt* for a biography.)

1930   Über die Vollständigkeit eines gewissen Systems der Arithmetik ganzer Zahlen, in
       welchem die Addition als einzige Operation hervortritt
       Sprawozdanie z I Kongresu matemtyków krajów slowianskich, Warszawa 1929
       (*Comptes-rendus du I Congrès des Mathématiciens des Pays Slaves, Varsovie, 1929*),
       pp. 92–101, and supplement, p. 395. Translated as "On the completeness of a certain
       system of arithmetic of whole numbers in which addition occurs as the only operation",
       History and Philosophy of Logic, vol. 12, 1991, pp. 225–233.

PURKERT, Walter, and Hans Joachim ILGAUDS

Georg Cantor, 1845-1918

Birkhauser Verlag. (A biography in German.)

PUTNAM, Hilary

1967   Mathematics without foundations
       The Journal of Philosophy, vol. LXIV. Reprinted in *Putnam, 1979a*, pp. 43–59.

1979   What is mathematical truth?
       In Putnam, *Mathematics, Matter, and Method*, Cambridge U. Press, 2nd edition,
       pp. 60–78.

REID, Constance

1970   *Hilbert*
       Springer–Verlag.
       (A biography with much information about other mathematicians of Hilbert's time.)

1986   The autobiography of Julia Robinson
       The College Math. J., vol. 17, pp. 3–21.

RESNIK, Michael

1981   Mathematics as a science of patterns: ontology and reference
       Nous, vol. 16, pp. 529–550.

RESNIK, Michael and D. KUSHNER

1987   Explanation, independence, and realism in mathematics
       British Journal for the Philosophy of Science, vol. 38, pp. 141–158.

RICE, H. Gordon

1953   Classes of recursively enumerable sets and their decision problems
       Transactions of the American Mathematical Society, vol. 74, pp. 358–366.

RICHMAN, F., ed.

1981   *Constructive Mathematics, Proceedings of the New Mexico State University Conference*
       Lecture Notes in Mathematics no. 873, Springer-Verlag.

ROBINSON, John M.

1968   *An Introduction to Early Greek Philosophy*
       Houghton Mifflin, New York.

ROBINSON, Julia

(See also *Davis, Putnam, and Robinson.* For a biography see *Reid, 1986* and *Smorynski.*)

1949　Definability and decision problems in arithmetic
　　　The Journal of Symbolic Logic, vol. 14, pp. 98–114.

1969　Diophantine decision problems
　　　In *Studies in Number Theory,* W. J. LeVeque ed., Mathematical Association of America,
　　　pp. 76–116.

ROBINSON, Raphael

1950　An essentially undecidable axiom system
　　　Proceedings of the International Congress of Mathematics, vol. 1, pp. 729–30.

ROGERS, Hartley

1967　*Theory of Recursive Functions and Effective Computability*
　　　McGraw–Hill.

ROSSER, J. B.

1936　Extensions of some theorems of Gödel and Church
　　　Journal of Symbolic Logic, vol. 1, pp. 87–91, reprinted in *Davis, 1965,* pp. 231–235.

1984　Highlights of the history of the λ-calculus
　　　Annals of the History of Computing, vol. 6, no. 4, pp. 337–349.

RUSSELL, Bertrand

(See also *Whitehead and Russell* and *Frege, 1980.*)

1925　Mathematics and the metaphysicians
　　　In *Mysticism and Logic,* Longmans, Green and Co., pp. 74–96.

1967　*The Autobiography of Bertrand Russell*
　　　Allen & Unwin.

ST. GEORGE STOCK

1908　*Stoicism*
　　　Constable, London.

SAWYER, W. W.

1946　*Mathematician's Delight*
　　　Pelican Books.

SHEPHERDSON, J. C., and H. E. STURGIS

(See also *Bezboruah and Shepherdson* and *Fröhlich and Shepherdson.*)

1963　Computability of recursive functions
　　　The Journal of the Association of Computing Machinery, vol. 10, pp. 217–255.

SHOENFIELD, Joseph R.

1967　*Mathematical Logic*
　　　Addison–Wesley.

SKOLEM, Thoralf

1919　Untersuchungen über die Axiome des Klassenkalküls und über Produktations- und
　　　Summationsprobleme, welche gewisse Klassen von Aussagen betreffen
　　　Skrifter, Videnskabsakademiet i Kristiania. Reprinted in *Skolem, 1970,* pp. 67–101.

1923　Begründung der elementaren Arithmetik durch die rekurrierende Denkweise ohne
　　　Anwendung scheinbarer Veränderlichen mit unednlichem Ausdehnungsbereich
　　　Videnskapsselskapets skrifter, I. Matematisk-naturvidenskabelig klasse, no. 6.
　　　Translated as "The foundations of elementary arithmetic established by means of the

recursive mode of thought, without the use of apparent variables ranging over infinite domains" in *van Heijenoort, 1967b*, pp. 303–333.

1931   Über einige Satzfunktionen in der Arithmetik
        Skrifter utgitt av Det Norske Videnskaps-Akademi i Oslo, I. Mat.-natur. kl., no. 7, pp. 1–28. Reprinted in *Skolem, 1970*, pp. 281–306.

1947   The development of recursive arithmetic
        Dixième Congres des Math. Scand. Copenhagen, 1946, pp. 1–6. Reprinted in *Skolem, 1970*, pp. 499–514.

1970   *Selected Works in Logic*
        Universitetsforlaget, Oslo.
        (Contains a biography and historical discussion of his works.)

SKORUPSKI, John

2005   Later empiricism and logical positivism
        In *The Oxford Handbook of Philosophy of Mathematics and Logic*, ed. Stewart Shapiro, Oxford University Press, pp. 51–80.

SLUGA, Hans

1980   *Gottlob Frege*
        Routledge & Kegan Paul.

SMORYNSKI, C.

1986   Julia Robinson, *in memoriam*
        The Mathematical Intelligencer, vol. 8, pp. 77–79.

SOARE, Robert I.

1987   *Recursively Enumerable Sets and Degrees*
        Springer–Verlag.

SPENCER, Joel

1983   Large numbers and unprovable theorems
        American Mathematical Monthly , vol. 90, no.10, pp. 669–675.

STEIN, Dorothy

1985   *Ada: A Life and a Legacy*
        MIT Press.

STEINER, Mark

2000   Penrose and Platonism
        In *Grosholz and Breger*, pp. 133–141.

SWART, E. R.

1981   The philosophical implications of the four-color problem
        The American Mathematical Monthly, vol. 87, no. 9, pp. 697–707.

TARSKI, Alfred

        (See also *Banach and Tarski* and *Doner and Hodges.*)

1931   Sur les ensembles définissables de nombres réels. I.
        Fundamenta Mathematicae, vol. 17, pp. 210–239. Translated by J. H. Woodger as "On definable sets of real numbers" in *Tarski, 1983*.

1933   The concept of truth in formalized languages
        In *Tarski, 1983*, which has its publication history.

1983   *Logic, Semantics, Metamathematics*
        2nd edition, ed. J. Corcoran, Hackett Publ.

TARSKI, Alfred, Andrzej MOSTOWSKI, and Raphael ROBINSON
    1953   *Undecidable Theories*
          North-Holland.

THIEL, Christian
    1977   Leopold Löwenheim: life, work, and early influence
          In *Logic Colloquium 76*, eds. R. Gandy and M. Hyland, North-Holland.

TILLYARD, E. M. W.
    1943   *The Elizabethan World Picture*
          Chatto and Windus.

TROELSTRA, A. S., and D. van DALEN
    1988   *Constructivism in Mathematics*
          North-Holland.

TURING, Alan M.
    (See *Hodges* for a biography.)
    1936   On computable numbers, with an application to the Entscheidungsproblem,
          Proceedings of the London Mathematical Society, ser. 2, vol. 42, 1936-7, pp. 230-265;
          corrections, ibid, vol. 43, 1937, pp. 544-546. Reprinted in *Davis, 1965*, pp. 115-153.
    1939   Systems of logic based on ordinals
          Proceedings of the London Mathematical Society, ser. 2, vol. 45, pp. 161-228.
          Reprinted in *Davis, 1965*, pp. 155-222.
    1950   Computing machinery and intelligence
          Mind, vol. 59, pp. 433-460. Reprinted in *Perspectives on the Computer Revolution*,
          ed. Pylyshyn, Prentice-Hall, 1970.

VAN DANTZIG, D.
    1956   Is $10^{10^{10}}$ a finite number?
          Dialectica, vol. 9, pp. 273-277.

VAN DER WAERDEN, B. L.
    1930   *Moderne Algebra, Volume 1*
          Springer, Berlin. 2nd edition, 1937, translated as *Modern Algebra,* Frederick Ungar
          Publ., New York, 1949.
    1930   Eine Bemerkung über die Unzerlegbarkeit von Polynomen
          Mathematische Annalen, vol. 102, pp. 738-739.

VAN HEIJENOORT, Jean
    1967a Gödel's Theorem
          In *Edwards*, vol. 3, pp. 348-357.
    1967b *From Frege to Gödel: A Source Book in Mathematical Logic*
          J. van Heijenoort ed., Harvard University Press.
          (Contains original papers mentioned in the timeline, with synopses and commentaries.)

VON NEUMANN, John
    1927   Zur Hilbertischen Beweistheorie
          Mathematische Zeitschrift, vol. 26, pp. 1-46. Reprinted in von Neumann, *Collected*
          Works, Pergamon Press, 1961, pp. 256-300.

WANG, Hao
    1974   *From Mathematics to Philosophy*
          Routledge and Kegan Paul.

WEINSTEIN, Scott
>   1983   The intended interpretation of intuitionistic logic
>   *Journal of Philosophical Logic*, vol. 12, no. 2, pp. 261–270.

WHITEHEAD, Alfred North, and Bertrand RUSSELL
>   (See also *Grattan-Guinness* and *Russell.*)
>   1910–13 *Principia Mathematica*
>   In 3 volumes, Cambridge University Press.

WIGNER, Eugene
>   1967   The unreasonable effectiveness of mathematics in the natural sciences
>   In Wigner, *Symmetries and Reflections*, Indiana University Press, pp. 222–237.

WILDER, R. L.
>   1944   The nature of mathematical proof
>   *The American Mathematical Monthly*, vol. 51, pp. 309–323.

WITTGENSTEIN, Ludwig
>   1978   *Remarks on the Foundations of Mathematics*
>   Revised version, eds. G. H. von Wright, R. Rhees, and G. E. M. Anscombe, Basil
>   Blackwell and M.I.T. Press.

YESSENIN-VOLPIN, A. S.
>   1970   The ultraintuitionistic criticism and antitraditional program for the foundations
>   of mathematics
>   In *Intuitionism and Proof Theory*, eds. Kino, Myhill, and Vesley, North-Holland.

ZERMELO, Ernst   (See also *Moore, 1982*)
>   1904   Beweis, das jede Menge wohlgeordnet werden kann
>   *Mathematische Annalen,* vol. 59, pp. 514–516.
>   Translated as "Proof that every set can be well-ordered" in *van Heijenoort, 1967b,*
>   pp. 139–141.

ZYGMUNT, Jan
>   1991   Mojsesz Presburger: life and work
>   *History and Philosophy of Logic*, vol. 12, pp. 211—223.

Glossary and Index of Notation

General

§ (denotes section of this book)

∎ (end of proof)

$\equiv_{\text{Def}}$ (introducing a term by definition)

$\Rightarrow$ (direction of proof)

$\Leftarrow$ (direction of proof)

Nn, *see* NNNS

NNNS (natural number notation system), 267

Functions and operators

f, g, h (convention on lower case Roman letters denoting total functions), 124

φ, ψ (convention on Greek letters denoting partial functions), 124

$f: X \to Y$, 23

$\mapsto$ 24

$\downarrow$ (is defined), 124

$\not\downarrow$ (is not defined), 124

$\simeq$ (agree on input), 124

$x \mid y$ (x divides y), 101

β-function, 190

C_A (characteristic function of a set), 97

C_R (characteristic function of a relation), 98

$E(x, y)$ (equality), 96

$Even(n)$ (evens), 97

φ_c^4 (characteristic function of the universal computation predicate), 133

J (pairing function), 40, 103

K, L (unpairing functions), 103

λx 24

$lh(x)$ (length), 102

$n!$ 91

max (maximum), 96

min (min-operator), 99, 122

μx (μ-operator), 122

Odd 95

P (predecessor), 95

$p(n)$ (nth prime), 102

P_k^i (projection on the ith coordinate), 92

p_n (nth prime), 101

p. r. (abbreviation for "partial recursive"), 124

$prf(n)$ (n codes a proof), 164, 186

ψ_m 110

ψ (the Ackermann-Péter function), 112

$rem(x, y)$, 190

S (successor), 92

sg (signature), 95

$\overline{sg}$ (zero test), 95

Z (zero function), 92

$(x)_n$ 102

$(x)_{n, m}$ 102

$\langle \,, \dots, \rangle$ (coding function), 102

351

Index